Principles of
Organometallic
Chemistry

P. POWELL

Principles of Organometallic Chemistry

Second edition

London   New York

CHAPMAN AND HALL

First published in 1968 by Chapman and Hall Ltd
11 New Fetter Lane, London EC4P4EE
Second edition 1988
Published in the USA by Chapman and Hall
29 West 35th Street, New York NY 10001
© 1988 P. Powell
Printed in Great Britain by
J.W. Arrowsmith Ltd., Bristol

ISBN 0 412 27580 5 (hardback)
ISBN 0 412 27590 2 (paperback)

British Library Cataloguing in Publication Data

Powell, P.
 Principles of organometallic chemistry.
 —2nd ed.
 1. Organometallic compounds
 I. Title
 547'.05   QD411

 ISBN 0 412 27580 5
 ISBN 0 412 27590 2 Pbk

Library of Congress Cataloging in Publication Data

Powell, P. (Paul), 1936—
 Principles of organometallic chemistry / P. Powell.—2nd ed.
 p.cm.
 Rev. ed. of: Principles of organometallic chemistry / G.E. Coates ...[et al.].. 1968.
 Includes bibliographies and index.
 ISBN 0 412 27580 5. ISBN 0 412 27590 2 (pbk.)
 1. Organometallic chemistry. I. Principles of organometallic chemistry. II. Title.
 QD411.P64 1988   87—17193
 547'.05—dc 19   CIP

Contents

Contents

Contents

Foreword

The teaching of chemistry to undergraduate students presents course-content problems which grow ever more difficult as huge amounts of new material are added yearly to the overall body of the subject.

Some fifty years ago a student graduating with a good quality degree in chemistry could be expected to have a reasonable grasp of most areas, including a fair number of recent developments. Since then, the explosion of knowledge has made any such expectation quite unrealistic. Teachers of the subject continue, however, to need texts to which they can refer students. The excellent but mammoth compendia are clearly unsuitable for anything but reference.

Dr Powell's book provides a suitable anchor for students aiming at a useful grasp of organometallic chemistry, especially that of the transition elements. Applications of chemistry to industry often feature negligibly in undergraduate courses; the chapter on industrial processes should help to remedy this unfortunate situation.

The growth of organometallic chemistry over the past 40–50 years has been immense. The present book can trace its ancestry to a 'slim volume' published in 1955, almost the whole of the very short chapter on transition metals having been added in the proof stage in 1954. The third edition, written by three authors, devoted an entire volume to transition metals. After the publication of this two volume work, its authors decided that the book was no longer suitable for most students, and that a text aimed at undergraduate students should be written. Dr Powell was one of the four authors of the resulting book (1968), and the present volume stems from it.

G.E. Coates, M.L.H. Green and K. Wade

Preface

In the 20 years since 'Principles' first appeared the field of organometallic chemistry has continued to expand very rapidly. During the 1950s and 1960s the main theme was the preparation and structural characterization of new compounds, especially those of the transition elements. The last two decades have seen the development of this theme and many unexpected materials such as carbyne complexes and cluster compounds have been discovered. There is now increasing emphasis on the application of organometallics of all types in organic synthesis, both in the laboratory and on an industrial scale.

These developments, although very exciting, present problems to anyone rash enough to attempt to write an undergraduate text. Such a book should not be too long and should present a balanced view of the field without becoming completely indigestible through a mass of detail. The excellent reference work, 'Comprehensive Organometallic Chemistry', published in 1982, consists of 8 tomes and a large index volume. Unfortunately, as Professor Coates has remarked, "though the content of science has grown so much, it is unlikely that the effective capacity of students' brains has increased substantially over the past 15–20 years."

It has therefore been necessary to be very selective in the topics discussed. Emphasis is laid on structures, bonding, methods of preparation and general reactions of organometallic compounds. An elementary knowledge of thermodynamics, bonding theory and the use of spectroscopic methods of structure determination is assumed. The Periodic Groups provide a logical framework for treating the compounds of the typical elements. In the area of the transition elements, however, the chemistry has been classified in terms of the electron number of the principal ligand. The alternative approach, classification by reaction type, can have the disadvantage that examples have to be drawn from otherwise unrelated areas, which can prove confusing to the beginner.

The relevance of organometallic chemistry to industry has been indicated throughout. Thanks are due to several firms including Associated Octel, Bayer AG, British Petroleum plc, Dow Corning, Esso and Thomas Swan & Co. Ltd. for providing information in this connection. A chapter describes studies of the mechanisms of homogeneously catalysed reactions. A few of these reactions are used in industry. In other cases such mechanistic studies help to clarify how the related heterogeneous systems, which are used in practice, may work.

Preface

Some modern uses of organometallic compounds in specific organic synthesis are indicated. Examples from natural product chemistry have mostly been omitted, because at undergraduate level they may divert attention from the main point.

Study problems have been included at the end of chapters 3–12. A number of these derive from the finals papers of various UK universities. In these cases the name of the university is given at the end of the question, which has been reproduced by kind permission of that institution. A booklet containing answers to these problems may be obtained by writing directly to the publishers.

Undergraduate students have little time to consult research papers or even review articles. Consequently only a few references to textbooks and recent reviews are given. Sources of further information are indicated, however, for those who need them.

I am most grateful to Professor J.A. Connor, Professor T. Edmonds, Dr P.I. Gardner, Dr M.L.H. Green, Dr P.G. Harrison and Dr D.M.P. Mingos for their helpful suggestions during the preparation of the manuscript. I should like especially to thank Professor G.E. Coates who read the complete text and made many valuable comments. I would also like to thank the staff at Chapman and Hall Ltd for their help and guidance during the production of the book.

P. Powell
Royal Holloway and Bedford New College,
University of London
January 1988

Periodic table of
the elements

The periodic table *overleaf* has been included to give atomic weights for use in the problems at the ends of chapters 3–12. These weights have been taken from the IUPAC Inorganic Chemistry Division Commission on Atomic Weights and Isotopic Abundances published in 1986 in *Pure and Applied Chemistry*, **58**, 1677–92. Brackets indicate that the weight of the commonest nuclide has been given for a non-naturally occurring element.

Periodic table of elements

Period	Group Ia	Group IIa	Group IIIa	Group IVa	Group Va	Group VIa	Group VIIa	Group VIII		
1 1s	1 H 1.01									
2 2s 2p	3 Li 6.94	4 Be 9.01								
3 3s 3p	11 Na 22.99	12 Mg 24.31								
4 4s 3d 4p	19 K 39.10	20 Ca 40.08	21 Sc 44.96	22 Ti 47.88	23 V 50.94	24 Cr 52.00	25 Mn 54.94	26 Fe 55.85	27 Co 58.93	28 Ni 58.69
5 5s 4d 5p	37 Rb 85.47	38 Sr 87.62	39 Y 88.91	40 Zr 91.22	41 Nb 92.91	42 Mo 95.94	43 Tc (99)	44 Ru 101.07	45 Rh 102.91	46 Pd 106.42
6 6s (4f) 5d 6p	55 Cs 132.91	56 Ba 137.32	57* La 138.91	72 Hf 178.49	73 Ta 180.95	74 W 183.85	75 Re 186.21	76 Os 190.20	77 Ir 192.22	78 Pt 195.08
7 7s (5f) 6d	87 Fr (223)	88 Ra (226)	89** Ac (227)							

*Lanthanide series 4f	58 Ce 140.12	59 Pr 140.91	60 Nd 144.24	61 Pm (145)	62 Sm 150.36	63 Eu 151.97	64 Gd 157.25	65 Tb 158.93
**Actinide series 5f	90 Th 232	91 Pa 231	92 U 238	93 Np (237)	94 Pu (244)	95 Am (243)	96 Cm (247)	97 Bk (247)

Group Ib	Group IIb	Group IIIb	Group IVb	Group Vb	Group VIb	Group VIIb	Group 0
						1 H 1.01	2 He 4.00
		5 B 10.81	6 C 12.01	7 N 14.01	8 O 16.00	9 F 19.00	10 Ne 20.18
		13 Al 26.98	14 Si 28.09	15 P 30.97	16 S 32.07	17 Cl 35.45	18 Ar 39.95
29 Cu 63.55	30 Zn 65.39	31 Ga 69.72	32 Ge 72.61	33 As 74.92	34 Se 78.96	35 Br 79.90	36 Kr 83.80
47 Ag 107.87	48 Cd 112.41	49 In 114.82	50 Sn 118.71	51 Sb 121.75	52 Te 127.60	53 I 126.90	54 Xe 131.29
79 Au 196.97	80 Hg 200.59	81 Tl 204.38	82 Pb 207.20	83 Bi 208.98	84 Po (~210)	85 At (~210)	86 Rn (~222)

66 Dy 162.50	67 Ho 164.93	68 Er 167.26	69 Tm 168.93	70 Yb 173.04	71 Lu 174.97
98 Cf (251)	99 Es (252)	100 Fm (257)	101 Md (258)	102 No (259)	103 Lr (260)

1

General survey

1.1 Introduction

In this book we are concerned with the properties of compounds which contain metal—carbon bonds. Traditionally organometallic chemistry includes the carbon compounds of the metalloids boron, silicon and arsenic, but excludes those of phosphorus and of other more electronegative elements. Metal carbonyls are discussed, but not cyanides or carbides, which are more usefully considered in conjunction with inorganic rather than organometallic compounds.

1.2 Historical background

Whereas some organometallic compounds have been known for well over a hundred years (e.g. the alkyls of zinc, mercury and arsenic and Zeise's salt, the first transition metal compound, $K[Pt(C_2H_4)Cl_3]$), the development of the subject, especially of the organometallic chemistry of the transition elements, is much more recent. The study of organometallic compounds has often contributed significantly both to chemical theory and practice. Thus the preparation and investigation of the properties of ethylzinc iodide and of diethylzinc led Frankland (1853) to make the first clear statement of a theory of valency, in which he suggested that each element has a definite limiting combining capacity. From the more practical standpoint, the discovery of the organomagnesium halides (Grignard reagents) in 1900 provided readily handled and versatile intermediates for a variety of organic and organometallic syntheses. An industrially applicable method for the preparation of organosilicon halides, which are intermediates in silicone manufacture, from silicon and organic halides was discovered in the 1940s. Again the study of aluminium alkyls has led to their use in catalysts for the large scale polymerization and oligomerization of olefins.

The chance synthesis of ferrocene, $(\eta\text{-}C_5H_5)_2Fe$ (1951) and the determination of its structure (p. 279) in the following year, opened up a field of research of hitherto unforseen variety, which has contributed greatly to our understanding of chemical bonding. The rapid subsequent development of this area owes much to new methods of structure determination especially nuclear magnetic resonance spectroscopy and X-ray crystallography. Organotransition metal complexes are now finding wide application as reagents for specific organic syntheses. They are also involved as intermediates in many catalytic processes, some of which are used for the large scale conversion of carbon monoxide, hydrogen, alkenes and other small molecules into useful organic chemicals.

1.3 Properties

Most organometallic compounds resemble organic rather than inorganic compounds in their physical properties. Many possess discrete molecular structures, and hence exist at ordinary temperatures as low melting crystals, liquids or gases (see Table 1.1). Commonly they are soluble in weakly polar organic solvents such as toluene, ethers or dichloromethane. Their chemical properties vary widely and their thermal stability, for example, depends markedly on their chemical composition. Thus tetramethylsilane (Me_4Si) is unchanged after many days at 500°C, whereas tetramethyltitanium decomposes rapidly at room temperature. Similarly there are wide differences in their kinetic stability to oxidation; some (e.g. Me_4Si, Me_2Hg, $*Cp_2Fe$) are not attacked at room temperature by the oxygen in air, whereas others (e.g. Me_3B, Me_2Zn, Cp_2Co) are spontaneously inflammable.

1.4 Classification of organometallic compounds by bond type

Organometallic compounds may conveniently be classified by the type of metal—carbon bonding which they contain. Carbon is a fairly electronegative element (2.5 on the Pauling scale) and hence might be expected to form ionic bonds only with the most electropositive elements, but to form electron-pair covalent bonds with other elements. The Periodic Table may be divided into very approximate regions into which the various types of organometallic compound predominantly fall (Fig. 1.1). As is usually found in inorganic chemistry, such a classification is far from rigid and the regions overlap considerably. It will be noticed that these regions are very similar to those observed in the classification of hydrides into (a) ionic, (b) volatile covalent, (c) hydrogen-bridged and (d) metal-like. As we shall see later, the organic compounds of the d-block transition elements often involve not only σ- but also π- or δ-bonding which is not commonly found amongst compounds of the main group elements. In detailed discussion of their chemistry, therefore, d-block transition metal organometallic compounds are better taken separately from those of the main group elements. Moreover, the chemistry of organic derivatives of the transition elements may be dominated more by the ligand (especially when it occupies in effect several co-ordination positions) rather than by the Periodic Group.

1.4.1 Ionic organometallic compounds

Organometallic compounds containing metal ions generally are formed only by the most electropositive elements. Thus methylpotassium crystallizes in a close

---

$*Cp = \eta\text{-}C_5H_5$

2

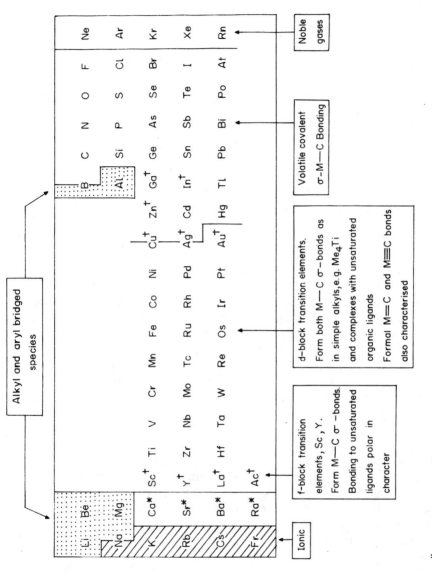

* The nature of the M—C bonding is not clearly established
† The elements also form alkyl and aryl bridged structures.

Fig. 1.1 Types of organometallic compound and the Periodic Table.

packed hexagonal, nickel arsenide structure, in which isolated methyl anions and metal cations are present (p. 39). The formation of ionic compounds is especially favoured when the hydrocarbon anion may be stabilized, for example by delocalization of the negative charge over several carbon atoms in an aromatic or unsaturated system. This applies to $Ph_3C^-Na^+$ or $C_5H_5^-Na^+$. The $C_5H_5^{\cdot}$ radical readily accepts an electron giving rise to a $C_5H_5^-$ anion in which six π-electrons form a delocalized, aromatic system, such as is present in benzene. With cations of electropositive metals, these stable anions can form salts which have a high degree of ionic character. In sodium ethynyl, $Na^+\bar{C}{\equiv}CH$, the negative charge is stabilized mainly on account of the greater electronegativity of sp relative to sp^3 hybridized carbon atoms.

1.4.2 Bridge bonding; cluster compounds

The cations of the light electropositive elements such as Li, Be, Mg and Al, however, are generally too strongly polarizing to coexist with polarizable carbanions in ionic structures. Strongly polarizing cations are those with a high charge/radius ratio. Aggregation of electron deficient monomer units such as LiMe, $MgMe_2$ or $AlPh_3$ can take place through bridging alkyl or aryl groups to form oligomers or polymers $(LiMe)_4$, $(Me_2Mg)_n$ or $(Ph_3Al)_2$ (Chapter 3). Similar bridge formation by hydrogen occurs in boranes. The structures of higher boranes are based on clusters of boron atoms. A wide range of carboranes also exists, which derive from boranes by replacement of BH_2 or BH^- by CH (Chapter 11). Cluster formation is also a feature of the chemistry of d-block transition elements. Many of their cluster compounds are carbonyl complexes e.g. $Fe_3(CO)_{12}$ and $Fe_5C(CO)_{15}$. The bonding in such aggregates cannot be explained in terms of classical two centre-two electron bonds. It is best discussed in terms of delocalized molecular orbitals.

1.4.3 Covalent two-electron bonds

The simplest type of metal–carbon link consists of an essentially covalent single two-electron bond M—C. The polarity of such a bond depends on the electronegativity of the metal M and to a lesser extent of the organic group. For example, the B—C bond in Me_3B (electronegativity difference $x_C - x_B = 2.5 - 2.1 = 0.4$) is less polar in character than the Al—C bond in the monomeric form of Me_3Al (electronegativity difference $x_C - x_{Al} = 2.5 - 1.6 = 0.9$).

(a) BOND ENERGIES. The bond energies for M—C bonds in some derivatives of main group elements are plotted against atomic number in Fig. 1.2. This graph shows that within any one main group of the Periodic Table; the metal–carbon bond energy falls with increase in atomic number. In general $\bar{D}(M—Ph) > \bar{D}(M—Me) > \bar{D}(M—Et)$. Metal–hydrogen bond energies follow a similar trend. While binary alkyls and aryls, MR_n, are known for all the main group elements

4

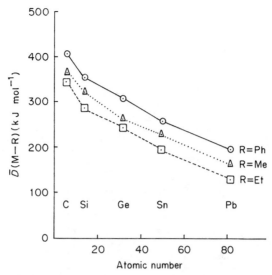

Fig. 1.2 Mean bond dissociation enthalpies in methyl, ethyl and phenyl derivatives of Group IV elements at 298.15 K.

except the noble gases, analogous derivatives of *d*-block transition elements generally cannot be isolated at all (e.g. Et_4Ti or Et_2Ni) or decompose readily (e.g. Me_4Ti or Me_6W). This lability is due to kinetic factors and not to any inherent weakness of the metal–carbon bonds. By suitable selection of the organic groups and/or by inclusion of supporting ligands, however, it has proved possible to prepare alkyl and aryl complexes which are reasonably robust. The factors which determine the resistance of organometallic compounds to thermal decomposition are discussed below (p. 9).

In spite of a lack of data, the indications are that carbon forms σ-bonds to transition elements of comparable strength with those it forms with main group elements. In contrast to the main groups, however, bond energies *increase* down any group, so that the strongest bonds are formed by elements of the Third Transition Series. This important difference can probably be explained in terms of the extent of overlap between the orbitals of the central atom and the $2s/2p$ orbitals of carbon. In the main groups, the best overlap with C $(2s/2p)$ is achieved by elements in the same row as carbon. Down any group the valence *s* and *p* orbitals become more diffuse so that their overlap with the compact orbitals of carbon becomes progressively weaker. In the elements of the First Transition Series, however, the $3d$ orbitals are strongly contracted and do not interact as strongly with carbon orbitals as do the larger $4d$ orbitals present in atoms of the Second Series. Even better overlap is thought to occur when metal $5d$ orbitals are involved.

For a methyl derivative Me_nM the bond disruption enthalpy

$$Me_nM(g) = nMe(g) + M(g) + \Delta H^{\ominus}$$

Table 1.1. Mean metal—carbon bond dissociation enthalpies $\bar{D}(M-Me)$/kJ mol⁻¹ and boiling points/°C in parentheses for compounds Me$_n$M*.

Me₂M	$\bar{D}$	(bp)	Me₃M	$\bar{D}$	(bp)	Me₄M	$\bar{D}$	(bp)	Me₃M	$\bar{D}$	(bp)
Me₂Be	–	(220)[a]	Me₃B	373.9	(–22)	Me₄C	367	(10)	Me₃N	312	(3)
Me₂Mg	–	[b]	Me₃Al	283	(126)	Me₄Si	320	(27)	Me₃P	286	(40)
Me₂Zn	186	(44)	Me₃Ga	256	(56)	Me₄Ge	258	(43)	Me₃As	238	(52)
Me₂Cd	149	(106)	Me₃In	170	(136)	Me₄Sn	226	(77)	Me₃Sb	224	(79)
Me₂Hg	130	(93)	Me₃Tl	–	(147)[c]	Me₄Pb	161	(110)	Me₃Bi	151	(110)

$n\bar{D}(M-Me) = \Delta H_f^{\ominus}(M,g) + n\Delta H_f^{\ominus}(Me,g) - \Delta H_f^{\ominus}(Me_nM,g)$

*M—C Bond lengths for Me₄M are where M = C, 1.54 Å; Si, 1.87 Å; Ge, 1.94 Å; Sn, 2.14 Å; Pb, 2.30 Å.
[a]Extrapolated sublimation temperature. [b]Involatile. [c]Extrapolated boiling point.
$\bar{D}$(Ta—Me) for Me₅Ta, 261 kJ mol⁻¹; $\bar{D}$(W—Me) for Me₆W, 161 kJ mol⁻¹.
The dissociation energies quoted are generally ca. ± 5 kJ mol⁻¹.

is given by

$$\Delta H^\ominus = n\bar{D}(\text{M}\!-\!\text{Me}) = \Delta H_f^\ominus (\text{M}, g) + n\Delta H_f^\ominus (\text{Me}, g) - \Delta H_f^\ominus (\text{Me}_n\text{M}, g)$$

In general, $\Delta H_f^\ominus(\text{Me}_n\text{M}, g)$ is small compared with the enthalpy of atomization of the element, $\Delta H_f^\ominus(\text{M}, g)$, so that the latter term is dominant. It is found that within a series of compounds with given n, $\Delta H^\ominus$ and hence $\bar{D}(\text{M}\!-\!\text{Me})$ qualitatively follow the trend in atomization enthalpy for the elements, $\Delta H_f^\ominus(\text{M}, g)$. The strongest bonds to carbon are thus formed by those elements which are most strongly bound in their standard states (Fig. 1.4).

While the strengths of single metal—carbon bonds vary from strong (e.g. B—C) to rather weak (e.g. Pb—C), they are of a similar order of magnitude to the strengths of single bonds C—X (X = C, N, O, Cl, S etc.) which are present in familiar organic compounds.

(b) MULTIPLE BONDING TO CARBON.   Carbon forms strong multiple bonds with itself and with nitrogen and oxygen. Multiple bonding between carbon and other main group elements, however, (except S and Se) is uncommon. In phosphorus yields (p. 136), $R_3P\!=\!CH_2$, the π-component probably includes some P(3d) character, as in the isoelectronic phosphine oxides, $R_3P\!=\!O$. Otherwise there are only curiosities such as the phospha- and arsabenzenes (p. 144) and the silaethenes (p. 124). Except at high temperatures in the gas phase, compounds such as $R_2C\!=\!SiR_2'$ do not exist as monomers unless their aggregation to polymers or cyclic oligomers is prevented, perhaps by introducing very bulky substituents.

Transition elements, however, form complexes in which formal metal–carbon double or triple bonds are present. Examples are $(OC)_5W\!=\!C(OMe)Me$ (p. 233) and $(Bu^tO)_3W\!\equiv\!CEt$ (p. 377). The π-components of these bonds are thought to arise through interaction between metal $(n\text{-}1)d$ and carbon $2p$ orbitals.

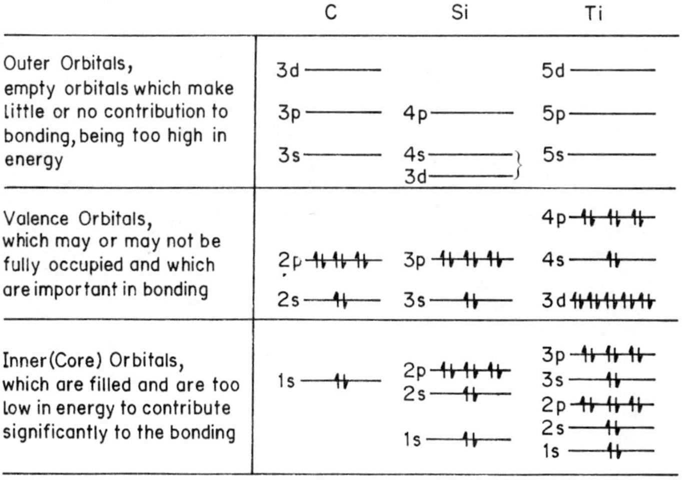

	C	Si	Ti
Outer Orbitals, empty orbitals which make little or no contribution to bonding, being too high in energy	3d ——— 3p ——— 3s ———	4p——— 4s——— } 3d——— }	5d ——— 5p——— 5s———
Valence Orbitals, which may or may not be fully occupied and which are important in bonding	2p ⇅⇅⇅ 2s —⇅—	3p ⇅⇅⇅ 3s —⇅—	4p ⇅ ⇅ ⇅ 4s —⇅— 3d ⇅⇅⇅⇅⇅
Inner (Core) Orbitals, which are filled and are too low in energy to contribute significantly to the bonding	1s —⇅—	2p ⇅⇅⇅ 2s —⇅— 1s —⇅—	3p ⇅⇅⇅ 3s —⇅— 2p ⇅⇅⇅ 2s —⇅— 1s —⇅—

Fig. 1.3 Division of orbitals into outer, valence and core orbitals.

(c) COMPLEXES OF UNSATURATED HYDROCARBONS. The *d*-block transition elements form an enormous range of complexes with unsaturated hydrocarbons. This is the chief feature which distinguishes their organometallic chemistry from that of the main groups. Virtually all these compounds have been prepared and characterized since 1951, the year in which ferrocene was reported. Some examples of typical complexes are illustrated in Fig. 6.1. In these compounds the metal–carbon bonding is essentially covalent in character. It arises by interaction between the π-electron system of the organic ligand and the metal valence orbitals, n*s*, n*p* and (n-1)*d*. It is only with *d*-block transition elements that the *d* orbitals have suitable energy for bond formation and also contain electrons (Fig. 1.3). Moreover because of their shape, metal *d* orbitals are well suited to interact with the π-molecular orbitals of unsaturated organic ligands. These interactions can give rise to components of σ-, π- and δ-symmetries with respect to the metal–ligand axis (p. 199).

Complex formation between unsaturated organic species and main group

Table 1.2. Summary of types of bonds in organometallic compounds

Ionic	Bridge bonding, Clusters
The most electropositive elements, including lanthanides. Favoured where organic anion is stable.	Multicentre bonds involving carbon. Carbon atom associated with two or more metal atoms in multicentre m.o.s. Electropositive elements where cation would be strongly polarizing (Chapter 3). Carboranes. Some transition metal clusters, especially carbonyls (Chapter 11).

COVALENT σ-Bonding (single M—C bonds)	σ- and π-Bonding (multiple M—C bonds)
Essentially covalent M—C bonds occurs with all except the most electropositive Group IA and Group IIA elements.	In main groups multiple bonds often labile to oligomerization or polymerization e.g. $R_2C{=}SiR_2'$. Some exceptions e.g. ylides.
Both main group and transition elements. Compounds of transition elements labile to thermal decomposition unless low energy pathways blocked.	*d*-Block transition elements form alkylidene and alkylidyne complexes which contain formal double and triple metal–carbon bonds (Chapter 7).

COMPLEXES WITH UNSATURATED ORGANIC LIGANDS

Characteristic of *d*-block transition elements. Covalent interaction between 2*p* orbitals of ligand and valence orbitals, (n-1)*d*, n*s* and n*p* of metal (Chapter 6). Involvement of metal *d*-orbitals important. 18-Electron configuration of metal commonly associated with kinetic stability.

elements is very much less common. The cyclopentadienyls comprise the largest group; in some of these such as Cp_2Mg the bonding seems to be appreciably ionic. There are also a few complexes of heavy metals with arenes (p. 93).

The lanthanides and actinides also form a wide range of complexes with unsaturated organic ligands including allyls, cyclopentadienyls and cyclooctatetraene derivatives. The bonding is strongly polar. Coordination numbers are determined largely by steric requirements around the metal centre.

1.5 The 'stability' of organometallic compounds

When discussing the stability of a compound one must be quite clear with what type of stability one is concerned. Loose description of a compound as 'stable' may refer to thermal stability, or to resistance to chemical attack, especially to oxidation or hydrolysis. Many organometallic compounds are readily oxidized in air but may be kept indefinitely under an inert atmosphere. Nitrogen gives adequate protection in many cases, but there are some complexes which react even with this normally rather inactive gas. The first dinitrogen complex, $[Ru(NH_3)_5N_2]^{2+}$ was discovered inadvertently by reducing $[Ru(NH_3)_5(H_2O)]^{3+}$ under a nitrogen atmosphere. In these cases argon affords suitable protection.

Very reactive species such as the silaethene, $Me_2Si{=}CHMe$ which in most situations would have only a transient existence, can be generated in solid inert gas matrices at very low temperatures (4–20 K). Under these conditions the individual molecules are trapped in isolation from each other, so that they can be studied spectroscopically. Above 45 K, however, diffusion through the matrix is sufficiently rapid for dimerization to occur. Silaethene monomers have also been detected by mass spectrometry in the gas phase at low pressures and high temperatures. The entropy term then favours their formation. The environment is therefore critical in determining whether or not a compound can be isolated or studied. Not only thermodynamic, but also kinetic factors must always be taken into account.

1.5.1 Thermal stability

In Table 1.3 and Fig. 1.4 the heats of formation of a number of organometallic compounds are presented. The enthalpy (or strictly the free energy*) of formation of a compound gives a measure of its thermodynamic stability. The low heats of formation of the methyls of the first short period elements (notably Me_4C and Me_3N) are largely a consequence of the high binding energies of the elements in their standard states (298 K and 1 atm). These data confirm that whereas some organometallic compounds are thermodynamically stable at room temperature

---

*Few standard entropy data for organometallic compounds are available; consequently $\Delta H_f^{\ominus}$ rather than $\Delta G_f^{\ominus}$ values are used to indicate thermodynamic stabilities. The omission of the entropy term must be borne in mind when considering the data.

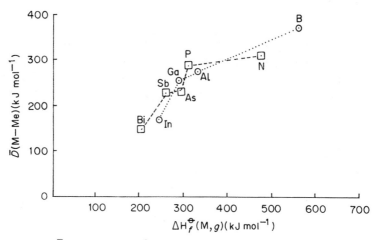

Fig. 1.4 Plots of $\bar{D}(M\text{—Me})$ vs $\Delta H_f^{\ominus}(M,g)$ for methyls Me_3M.

(e.g. Me_4Si, Me_3B) with respect to decomposition to their constituent elements, others, notably those of the B-elements of the third Long Period (viz. Me_2Hg, Me_3Tl, Me_4Pb) are unstable to such decomposition, i.e. they are endothermic compounds.

All such endothermic compounds and many more are thermodynamically unstable to reactions such as

$$R_nM \longrightarrow \text{Hydrocarbons} + \text{metal}$$

Table 1.3. Enthalpies of formation of some organometallic compounds

Compound*	$\Delta H_f^{\ominus}$ (kJ mol^{-1})	Compound	$\Delta H_f^{\ominus}$ (kJ mol^{-1})
EtLi (c)	-59	Me_3P (g)	-101
MeMgI (ether)	-288	Me_3As (g)	13
Me_2Zn (g)	50	Me_3Sb (g)	32
Me_2Cd (g)	106	Me_3Bi (g)	194
Me_2Hg (g)	94	Ph_3P (c)	218
Me_3B (g)	-123	Ph_3Bi (c)	601
Me_3Al (g)	-81	Cp_2TiMe_2 (c)	54
Me_3Ga (g)	-42	Cp_2ZrMe_2 (c)	-44
Me_3In (g)	173	Cp_2MoMe_2 (c)	283
Me_4Si (g)	-245	Cp_2WMe_2 (c)	285
Me_4Ge (g)	-71	Cp_3Y (c)	-45
Me_4Sn (g)	-19	Cp_3La (c)	36
Me_4Pb (g)	136	$(C_6Me_6)Cr(CO)_3$ (c)	-671

See also pp. 164, 280 for transition metal complexes.
*(g) denotes gaseous state; (c) denotes crystalline state.
The enthalpies quoted are generally ca. ± 5 kJ mol^{-1}.

An example is

$$Me_4Pb(g) = Pb(s) + 2C_2H_6(g); \quad \Delta H^{\ominus} = -306 \, kJ \, mol^{-1}$$

Why is it, then, that many of these compounds can be isolated, and in some cases show considerable resistance to thermal decomposition? The reason for this is that in addition to a favourable free energy change for the decomposition, a reaction path of sufficiently low activation energy must also be available for it to occur at a measurable *rate*. In other words, the decomposition of isolable yet thermodynamically unstable compounds may be *kinetically* controlled. In order for decomposition to proceed, the activation energy, or energies for a multi-stage process (strictly free energy of activation) must be sufficiently low. Low activation energies imply a low energy pathway for the rate controlling step in the decomposition. Such a step may involve breaking of the M—C bond to form M$\cdot$ and R$\cdot$ radicals (homolytic dissociation). This occurs, for example, in the thermolysis of tetramethyllead (p. 98). In such cases the activation energy will depend qualitatively on the strength of the metal–carbon bond to an extent related to the degree of bond-breaking in the transition state. For example in the Group IVB methyls, as the strengths of the M—C bonds decrease down the Group, thermal decomposition will become increasingly favoured on both kinetic and thermodynamic grounds.

Another common pathway for decomposition involves β-hydrogen transfer (p. 217).

It is potentially reversible, but if the metal hydride is unstable to further decomposition, the process becomes irreversible. If the metal alkyl does not possess a β-hydrogen atom this pathway is suppressed. $Ti(CH_2Ph)_4$, for example, is stable at room temperature, whereas $Ti(CH_2CH_3)_4$ has never been isolated. Other mechanisms such as α-hydrogen transfer, however, may then take over (p. 220). Both α- and β-hydrogen transfer are aided by the presence of low-lying empty orbitals on the metal atom.

1.5.2 Stability to oxidation

All organometallic compounds are thermodynamically unstable to oxidation, the driving force being provided by the large negative free energies of formation of metal oxide, carbon dioxide and water. Many are also kinetically unstable to oxidation at or below room temperature. Nearly all the methyls of the main group elements are rapidly attacked, although Me_2Hg and the derivatives of the Group IVB elements are inert. Many (e.g. Me_2Zn, Me_3In, Me_3Sb) are spontaneously inflammable in air. Kinetic instability to oxidation may be associated with the presence of empty low-lying orbitals, e.g. $5p$ in Me_3In, or of a non-bonding pair of electrons e.g. Me_3Sb. In contrast, the Group IVB alkyls possess neither of these features and behave as saturated compounds.

Most transition metal derivatives too are sensitive to oxygen and it is customary and often obligatory to handle them under an inert atmosphere of nitrogen or argon.

1.5.3 *Stability to hydrolysis*

Hydrolysis of an organometallic compound often involves nucleophilic attack by water and hence is facilitated by the presence of empty low-lying orbitals on the metal atom. In agreement with this, the organic derivatives of the elements of Groups IA and IIA and of Zn, Cd, Al, Ga and In are readily hydrolysed. The *rate* of hydrolysis is dependent on the polarity of the M—C bond; where this is high (e.g. Me_3Al), rapid attack by water occurs, whereas Me_3B is unaffected by water at room temperature, even though an empty $2p$ orbital is present on the boron atom. The alkyls and aryls of Group IVB and VB elements, however, are kinetically inert to hydrolysis by water. In these compounds the metal atom is surrounded by a filled shell of eight electrons, so that nucleophilic attack is no longer favoured. The majority of the neutral organic derivatives of transition elements are inert to hydrolysis. Organo-lanthanides, however, are extremely susceptible, on account of the polar character of the bonding, the large size of the central atom and the presence of many low lying empty orbitals, all of which aid coordination of a nucleophile.

1.5.4 *General features relating to stability; filled shells of electrons*

It is well known that much of carbon chemistry is controlled by kinetic factors; for example diamond would change spontaneously into graphite and ethyne into benzene at room temperature and atmospheric pressure if thermodynamics were controlling. In addition, all organic compounds are thermodynamically unstable to oxidation and exist in the presence of air only because no suitable low energy oxidation mechanism is available.

This 'kinetic stability' of carbon compounds has a variety of causes, notably the *full* use of the four valence orbitals (sp^3) in carbon, leading to the common maximum coordination number of four (exceptions e.g. Me_4Li_4, $(Me_3Al)_2$ in which the coordination number rises to 5–7 are discussed in Chapter 3) and the high energy of empty antibonding orbitals into which electrons could be donated in the case of nucleophilic attack.

Expansion of the coordination number above four commonly occurs in compounds of silicon and of other main group elements of the 2nd and later periods, when strongly electronegative groups such as halogen atoms are attached to the metal atom (e.g. SiF_6^{2-}). The kinetic lability of $SiCl_4$ to hydrolysis, in contrast to CCl_4, may be associated with the presence of empty relatively low lying ($3d$?) orbitals in the former which can accept electrons from water in nucleophilic attack. Where such electron-attracting groups are absent, however, as in Me_4Si, there is generally no tendency for the covalency of the metal atom to expand above four, so that Me_4Si is *kinetically* inert to thermal decomposition, oxidation and hydrolysis at room temperature.

Thus kinetic stability of organometallic compounds may be associated with a closed shell of electrons, often of essentially spherical symmetry, around the metal atom.

For compounds of the transition elements, however, empty valence shell ns, np or (n-1)d orbitals are often available and this can markedly decrease their kinetic stability. This accounts for the ready thermal decomposition of many binary alkyls by α- or β-hydrogen transfer referred to above. If all the valence orbitals are fully used and occupied a closed shell results, so that such facile hydrogen transfer is inhibited. Generally a closed shell for transition elements consists of 18 electrons i.e. ns^2, np^6, $(n-1)d^{10}$. These additional electrons can be supplied by 'spectator' ligands such as cyclopentadienyl, which themselves are not readily displaced. Whereas WMe_6 decomposes below room temperature, sometimes explosively, Cp_2WMe_2 can be sublimed without decomposition at $120°C/10^{-3}$ mm. In the former the tungsten atom has a configuration of only 12 electrons, whereas in the latter the closed shell of 18 electrons is attained. Transition metal atoms are, like atoms of Group IA and IIA elements, inherently electron deficient, and this electron deficiency must be satisfied if thermally stable organometallic compounds are to be formed. The 18-electron rule and related ideas are discussed further in Chapters 5 and 6.

1.6 References

On account of limited space it is possible to list only a few relevant books and review articles at the end of each chapter. Most of those chosen have appeared since about 1979. Readers who require detailed access to the literature, including original research papers are referred to 'Comprehensive Organometallic Chemistry'. This contains an exhaustive but quite readable treatment of the subject up to and including the year 1980. Some of the texts listed below also include fairly extensive bibliographies. In particular, the hardback edition of Haiduc and Zuckerman's book contains 111 pages of references to the secondary literature up to 1984 (books and reviews only).

Texts on organometallic chemistry

Atwood, J.D. (1985) *Mechanisms of Inorganic and Organometallic Reactions*, Brooks/Cole, CA.

General survey

Collman, J.P., Hegedus, L.S., Norton, J.R. and Finke, G. (1987) *Principles and Applications of Organotransition Metal Chemistry*, 2nd edn, University Science Books, Mill Valley, CA.

Davies, S.G. (1982) *Organotransition Metal Chemistry: Applications to Organic Synthesis*, Pergamon Press, Oxford.

Haiduc, I. and Zuckerman, J.J. (1985) *Basic Organometallic Chemistry*, de Gruyter, Berlin.

Lukehart, C.M. (1985) *Fundamental Transition Metal Organometallic Chemistry*, Brooks/Cole, CA.

Parkins, A.W. and Poller, R.C. (1986) *An Introduction to Organometallic Chemistry*, Macmillan, London.

Pearson, A.J. (1985) *Metallo-organic Chemistry*, Wiley, Chichester.

Previous general texts on organometallic chemistry

These books still convey something of the excitement which was aroused as organometallic chemistry grew rapidly during the period 1950–1970.

Coates, G.E. (1960) *Organometallic Compounds*, 2nd edn, Methuen, London.

Coates, G.E., Green, M.L.H. and Wade, K. (1967) *Organometallic Compounds*, (*The main Group Elements*, Vol. I), (1968). (*The Transition Elements*, Vol. II), 3rd edn, Methuen, London.

Fischer, E.O. and Werner, H. (1966) *Metal-Complexes*, (*Complexes with Di- and Oligo-Olefinic Ligands*, Vol. I), Elsevier, Amsterdam.

Herberhold, M. (1972) *Metal-Complexes*, (*Complexes with Mono-Olefinic Ligands*, Vol. II), Elsevier, Amsterdam.

Preparation, properties and reactions of organometallic compounds

Hartley, F.R. and Patai, S. (eds) *The Chemistry of the Metal—Carbon Bond*, (1982) Vol. I, (1984) Vol. II, (1985) Vol. III, Wiley/Interscience, Chichester.

Wilkinson, G., Stone, F.G.A. and Abel, E.W. (eds) (1982) *Comprehensive Organometallic Chemistry*, Vols. 1–9, Pergamon Press, Oxford. Detailed coverage up to about 1980, arranged element by element. An invaluable reference work.

Other useful texts

Carruthers, W. (1987) *Some Modern Methods of Organic Synthesis*, 3rd edn, Cambridge University Press, Cambridge.

Greenwood, N.N. and Earnshaw, A. (1984) *Chemistry of the Elements*, Pergamon Press, Oxford. An excellent book which provides suitable background reading. It includes numerous references to the chemical literature.

Shriver, D.F. (1969) *The Manipulation of Air-Sensitive Compounds*, McGraw-Hill, New York.

Thayer, J.S. (1975) Organometallic chemistry; an historical perspective. *Adv. Organomet. Chem.*, **13**, 1.

2

Methods of formation of metal—carbon bonds of the main group elements

2.1 The reaction between a metal and an organic halogen compound

$$2M + nRX = R_nM + MX_n \text{ (or } R_xMX_y; x + y = n)$$

The reaction between a metal and an organic halogen compound provides a fairly general route to organometallic compounds. One of the earliest organometallic compounds ever to be prepared, ethylzinc iodide, was made in this way from zinc/copper couple and ethyl iodide.

2.1.1 *Thermochemical considerations*

It is useful to examine the scope of the reactions between metals and organic halides in the light of their thermochemistry. Some typical values of the heats of reaction between methyl halides and main group elements are given below (Figs. 2.1, 2.2). Within one series of reactions involving compounds of the same group of the Periodic Table where the stoichiometry is the same for each element, the neglect of the entropy term should not, to a first approximation, seriously alter the gross pattern of changes within the Group.

These data show that, apart from the elements C, N and O and the B-elements of the last row of the Periodic Table (Hg, Tl, Pb and Bi), the reaction between an element and methyl chloride to yield the chloride and methyl of the element is exothermic, and therefore might be expected to provide a method of synthesizing the methyls. In nearly all the examples listed where a negative value of ΔH is found, such reactions have been observed and many are important synthetic routes. The mechanisms of the reactions and the conditions under which they occur, however, vary greatly with different elements. Kinetic as well as thermodynamic factors are thus significant in controlling the reactions.

Figure 2.1 shows that the main driving force for these reactions is, in general, the strongly exothermic formation of the metal halide. This is particularly striking for the Group IIB elements, Zn, Cd and Hg, where the methyls are in each case

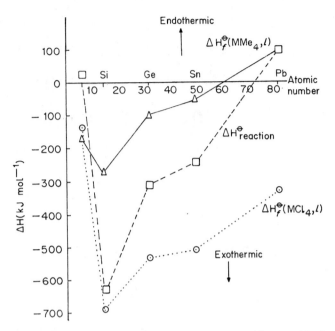

Fig. 2.1 Thermochemistry of reactions $2M + 4CH_3Cl_{(g)} \rightarrow MCl_{4(l)} + M(CH_3)_{4(l)}$. (In fact mixed methyl chlorides Me_nMCl_{4-n} are the thermodynamically favoured products.)

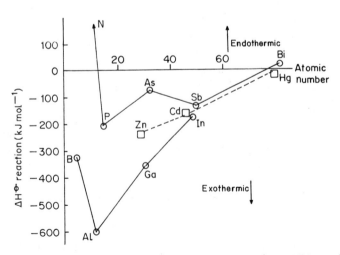

Fig. 2.2 Thermochemistry of reactions $2M + nCH_3Cl_{(g)} \rightarrow MCl_{n(c)} + MMe_{n(l)}$. (a) In fact mixed methyl chlorides Me_mMCl_{n-m} are the thermodynamically favoured products. (b) For B, N, P, As, points refer to $MCl_{n(l)}$.

endothermic compounds. Although Cd and Hg react directly with methyl iodide only in the presence of light, which initiates homolytic dissociation of the organic halide into ·CH$_3$ and I·, zinc as zinc/copper couple reacts readily to yield methylzinc iodide.

In the B-elements the strength of the M—C bonds falls sharply on going from the elements of the second long Period (e.g. Sn, Sb) to those of the third long Period (e.g. Pb, Bi). The alkyls of Hg, Tl and Bi are all strongly endothermic compounds and their unfavourable heats of formation are not outweighed by the formation of the exothermic halides when the elements are treated with alkyl halides.

Calculated heats of reaction of methyl halides with metals suggest that CH$_3$F should react more exothermically than the other halides. In practice, however, methyl fluoride reacts by far the most reluctantly. Although the detailed mechanism of attack of a methyl halide on a metal may well vary with the metal concerned, the activation energy will almost certainly be related to the strength of the carbon—halogen bond which is breaking. The C—Hal bond energies fall from C—F (ca. 440), C—Cl (ca. 330), C—Br (ca. 276) to C—I (ca. 238 kJ mol^{-1}), whereas the relative reactivities (rates) of organic halides to metals usually rise: RF ≪ RCl < RBr < RI. This is another example of reactions which are thermodynamically very favoured, but which are controlled by kinetic factors.

2.1.2 *Applications*

The reaction between metals and organic halides is often suitable for the synthesis of organometallic compounds of the most electropositive elements. On a laboratory scale, the derivatives of Li, Mg and Al, which are very important in synthetic work, are usually prepared in this way. Some illustrations of the method are given briefly here; more detailed discussion of the preparation and properties of the compounds involved appears in Chapter 3.

Organolithium compounds may be prepared by the action of lithium metal on alkyl or aryl halides, in ether or hydrocarbon solvents,

$$2Li + RX \longrightarrow RLi + LiX$$

Chlorides are preferred to bromides in the alkyl series, and with the important exception of methyl iodide, iodides cannot be used since they react too rapidly with the organolithium compound,

$$RLi + RI \longrightarrow R—R + LiI \quad \text{(Wurtz Coupling)}.$$

This coupling is less serious in the aryl series, so bromides or iodides can often be used successfully.

Magnesium reacts with alkyl and aryl halides in ethers to form Grignard reagents:

$$RX + Mg \longrightarrow RMgX$$

For R = alkyl, chlorides are used whenever possible. They are cheaper than the bromides and iodides and although they react more slowly with magnesium, side reactions such as Wurtz coupling or olefin elimination are not so serious. Yields of Grignard reagents from primary alkyl chlorides or bromides are usually better than 80%. Secondary and tertiary halides, especially bromides and iodides, are not so satisfactory. With *tert*-butyl bromide, for example, much butene is evolved by elimination and only low yields of *tert*-butylmagnesium bromide are obtained. On the basis of many unsuccessful attempts, it was once thought that Grignard reagents could not be prepared from alkyl fluorides. In 1971 it was found that in the presence of a small quantity of iodine (ca. 4%) in tetrahydrofuran, good yields of RMgF (R = e.g. Me, C_6H_{13}) could sometimes be obtained. Tetrahydrofuran coordinates to magnesium better than the usual solvent, diethyl ether. Aryl fluorides, however, still do not react under these conditions.

A useful method for producing very finely divided, reactive metals has been devised by Rieke. An anhydrous metal halide is reduced by potassium in boiling tetrahydrofuran. The black slurry of magnesium thus prepared gives good yields of Grignard reagents with alkyl- and even with aryl fluorides. Zinc and cadmium alkyls can also be made in this way from alkyl bromides and iodides. Active metals can also be made by the 'metal atom' technique, of which a fuller description is given on p. 312. In one such experiment magnesium was heated under high vacuum and the vapour cocondensed with ethyl bromide at −196°C. On warming to room temperature reaction occurred to give unsolvated EtMgBr. In organometallic chemistry the method has most usefully been applied to the synthesis of endothermic complexes of transition elements, some of which are not accessible by other routes.

The formation of organometallic compounds from bulk metals and organic halides can sometimes be aided by ultrasonic irradiation. The human ear is sensitive to frequencies between 16 Hz and 16 kHz. Beyond 16 kHz lies the region of ultrasound. The upper practical limit is about 1 MHz, 20–50 kHz being most commonly used. Ultrasonic radiation probably acts by scrubbing the surface of the metal, so that a clean metal-solvent interface is maintained. Zinc normally requires activation as a zinc-copper couple, for example, before it will enter into reaction with organic halides. The Reformatsky reaction between ethyl bromoacetate and aldehydes or ketones in the presence of activated zinc (p. 59) normally requires heating to 80°C for several hours. Under sonic irradiation it is essentially complete within 5–30 minutes at room temperature. Organozinc reagents can be prepared simply by irradiating haloalkanes, lithium and zinc bromide for ten minutes in a mixture of toluene and tetrahydrofuran. Similar acceleration has been observed in the formation of organolithium, magnesium and copper compounds.

Aluminium reacts with a limited number or aryl and alkyl halides, forming $R_3Al_2X_3$. These reactions are considered in Chapter 3.

When volatile organic halides are passed over silicon, mixed with a metal such as copper and heated at 250–400°C, organosilicon compounds, notably R_2SiCl_2, are produced.

$$Si + 2RCl = R_2SiCl_2 \quad (R = \text{alkyl or aryl})$$

The chlorides are almost always used; methyl fluoride and methyl iodide do not react according to the equation given above. This reaction is often called the 'Direct' synthesis and, with its applications, is discussed in Chapter 4. It is not restricted to silicon chemistry, however, although it is in this field that its main importance lies. Similar reactions occur between alkyl halides and metals and metalloids such as Ge, Sn, P, As and Sb. As long ago as 1870 Cahours observed that arsenic reacts with methyl iodide in a sealed tube,

$$4MeI + 2As = Me_4As^+I^- + AsI_3$$

More recently it has been shown that the following reaction occurs when methyl bromide is passed over a heated mixture of arsenic and copper

$$3MeBr + 2As = MeAsBr_2 + Me_2AsBr$$

2.1.3 Alloy methods

The alkyls and aryls of the heaviest elements (e.g. Hg, Tl, Pb, Bi) are endothermic compounds. Even the reactions between the element and organic halide may be endothermic, in spite of a high heat of formation of the metal halide tending to drive the reaction to the right. In these cases the direct synthesis can still be used by taking an alloy of the metal with an alkali metal such as sodium, instead of the free metal. Here the formation of the strongly exothermic sodium halide provides the driving force. Alkyls and aryls of Hg, Tl, Sn, Pb and Bi can be made in this way.

$$2Hg(l) + 2MeBr(l) = Me_2Hg(l) + HgBr_2(s) \quad \Delta H^\ominus = -1 \text{ kJ mol}^{-1}$$

$$HgBr_2(s) + 2Na(s) = Hg(l) + 2NaBr(s) \quad \Delta H^\ominus = -551 \text{ kJ mol}^{-1}$$

$$Hg(l) + 2MeBr(l) + 2Na(s) = Me_2Hg(l) + 2NaBr(s) \quad \Delta H^\ominus = -552 \text{ kJ mol}^{-1}$$

The alloy method is used in the manufacture of tetramethyl- and tetraethyllead, antiknock additives to petrol (p. 100). Similarly bismuth aryls were first made by Michaelis in 1877 by the reaction of powdered bismuth–sodium alloy with aryl halides,

$$3ArX + Na_3Bi = Ar_3Bi + 3NaX$$

2.2 Metal exchange: the reaction between a metal and an organometallic compound of another metal

$$M + RM' = M' + RM$$

This method depends on the difference in the free energies of formation of the two species RM' and RM. Owing to the lack of $\Delta G^\ominus$ values, enthalpies of formation are used to indicate the thermodynamic feasibility of such reactions. Endothermic or weakly exothermic organometallic compounds should be the most versatile reagents RM′ in such reactions. Such compounds are found among the heavy B elements (Hg, Tl, Pb and Bi). Of these the air-stable and readily prepared (though toxic) organomercury compounds are the most widely used e.g.

(i) $\qquad Zn(s) + Me_2Hg(l) = Me_2Zn(l) + Hg(l) \quad \Delta H^{\ominus} = -39\,kJ\,mol^{-1}$

(ii) $\qquad Cd(s) + Me_2Hg(l) = Me_2Cd(l) + Hg(l) \quad \Delta H^{\ominus} = +8\,kJ\,mol^{-1}$

This shows that the reaction with zinc goes essentially to completion whilst an equilibrium mixture is observed in the cadmium system. It is therefore inadvisable to handle volatile organocadmium compounds in vacuum apparatus containing mercury, as the reverse of reaction (ii) can occur.

Other elements which react satisfactorily with dialkyl and diaryl mercury compounds include the alkali metals, the alkaline earths, Al, Ga, Sn, Pb, Sb, Bi, Se and Te (see Chapters 3, 4). In some cases the formation of a metal amalgam assists the reaction. The transition elements, when they react, give metallic mercury and a mixture of hydrocarbons. An unstable organotransition metal species may be formed initially, only to decompose. The reactions,

$$P(white) + Ph_3Bi(c) = Bi(s) + Ph_3P(c) \quad \Delta H^{\ominus} = -272\,kJ\,mol^{-1}$$

$$As(s) + Ph_3Bi(c) = Bi(s) + Ph_3As(c) \quad \Delta H^{\ominus} = -180\,kJ\,mol^{-1}$$

$$Sb(s) + Ph_3Bi(c) = Bi(s) + Ph_3Sb(c) \quad \Delta H^{\ominus} = -161\,kJ\,mol^{-1}$$

are similar. The enthalpy (or more strictly, free energy) of formation of the organoelement compound is the controlling factor. The order of displacement of one element by another in such reactions has sometimes been related to the electrochemical series. An element high in this series will displace another element from its organometallic derivatives just as from its salts. Such a rule does hold for the metallic elements but it is based upon an artificial comparison. As the above examples show, however, where non-metals or metalloids are concerned, it is better and more meaningful to consider $\Delta H_f^{\ominus}$ rather than any position in the electrochemical series.

The transfer of alkyl or aryl groups from one element to another may not necessarily be complete. Thus the aryls of Groups IVB and VB react with alkali metals in ether solvents or in liquid ammonia,

$$Ph_4Sn + 2Na \xrightarrow[NH_3]{liq.} Ph_3SnNa + NaPh \text{ (attacks solvent)}$$

$$Ph_3P + 2Na \xrightarrow{THF} Ph_2PNa + NaPh \text{ (attacks solvent)}$$

2.3 Reactions of organometallic compounds with metal halides

$$RM + M'X = RM' + MX$$

These provide another general method for the preparation of organo–element

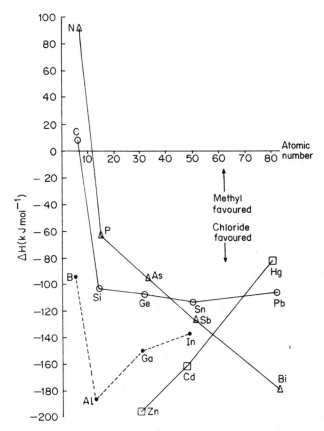

Fig. 2.3 Plot of $1/n \{\Delta H_f^{\ominus}(\mathrm{MCl}_n(\mathrm{SS}) - \Delta H_f^{\ominus}(\mathrm{MMe}_{n2})l)\}$ against atomic number. (SS = standard state at 298.15°C/1 atm.)

compounds; they are probably the most widely used and versatile of all laboratory methods. In order to understand the direction in which any such reaction will go, consider the thermochemical data given diagrammatically in Fig. 2.3. Here the difference between the enthalpies of formation of the chloride and methyl of a given element is plotted against atomic number. For the elements low on the diagram (e.g. Zn, Al) the chloride is especially favoured with respect to the methyl. On descending Group VB (N to Bi) the trend is towards increasing relative stability of the chloride. This is also the general trend of increasing electropositive character of the elements. There is a good correlation between $\Delta H_f^{\ominus}(\mathrm{MCl}_n)$ $-\Delta H_f^{\ominus}(\mathrm{MMe}_n)$ and the Pauling electronegativity values of M. Starting from Pauling's definition, which is based upon bond energies, it can be shown that these two quantities are directly proportional. In these exchange reactions, therefore, the effect is to combine the halogen with the more electropositive

21

element and the organic radical with the more electronegative element of the pair.

By far the most widely used reagents in such syntheses are the organic derivatives of Li and Mg. These react rapidly and exothermically with halides of more electronegative elements. e.g.

$$2RMgX + CdCl_2 = R_2Cd + 2MgXCl \quad \text{(see p. 60)}$$

Organolithium compounds are more reactive than Grignard reagents and in certain cases allow more complete substitution. For example, Grignard reagents substitute only two of the chlorine atoms in thallium(III) chloride when diethyl ether is the solvent, whereas lithium alkyls substitute all three:

$$Et_3Tl \xleftarrow[\text{ether}]{\text{EtLi}} TlCl_3 \xrightarrow[\text{ether}]{\text{EtMgCl}} Et_2TlCl$$

Similarly $(Et_3P)_2PtBr_2$ yields the dimethyl derivative $(Et_3P)_2PtMe_2$ with methyl lithium, whereas the Grignard reagent gives only the monomethyl complex $(Et_3P)_2PtMeBr$. The preparation of alkyl and aryl complexes of transition elements by this general method is further discussed in Chapter 7.

Another factor which influences the choice between organolithium and Grignard reagents is the hydrolytic stability of the product. Products from a Grignard reaction are most conveniently isolated by hydrolysis of the reaction mixture with dilute hydrochloric acid, saturated aqueous ammonium chloride or water to destroy any excess Grignard reagent and to dissolve magnesium salts, followed by separation of the organic layer, drying and removal of the solvent. Omission of the hydrolysis step leads to a rather intractable mixture of the desired product with sticky ether complexes of magnesium halides. The products are especially difficult to separate when bromides and iodides are present. Lithium alkyls, on the other hand, can be prepared in petroleum ether in which lithium halides are quite insoluble. Water-sensitive products can be isolated by filtration from the lithium halide and evaporation of the filtrate in a dry atmosphere.

2.4 Addition of metal hydrides to alkenes and alkynes

These reactions may be represented by the general equation:

$$M-H + \underset{/}{\overset{\backslash}{C}}=\underset{\backslash}{\overset{/}{C}} \rightleftharpoons M-\overset{|}{\underset{|}{C}}-\overset{|}{\underset{|}{C}}-H$$

The most important examples concern B, Al and Si. (Additions of non-metal hydrides N—H, P—H, O—H, S—H and H—Hal are not included here.) These reactions are not as simple as the equation suggests and several different mechanisms can operate, even in the reactions of hydrides of one element such as silicon.

Addition of metal hydrides to alkenes and alkynes

The addition of boron hydrides to alkenes and alkynes is known as hydroboration. It provides a simple laboratory route to boron alkyls and thence to a wide range of organic compounds (p. 66). It takes place under mild conditions at room temperature in an ether solvent.

The corresponding reaction of aluminium hydrides is important in industry. Although aluminium metal does not form AlH_3 by direct reaction with hydrogen, it does react in the presence of aluminium alkyl. Tri-isobutylaluminium, for example, is normally made in a single-stage process in which the metal reacts with liquid 2-methylpropene (isobutene) and hydrogen at $80-110°C/200\,atm$, in the presence of some preformed Bu^i_3Al. It loses isobutene when heated at $140°C/20\,mm$ but this reaction can be reversed by warming Bu^i_2AlH to $60-70°C$ with isobutene under mild pressure.

$$Bu^i_3Al \rightleftharpoons Bu^i_2AlH + Me_2C{=}CH_2$$

If propene and Pr^n_3Al are heated at $140-200°C$ under pressure, an insertion of the alkene into Al—C bonds occurs. This is followed by elimination of the dimer of propene, 2-methylpentene.

$$Pr^n_2Al{-}Pr^n + MeCH{=}CH_2 \rightleftharpoons Pr^n_2Al{-}CH_2CH{\overset{Me}{\underset{Pr^n}{\diagdown}}} \rightleftharpoons Pr^n_2Al{-}H + H_2C{=}C{\overset{Me}{\underset{Pr^n}{\diagdown}}}$$

With ethene, further chain growth occurs, leading to oligomers (p. 80).

Gallium hydrides also add to alkenes. For example, Et_2GaH adds 1-decene giving $Et_2GaC_{10}H_{21}$, and the dimer $(HGaCl_2)_2$ adds to terminal olefins,

$$(HGaCl_2)_2 + RCH{=}CH_2 \xrightarrow[\text{room temp.}]{\text{ether}} (RCH_2CH_2GaCl_2)_2$$

Hydrozirconation (p. 290) is related to these reactions. It is useful in organic synthesis.

2.4.1 Addition of Group IVB hydrides

The ease of addition of the hydrides (e.g. R_3MH, M = Si, Ge, Sn, Pb) to alkenes increases down Group IVB as the M—H bonds decrease in strength. Thus trialkylsilanes require heating to about $300°C$ under olefin pressures of at least $300\,atm$, UV irradiation, or catalysis by peroxides, certain metal salts or tertiary amines. Trialkylgermanes and trialkylstannanes add to activated terminal carbon–carbon double bonds at $120°C$ and $90°C$ respectively and a catalyst is not necessary, e.g.

$$Bu_3GeH + H_2C{=}CHCN \longrightarrow Bu_3GeCH_2CH_2CN$$

Tri-n-butyllead hydride is so reactive that it adds rapidly to alkenes and alkynes in ether solution without a catalyst even at $0°C$, e.g.

$$Bu_3PbH + PhC{\equiv}CH \longrightarrow trans\text{-}PhCH{=}CHPbBu_3$$

23

The main group elements

(a) HYDROSILATION. The addition of Si—H bonds to alkenes and alkynes is used industrially in the production of intermediates for silicone manufacture (p. 115).

$$X_3SiH + H_2C{=}CHY \longrightarrow X_3SiCH_2CH_2Y$$
$$X_3SiH + HC{\equiv}CH \longrightarrow X_3SiCH{=}CH_2 \longrightarrow X_3SiCH_2CH_2SiX_3$$

Addition may occur by thermal reaction under pressure, but proceeds better under UV irradiation or when catalysed. The thermal, photolytic or peroxide catalysed reactions involve free-radical chain mechanisms. Radicals are first produced from the silicon hydride,

$$X_3SiH \xrightarrow{hr} X_3Si^{\cdot} + H^{\cdot}$$

which then add to the alkene by various pathways,

$$Cl_3Si^{\cdot} + H_2C{=}CH_2 \longrightarrow Cl_3SiCH_2\dot{C}H_2 \longrightarrow Cl_3SiCH_2CH_2CH_2\dot{C}H_2$$

$$\nwarrow^{HSiCl_3} \qquad \searrow^{R\cdot} \qquad \text{Chain propagation}$$

$$Cl_3SiCH_2CH_3 + {}^{\cdot}SiCl_3 \qquad Cl_3SiCH_2CH_2R$$
$$\text{Chain transfer} \qquad \text{Chain termination}$$

The chain propagation step is often not very important as the silyl radicals are very efficient chain transfer agents.

   Catalysis by transition metal complexes such as H_2PtCl_6 is thought to involve an alkene complex of Pt and intermediates with covalent Pt—Si bonds. Addition of transition metal hydrides to olefins and the role of such reactions in catalytic processes are discussed in Chapter 12.

(b) HYDROSTANNATION. Addition of organotin hydrides to carbon–carbon double and triple bonds provides a route to intermediates which are valuable in organic synthesis (p. 228). When functional groups (E) are present, the Grignard method is not applicable. Addition is accelerated by free radical initiators such as azoisobutyronitrile, or by ultraviolet irradiation.

$$R_3SnH + H_2C{=}CHE \xrightarrow{radical} R_3SnCH_2CH_2E$$

$$(E = CONH_2, CH(OEt)_2, CH_2OH, CN)$$

Homolytic addition to substituted alkynes is mostly regiospecific. Formation of the E-isomer is favoured, although an E/Z mixture is often produced.

$$HC{\equiv}CCO_2R + R'_3SnH \xrightarrow{radical} R'_3Sn\diagup\hspace{-0.3em}\diagdown CO_2R + R'_3Sn\diagup\hspace{-0.3em}\diagup^{CO_2R}$$

Where a strongly polar substituent is present in the unsaturated compound, addition by a polar mechanism can occur.

2.5 Formation of metal–carbon bonds by other insertion reactions

The addition of metal hydrides to carbon–carbon double or triple bonds is a special case of a very general reaction type, which involves the addition of a species A—B to an unsaturated system X=Y (or X≡Y).

$$A—B + X=Y \longrightarrow A—X—Y—B$$

(A = Metal; B = $—C\lessgtr$, $—N\lessgtr$, $—P\lessgtr$, $—O—$, $—Hal$, $—Metal$, etc.). Formally similar additions of these bonds to species X=Y, where Y≠C (e.g. to C=O, C=N, etc.) are described briefly elsewhere (p. 45) as they do not lead to the formation of new M—C bonds.

Polar metal–carbon bonds, such as are found in the organometallic compounds of the most electropositive elements, can add to alkenes and alkynes, especially when the latter bear electron-withdrawing substituents, which either increase the polarity of the C=C or C≡C bond or can stabilize a negative charge in the transition state. Thus butyllithium adds to diphenylethyne in diethyl ether, though not in pentane, and *trans*-butylstilbene may be isolated after hydrolysis, showing that *cis*-addition occurs:

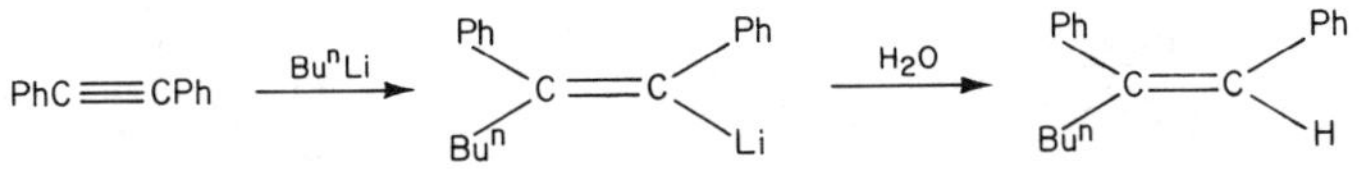

The polymerization of 1,3-butadiene is catalysed by butyllithium (p. 48). Oligomerization of ethene is effected by trialkylaluminiums (p. 80). Grignard reagents, however, do not normally add to alkenes or alkynes.

The addition of M—N, M—P, M—O and M—M bonds to alkenes and alkynes sometimes occurs, especially when these bonds are weak, as with the heavier elements such as Sn and Pb. Such reactions are favoured when the alkene or alkyne bears electron-attracting substituents such as —COOMe or —CN.

$$Et_3Sn—OMe + H_2C=C=O \longrightarrow Et_3SnCH_2CO(OMe)$$

$$Me_3Sn—SnMe_3 + C_2F_4 \longrightarrow Me_3SnCF_2CF_2SnMe_3$$

2.6 Reactions of diazo compounds

The use of diazo compounds in the formation of metal–carbon bonds is conveniently divided into two sections—the reactions of (a) aliphatic and (b) aromatic diazo compounds.

2.6.1 Aliphatic diazo compounds

Diazomethane and substituted diazomethanes (e.g. ethyl diazoacetate) react with many metal halides or metal hydrides under mild conditions by methylene

The main group elements

insertion:

$$SiCl_4 + CH_2N_2 \xrightarrow[\text{ether}]{-50°C} Cl_3SiCH_2Cl + N_2$$

$$HgCl_2 + CH_2N_2 \xrightarrow[-N_2]{\text{ether}} ClCH_2HgCl \xrightarrow[-N_2]{CH_2N_2} ClCH_2HgCH_2Cl$$

$$R_3SnH + N_2CHCOOEt \longrightarrow R_3SnCH_2COOEt + N_2$$

Replacement of all the M—Cl groups is often difficult or impossible.

Where the reactant is a strong electron-pair acceptor, the formation of polymethylene is a serious complication. In these cases the initial step involves coordination of the CH_2 group of diazomethane with the acceptor orbital e.g.

$$X_3B + H_2\bar{C}-N{\equiv}\overset{+}{N} \longrightarrow X_3\bar{B}-CH_2-\overset{+}{N}{\equiv}N$$

This is followed in fast subsequent steps by elimination of nitrogen and polymerization. Aluminium compounds usually behave similarly. At low temperatures, however, dialkylaluminium halides and diazomethane give, initially, dialkylhalomethylaluminium derivatives, which are stabilized as adducts with ether.

The methylene group can be transferred quantitatively, like a carbene, to cyclohexene. Norcarane, [4.1.0]-bicycloheptane, is formed.

2.6.2 Reaction of aromatic diazonium salts with metal and metalloid halides or oxides in aqueous solution

This method is applicable to a wide range of heavy metal and metalloid elements. Two conditions must be fulfilled for it to be applied successfully. First, as aqueous media are used, the product must not be susceptible to hydrolysis. This restricts its use to aryls of Hg, Tl and the Group IVB and VB elements. Secondly the element involved must be susceptible to electrophilic attack by the aryldiazonium ion ArN_2^+.

The best known and probably the most successful applications are the Bart reaction for the preparation of arylarsonic acids and the analogous Schmidt reaction in antimony chemistry,

$$PhN_2X + M(OH)_3 \xrightarrow[\text{aq. acetone}]{\text{alkaline}} PhMO(OH)_2 + N_2 + HX \quad (M = As, Sb)$$

Nesmeyanov has extended this reaction to the synthesis of aryl derivatives of the heavier B-elements, but the yields are often low. Often the double salts of the metal halide with the diazonium salt are first isolated and then decomposed by a metal such as copper or zinc, e.g.

$$ArNH_2 \xrightarrow[HCl]{NaNO_2} ArN_2{}^+Cl^- \xrightarrow{HgCl_2} ArN_2^+HgCl_3^- \xrightarrow{Cu/Cu^{2+}} ArHgCl$$

In this way an aromatic amine is converted regiospecifically into the corresponding mercurial. An alternative method of preparation, electrophilic mercuration of an arene, however, leads to a mixture of position isomers (p. 62).

2.7 Decarboxylation of heavy B-metal salts

The decarboxylation of the calcium and barium salts of aliphatic acids leads to ketones, e.g.

When certain salts of organic acids with heavy B-elements decarboxylate, however, metal–carbon bonds are formed. It is generally necessary for the organic acid to contain electron-withdrawing groups. Most examples come from Cu, Hg, Tl, Sn or Pb chemistry, e.g.

Contrast the effect of heat on mercury(II) benzoate, when substitution of the aromatic ring *ortho* to the carboxyl group occurs. Mercury(II) trichloroacetate decarboxylates so rapidly that it has not yet been isolated. Bis(trichloromethyl)mercury was obtained in good yield by the reaction of sodium trichloroacetate and mercuric halides in 1, 2-dimethoxyethane:

$$2Cl_3CCO_2Na + HgCl_2 \longrightarrow (Cl_3C)_2Hg + 2NaCl + 2CO_2$$

Similarly certain triphenyllead salts of organic acids with electron-withdrawing substituents decarboxylate on heating at about 160°C under reduced pressure, e.g.

$$Ph_3PbOCOCH_2COOEt \longrightarrow Ph_3PbCH_2CO_2Et + CO_2$$

A rather unusual reaction of this type is the decomposition of tributyltin formate, in which tributyltin hydride is produced:

$$Bu_3SnOCHO \xrightarrow[10\,mm]{170°C} Bu_3SnH + CO_2$$

Electrophilic mercuration and thallation of aromatic compounds, alkenes and alkynes are considered in Chapter 3.

2.8 References

Bremner, D. (1986) Chemical ultrasonics. *Chem. Brit.*, **22**, 633.

Deacon, G.B., Faulks, S.J. and Pain, G.N. (1986) Synthesis of organometallics by decarboxylation reactions. *Adv. Organomet. Chem.*, **25**, 237.

Eisch, J.J. (1981) *Organometallic Syntheses*, (*Non-transition metal Compounds, Experimental procedures* Vol. 2), Academic Press, New York.

Eisch, J.J. and Gilman, H. (1960) Preparation of organometallic compounds of the main group elements: a general review. *Adv. Inorg. Chem. Radiochem.*, 2.

Maslowsky, E. (1980) Synthesis, structure and vibrational spectra of organomethyl compounds. *Chem. Soc. Revs.*, **9**, 25.

Pilcher, G. and Skinner, H. (1982) in *The Chemistry of the Metal–Carbon Bond* (*Thermochemistry of organometallic compounds.* Vol. I, Ch. 2), (eds Hartley, F.R. and Patai, S.), Wiley/Interscience, Chichester.

Rieke, R.D. (1977) Preparation of active metals by reduction. *Acc. Chem. Res.*, **10**, 301.

Tel'noi, V.I. and Rabinovitch, I.B. (1980) Thermochemistry of organic derivatives of non-transition elements. *Russ. Chem. Revs.*, **49**, 603.

3 Organometallic compounds of elements of the first three periodic groups

3.1 Introduction

In this chapter the organic compounds of the first three Periodic Groups are discussed. The alkali metals, Groups IIA and IIB, boron, aluminium and Group IIIB are included. Scandium, yttrium and lanthanum are best considered with the lanthanides and are therefore covered in Chapter 13. Although copper, silver and gold are strictly transition elements, some of their organic chemistry in oxidation state $+1$ relates quite well to that of the main groups. It is therefore mentioned briefly here.

The heavy alkali metals K, Rb and Cs form essentially ionic derivatives. These are extremely reactive; caesium alkyls, for instance, are such strong bases that they can even deprotonate alkanes. Consequently their organic chemistry is less developed than that of the lighter elements, lithium and sodium. These small electropositive atoms enter into bridge bonding with organic groups; this property also extends in particular to beryllium and magnesium in Group II and to aluminium in Group III. Some copper and silver alkyls and aryls also possess this structural feature.

Organomagnesium compounds, especially the Grignard reagents, RMgHal, have played a key role in organic and organometallic synthesis for nearly 90 years. Of particular value is their ability to add to polar groups such as $>C=O$ and $-C\equiv N$. Organoberyllium compounds have not yet found any significant synthetic applications, largely because beryllium oxide, formed by their combustion, is toxic by inhalation as dust. They are generally less accessible, less stable and less reactive than the derivatives of magnesium and lithium. Their behaviour is less complicated by side-reactions than that of Grignard reagents. Even less is known of the organic chemistry of calcium, strontium and barium. Organozinc, cadmium and mercury compounds, however, have been known for over a century and are still quite widely used in synthesis.

The general reactivity of the organic compounds of Group II lies in the sequence $Ba > Sr > Ca > Mg > Be \sim Zn > Cd > Hg$. This order follows bond

29

Table 3.1. Electronegativity and reactivity of alkyls

Metal	Pauling electro- negativity	Alkenes	Ethers	RC≡N	$R_2C{=}O$	CO_2	RCOCl	H_2O	O_2
Ca	1.00	+	+[a]	+	+	+	+	+	+
Mg	1.31	−	−[b]	+	+	+	+	+	+
Be	1.57	−	−[b]	+	+	+	+	+	+
Zn	1.65	−	−[b]	+[c]	−[d]	−	+	+	+
Cd	1.69	−	−	−	−	−	−	+	+
Hg	2.00	−	−	−	−	−	−	−	−
Na	0.93	+	+	+	+	+	+	+	+
Mg	1.31	−	−[b]	+	+	+	+	+	+
Al	1.61	+	−[b]	+	+[e]	+	+	+	+
Si	1.90	−	−	−	−	−	−	−	−
Li	0.98	±	±[a]	+	+	+	+	+	+
Be	1.57	−	−[b]	+	−	+	+	+	+
B	2.04	−	−	−	−	−	−	−	+

[a]Rates vary greatly. [b]Does not cleave ethers; forms adducts. [c]R_2Zn does not react, RZnCl does. [d]Does not react by addition, reduction is the main reaction. [e]Complicated by reduction and enolization.

polarity as indicated by Pauling electronegativity values (Allred-Rochow electro-negativities do not fit very well). Similarly, reactivity decreases in passing from left to right across a given Period e.g. Li > Be > B and Na > Mg > Al > Si. These trends are summarized in Table 3.1, which shows the qualitative relationship between electronegativity (and hence M—C bond polarity) and the ability to react with some common functional groups.

The organic compounds of the elements of Group I to III possess vacant valence orbitals and hence are able to function as Lewis acids. They therefore form adducts with electron pair donors such as ethers, tertiary amines and phosphines. This coordination chemistry is discussed further on p. 82. In general the lower the electronegativity of the central atom, the stronger the Lewis acidity. If the Lewis base also has an active (acidic) hydrogen atom, the initially formed adduct may eliminate hydrocarbon e.g.

$$Me_2Zn + MeOH \longrightarrow \tfrac{1}{4}(MeZnOMe)_4 + CH_4$$

$$Me_2Zn + Ph_2NH \longrightarrow \tfrac{1}{2}(MeZnNPh_2)_2 + CH_4$$

The products, if monomeric, would still possess donor and acceptor sites; thus MeZnOMe monomer would have two oxygen lone pairs and two acceptor orbitals on zinc. Association therefore occurs to form oligomers or, in some cases, polymers. The structures of the methylzinc methoxide tetramer and of the dimer $(MeZnNPh_2)_2$ illustrate this well (p. 58). Hydrolysis of an alkyl is probably generally initiated by adduct formation. The hydrolyses of Me_3B and of Me_3Al are both strongly exothermic,

$$Me_3B(l) + \tfrac{3}{2}H_2O(l) = \tfrac{1}{2}B_2O_3(c) + 3CH_4(g); \quad \Delta H^{\ominus} = -289 \text{ kJ mol}^{-1}$$

$$Me_3Al(l) + \tfrac{3}{2}H_2O(l) = \tfrac{1}{2}Al_2O_3(c) + 3CH_4(g); \quad \Delta H^{\ominus} = -497 \text{ kJ mol}^{-1}$$

If the entropy terms are included, however, the free energy changes would be even more favourable, on account of the formation of three moles of gaseous methane. The fact that Me_3Al is hydrolysed with explosive violence, whereas Me_3B is not attacked by water at room temperature is due to kinetic rather than to thermodynamic factors.

As noted in Chapter 1, the ability to form complexes with unsaturated organic ligands is a property of d-block transition elements rather than of main group elements. Alkyls of the alkali metals and especially of aluminium are able to add to alkenes and hence to initiate oligomerization or polymerization (p. 80). Some of these reactions are used industrially on a very large scale. There is evidence for weak interactions between the π-electron systems of unsaturated hydrocarbons and these electropositive metal centres. This is illustrated by the structure of benzyllithium, although there is some controversy about the extent of covalent bonding between metal orbitals and the π-orbitals of the ligand. The cyclopentadienyl derivatives have a wide range of structures. In some of these, e.g. Cp_2Mg the rings are symmetrically bonded to the metal in a sandwich (p. 91). In others, e.g. solid CpTl the individual units are linked in chains through bridging C_5H_5 groups. While a case could perhaps be made that these compounds are ionic, this is difficult to envisage in the intriguing arene complexes of main group elements, such as $(C_6H_6)_2Ga_2Cl_4$ (p. 93). Formation of π-complexes between copper(I) and silver(I) and alkenes and arenes has been known for many years. Similar π-complexes are implicated in the electrophilic attack by Hg^{2+} or Tl^{3+} on alkenes or arenes (p. 62). The polymeric alkynyls of Cu(I), Ag(I) and Au(I) are associated by π-interactions.

3.2 Structures

A wide range of structural types is found among the organic derivatives of the elements of the first three groups. On the one hand the alkyls and aryls of the heavy alkali metals are essentially ionic in character. On the other, those of the least electropositive elements form covalently bonded monomers e.g. R_3B, R_2M (M = Zn, Cd, Hg). Intermediate between these are the derivatives of Li, Be, Mg and Al which are associated through alkyl or aryl bridges.

In contrast to diborane, B_2H_6, trimethylborane is monomeric in the vapour and also in the condensed states. The methyl groups are arranged trigonally in a plane about the boron atom. The boron atom has an empty $2p$ orbital and consequently acts as a Lewis acid, forming complexes with electron pair donors such as amines, phosphines or hydride ion. Hydrogen readily forms bridges between boron atoms, for instance in tetramethyldiborane or in diborane itself. Stable alkyl bridges, however, have not been found, although they are presumably present in the transition state when groups are exchanged between boron centres. It is thought that an important reason why Me$_3$B is monomeric is that the dimerization is sterically hindered.

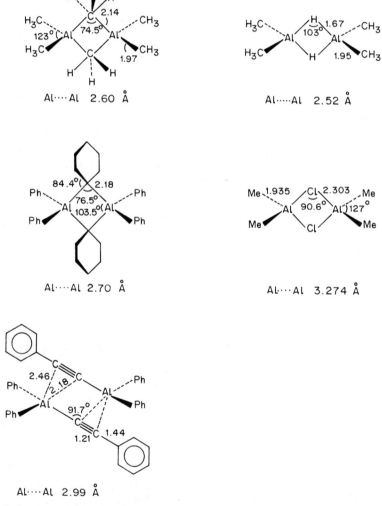

Fig. 3.1 Structures of some organoaluminium compounds.

The boiling point of trimethylaluminium (126°C) is rather higher than that of either Me_4Si (27°C) or of Me_3P (38°C) (Table 1.1). Trimethylaluminium crystallizes as a dimer (m.p. 15°C) and remains associated in the liquid state and also in the vapour well above the boiling point. The structure of the dimer, determined by X-ray diffraction, is shown in Fig. 3.1. As the aluminium atom is bigger than boron there is no steric problem; steric constraints do arise, however, when large alkyl groups are introduced. This appears in the enthalpy of dissociation; while $(Me_3Al)_2$ is still largely dimeric even at 150°C, tri-isobutylaluminium is appreciably dissociated into monomer at 25°C.

Monomeric trimethylaluminium has a trigonal planar (D_{3h})* structure. Electron diffraction on the vapour at 215°/30 mm gives a value of 1.957 Å for the Al—C bond length, very close to that in the dimer. Proton n.m.r. of Me_6Al_2 in cyclopentane solution shows two signals below $-55°C$ with intensity ratio 2:1, corresponding to terminal and bridging methyl groups respectively. As the temperature is raised these signals coalesce (at $-35°C$) and at room temperature there is only one averaged signal. The activation energy for exchange of methyl groups in solution (65.5 kJ mol^{-1}) is lower than the dissociation enthalpy into monomer either in the pure liquid (81.5 kJ mol^{-1}) or in the gas phase (85.7 kJ mol^{-1}). Exchange is thought to occur by dissociation to monomer units which remain trapped in a solvent cage.

The bonding in Me_6Al_2 has conventionally been discussed in terms of three-centre molecular orbitals. If the valence orbitals of Al are sp^3 hybridized, two of the hybrids can form the terminal Al—CH_3 bonds. These are classical electron pair bonds and account for two of the three valence electrons of each Al. The remaining sp^3 hybrids are directed in the Al_2C_2 plane and are well placed to interact with the Csp^3 orbitals of the bridging methyl groups. Each of these three centre bonding m.o.s contains two electrons, one from Al and one from a bridging methyl group (Fig. 3.2(a)).

An alternative description is shown in Fig. 3.2(b). Here the Al orbitals are sp^2 hybridized, and the 3-centre bonds arise by interaction of $Al(3p)$, $C(sp^3)$ and $Al(3p)$. This model is consistent with the wide Me^t—Al—Me^t angle which is observed, but it overemphasizes the Al—Al interaction. While the Al$\cdots$Al distance is short (2.60 Å, cf. sum of covalent radii 2.36 Å), it is not thought likely that Al$\cdots$Al bonding is an important source of stabilization in these dimers.

Alkoxy, halogen, hydrogen and even aryl groups all form stronger bridges to aluminium than alkyl groups. Some structures are illustrated opposite and also on p. 85. In the first alkoxides and halides oligomers arise by Lewis acid/base interactions. The bonding in aryl bridges can be described by similar schemes to those shown in Fig. 3.2(a), (b), with the addition of some π-bonding with the phenyl groups. Alkenyl groups bridge in a similar fashion. The alkynyl bridges in $[Ph_2AlC{\equiv}CPh]_2$, however, are of a different type. Here it seems that there is electron donation from the π-system of the alkyne to aluminium. Further

---

*Symmetry of AlC_3 skeleton, exclusive of hydrogen atoms.

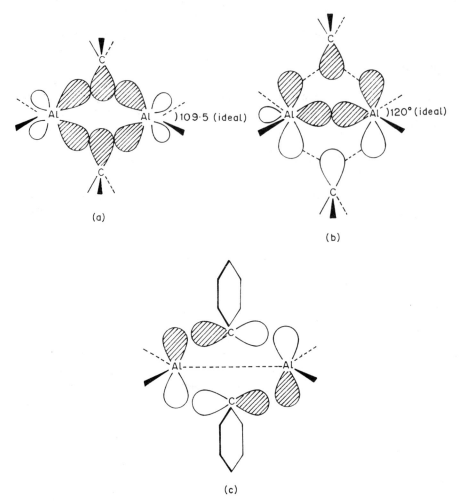

Fig. 3.2 Alternative descriptions of bridge bonding in Me$_6$Al$_2$. Bonding combinations only are shown. (a) sp^3 hybridization of Al orbitals. (b) sp^2 hybridization of Al orbitals. (c) π-bonding in Ph$_6$Al$_2$.

evidence for π-bonding between alkenes and aluminium centres is presented on p. 81.

In view of the very similar sizes of aluminium and gallium (covalent radii 1.18 Å and 1.26 Å respectively) it is rather surprising that Me$_3$Ga is monomeric in the vapour and liquid states and also in hydrocarbon solvents. Gallium is rather more electronegative than aluminium (Al, 1.61; Ga, 1.81) and the lower polarity of the Ga—C bonds may be decisive. Trivinylgallium, however, is dimeric. While the solid state structure of Me$_3$Ga is not known, trimethylindium and trimethylthallium crystallize as weakly associated tetramers, in which the planar monomer units are still easily distinguishable. Similar weak association may also occur in solid triphenylindium. Proton n.m.r. studies show that exchange of methyl groups between Me_3M molecules (M = Ga, In, Tl) occurs readily, presumably through methyl bridged dimers.

The Group IIB dialkyls and diaryls are all monomeric, having linear skeletons in which the metal is 2-coordinate. The change in bond length in the dimethyls is worth noting; Zn—C 1.929 Å; Cd—C, 2.112 Å; Hg—C, 2.094 Å, all ± 0.005 Å. The anomalous position of cadmium is reflected also in the boiling point sequence (p. 6) and is attributed to bigger van der Waals forces between Me_2Cd than between Me_2Hg molecules. Proton n.m.r. spectra of mixtures of Me_2Zn and Me_2Cd with each other or with Me_3Al indicate that alkyl exchange occurs readily. The methyl groups in Me_2Hg, however, exchange only slowly. In solid Ph_2Hg the two phenyl rings are coplanar. In the di-*o*-tolyl compound, however, they are twisted through 59° relative to each other, on account of interactions between the *ortho* substituents on different rings.

The dialkyls and diaryls of beryllium and magnesium are generally associated. Steric effects play an important part in determining the degree of association in beryllium compounds. Dimethylberyllium, which forms colourless crystals decomposing above ca. 200°C, is a long chain polymer in which the metal atoms are surrounded by four methyl groups in a nearly tetrahedral arrangement. It is soluble only in donor solvents which are able to break down the polymeric structure. It has an appreciable vapour pressure above 100°C (0.6 mm at 100°C and 760 mm at 220°C extrapolated), and the vapour contains mainly the monomer, with some dimer and trimer. Diethylberyllium, b.p. 93–5°/4 mm is dimeric in benzene, whereas the very hindered di-*tert*-butylberyllium is monomeric both in the vapour and in solution. The Be—C bonds in this linear monomer are 1.70 Å long (electron diffraction).

The magnesium dialkyls are colourless solids, which react readily with oxygen and water and which usually decompose without melting at high temperatures. The pathways followed in their thermal decomposition are typical of many alkyl derivatives of main group and of transition elements. Where a β-C—H is present, as in Et_2Mg, β-hydrogen transfer predominates with loss of alkene:

$$[Mg(CH_2CH_3)_2]_n \xrightarrow[210°]{170-} nMgH_2 + nH_2C{=}CH_2$$

$$\searrow 370°$$

$$nMg + nH_2$$

where such a function is absent, as in Me_2Mg, α-hydrogen transfer may occur:

$$\xrightarrow{>300°} n'MgC' + nH_2$$

$$[Mg(CH_3)_2]_n \xrightarrow[260°]{220-} n'MgCH_2' + nCH_4$$

Donor solvents tend to break down the polymers by coordination to the Lewis

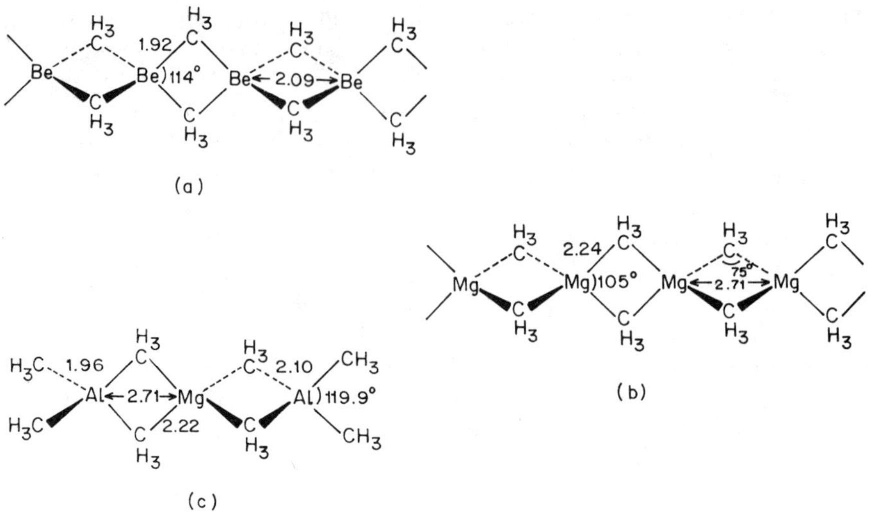

Fig. 3.3 Structures of (a) dimethylberyllium, (b) dimethylmagnesium and (c) Me_8Al_2Mg.

acidic metal centre. Thus Me_2Mg dissolves as a monomer in tetrahydrofuran but in diethyl ether some association persists at high concentrations (p. 53). Separate signals from bridging ($\delta -1.0$ to -1.5 p.p.m.) and terminal methyl groups ($\delta -1.69$ to -1.74 p.p.m.) are observed in the proton n.m.r. spectra of solutions of Me_2Mg in ether. A mixed derivative, $MgAl_2Me_8$, prepared from Me_3Al and Me_2Mg, contains methyl bridges between dissimilar metal atoms. (Fig. 3.3).

The highly polar character of organolithium compounds causes strong association. The geometry of the coordination sphere is determined essentially by steric effects, as in ionic structures, rather than by interaction of electron pairs. Even where lithium may appear to possess an octet configuration, it is not envisaged that the electrons are strongly held by the valence orbitals of the metal in covalent bonds.

Crystalline methyllithium consists of tetrameric aggregates (Fig. 3.4). Each carbon atom is bound to three hydrogen atoms and is equidistant from the three lithium atoms which are at the corners of a triangular face of the Li_4 tetrahedron. The formal coordination number of carbon is therefore seven. The structure consists of two interpenetrating and unequal tetrahedra, one Li_4 and one C_4. The Li_4C_4 skeleton can thus be represented as a distorted cube. The methyl carbon of one tetramer is only 2.36 Å from a lithium atom of a neighbouring tetramer. This interaction makes methyllithium involatile and sparingly soluble in hydrocarbons.

There are also tetramer units $(EtLi)_4$ in crystalline ethyllithium, although they associate in linear fashion rather than in three dimensions. These tetrameric clusters are retained in complexes of alkyl and aryllithiums with donor molecules.

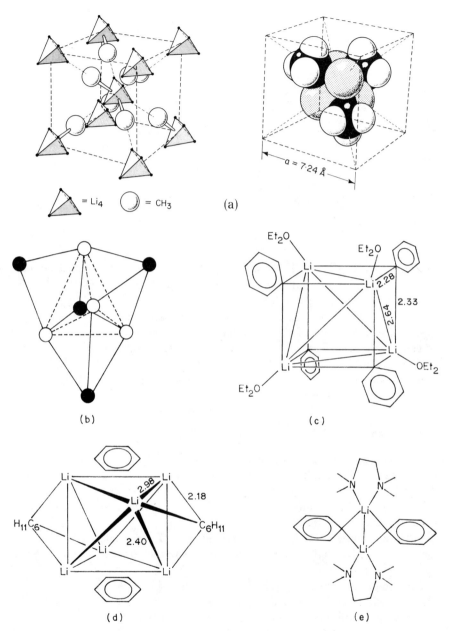

$= Li_4$ ⃝$= CH_3$ (a)

Et$_2$O

Et$_2$O

Li —— Li 2.28

2.33

2.64

Li

Li === Li

OEt$_2$

Et$_2$O

(b)   (c)

Li ═══ Li

2.98

2.18

Li

H$_{11}$C$_6$ C$_6$H$_{11}$

2.40

Li

Li ═══ Li

(d)   (e)

Fig. 3.4 (a) Unit cell of methyllithium. The (CH$_3$Li)$_4$ tetramer. (After E. Weiss and E.A.C. Lucken, *J. Organometal. Chem.* (1964) **2**, 197. (b) The Li$_4$C$_4$ skeleton of the tetramer viewed approximately along one of threefold axes. (c) Structure of (PhLiOEt$_2$)$_4$. (d) Structure of hexameric cyclohexyllithium (C$_6$H$_{11}$Li)$_6$.2C$_6$H$_6$. Cyclohexyl groups occupy six of the eight triangular faces of the octahedron – only two shown. (e) (PhLiTMEDA)$_2$.

Table 3.2. Aggregation of organolithiums, $(RLi)_n$ in solution

RLi	Solvent			
	Paraffins or cyclohexane	Benzene or toluene	Diethyl ether	Tetra-hydrofuran
MeLi	Insol	Insol	4	4
EtLi	6	6	4	4
Bu^{n}Li	4–6	6	4	4
Bu^{s}Li	4	4	–	–
Bu^{t}Li	4	4	4	4
PhLi	–	–	2	2
Menthyl Li	2	–	–	–

—Not known to author.

In $(MeLi)_4 TMEDA$ ($TMEDA = Me_2NCH_2CH_2NMe_2$ or tetramethylethylenedi-amine, 1,6-dimethyl-2,5-diazahexane), $(MeLi)_4$ units are bridged by TMEDA units. Bridging phenyl groups are present in the cluster $[PhLi \cdot OEt_2]_4$. Even in donor solvents such as tetrahydrofuran or diethylether, aggregation of organolithium species persists. The degree of aggregation, however, can vary with the solvent (Table 3.2). Steric factors are clearly important; in general the most bulky species are least associated. Cyclohexyllithium, however, can be crystallized from benzene as a solvate which has the structure shown in Fig. 3.4, which is based on a Li_6 octahedral cluster.

The closest Li—Li separations in these structures lie in the range 2.37–2.74 Å. The covalent radius of lithium is 1.34 Å although the ionic radius of Li^+ is only 0.76 Å.

Oligomers are present in the vapours above alkyllithiums. Mass spectroscopic studies show that ethyllithium vaporizes as a mixture of tetramer and hexamer. The bulk of the *tert*-butyl group reduces the interaction between tetramers even more, so that $(Bu^t Li)_4$ sublimes at 70°/0.1 mm; it is much more volatile than either methyl- or ethyllithium.

Lithium can interact with π-systems in alkenes, alkynes and arenes. This may account for the ability of lithium alkyls to initiate the polymerization of dienes (p. 48).

Lithium forms numerous bridged mixed alkyl and aryl derivatives with other elements. The structures of $LiAlEt_4$ and of $LiBMe_4$ are of interest on account of the contrasting nature of the bridge bonding. The former resembles Et_2Mg. The latter consists of planar sheets of Li atoms bridged by BMe_4 units. Neutron diffraction shows that the bridging to lithium occurs through the hydrogen atoms of the

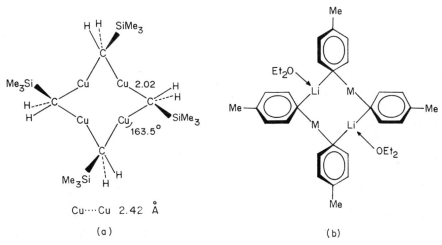

Fig. 3.5 (a) Structure of [Cu(CH$_2$SiMe$_3$)]$_4$. (b) Proposed structure of LiM(C$_6$H$_5$Me)$_2$ in ether (M = Cu, Ag, Au).

methyl groups and not through carbon (cf. transition metals, 'agostic hydrogen', p. 223). Both dibridged and tribridged configurations are present.

Much less is known about the structures of the alkyls of heavier alkali metals. That of methylsodium is based on (MeNa)$_4$ tetramers. MeM(M=K, Rb, Cs), however, possess ionic nickel arsenide structures with isolated methyl anions.

Copper, silver and gold form unstable organo derivatives which show some relation to those of the compounds discussed above. MeCu decomposes above −15°C and MeAg even at −50°C. The compounds [Cu(CH$_2$SiMe$_3$)]$_4$, m.p. 78–9°C and (CuPh)$_n$ (decomp. 100°) are more robust; the structure of the former has been determined by X-ray diffraction and consists of four coplanar copper atoms bridged by CH$_2$SiMe$_3$ groups. Phenyl copper probably has bridging phenyl groups. Alkyl or aryl-bridged tetramers are thought to be present in solutions of Li$_2$Cu$_2$R$_4$ (p. 88).

3.3 Alkali-metal derivatives

3.3.1 *Preparation of organolithium compounds*

The lithium alkyls are extremely reactive. They are very sensitive to oxygen and moisture. Dry apparatus and dry solvents must be used and air must be excluded by working under an atmosphere of dry oxygen-free nitrogen or argon. The following methods of preparation are available.

(a) LITHIUM METAL AND ALKYL HALIDE.   The most direct method of obtaining a solution of an organolithium reagent is reaction of lithium metal with an

The first three periodic groups

organic halide in a suitable solvent.

$$2Li + 2RX \rightarrow RLi + LiX$$

The physical state of the lithium is important and reaction does not start if the metal is too coated with corrosion product. Lithium is now supplied as a dispersion in mineral oil which can be weighed in air. The oil is removed by washing with hexane under an inert atmosphere, leaving the metal as a highly reactive, pyrophoric powder.

One of the most frequently used reagents is n-butyllithium, which can be prepared from chloro- or bromobutane in hexane, benzene or ether. Solutions in hexane (usually $3.6\,mol\,dm^{-3}\,BuLi$) are commercially available. Methyl chloride, bromide and iodide all react satisfactorily with lithium in diethyl ether. Normally alkyl iodides are unsuitable as they react too quickly with the organolithium compound: $RI + RLi \rightarrow R\!-\!R + LiI$. A similar problem arises with benzyl chloride, which gives 1, 2-diphenylethane. Aryl bromides and iodides, however, often give good yields, although the chlorides are often insufficiently reactive.

Organolithium reagents attack ethers, although in some cases (e.g. MeLi in diethyl ether) only very slowly. Because of such side reactions and the greater sensitivity of organolithiums to air and moisture, the use of Grignard reagents is usually more convenient, if they also perform the desired reaction (p. 44).

(b) METAL–HALOGEN EXCHANGE. The reversible reaction $RLi + R'X \rightleftharpoons R'Li + RX$ is generally fast at or below room temperature. The equilibrium favours the formation of the lithium compound of the more electronegative organic group ($C_{sp} > C_{sp^2} > C_{sp^3}$), which is the one which forms the more stable carbanion. Butyllithium can thus be used as a source of aryllithium compounds. Exchange is fastest with aryl iodides. Chlorides and fluorides react slowly and hydrogen–metal exchange (v.i.) often competes. A great advantage of metal–halogen exchange is its complete regiospecificity.

(c) METAL–HYDROGEN EXCHANGE (METALLATION).

$$R'H + RLi \rightleftharpoons R'Li + RH$$

or

$$R'H + R_2NLi \rightleftharpoons R'Li + R_2NH$$

Table 3.3. Acidity values (pK_a) of hydrocarbons

	pK_a		pK_a		pK_a
Me$_3$CH	47	PhH	39	Ph$_3$CH	32
Me$_2$CH$_2$	44	H$_2$C=CH$_2$	38.5	HC≡CH	24
MeCH$_3$	42	PhCH$_3$	37	Cyclo-pentadiene	18.5
CH$_4$	40	Ph$_2$CH$_2$	33.5	Fluorene	15

The reaction goes to the right if the hydrocarbon R'H is more acidic than RH (or R$_2$NH). In Table 3.3 the pK_a values of some hydrocarbons, a measure of their acidities, are listed. Phenylethyne, for example, reacts with either butyl- or phenyllithium

$$PhC≡CH + PhLi \longrightarrow PhC≡CLi + PhH$$

Metallation is believed to involve nucleophilic attack by the organolithium reagent on the 'acidic' hydrogen atom. The nucleophilic character of carbon bonded to lithium is enhanced by coordination of the lithium to a base. Organolithium compounds are therefore more reactive when dissolved in ethers than in hydrocarbons. Unfortunately ethers can themselves be metallated. This is especially a problem with the rather basic ether tetrahydrofuran, which is rapidly cleaved by butyllithium at room temperature.

This difficulty does not arise with tertiary amines since the C—H bonds adjacent to nitrogen are less susceptible to nucleophilic attack than those next to oxygen in ethers. TMEDA (p. 38) complexes strongly with alkyllithiums. The *n*-butyllithium chelate is monomeric and very soluble in hydrocarbons. It is a very strong metallating agent. It reacts with toluene at room temperature and more slowly, with benzene. Butyllithium in the absence of TMEDA does not normally attack benzene or toluene and can even be prepared in these solvents.

Aromatic compounds which bear electron attracting (−I) substituents are susceptible to metallation by butyllithium. On account of the inductive effect the acidity of the *ortho* hydrogen atoms in particular is enhanced. Methoxybenzene, for example, affords predominantly the *ortho*-lithium compound. In some cases the *ortho*-directing effect is increased through coordination of the substituent to lithium. The reagent is thus held near to the site of attack, (Fig. 3.6). In spite of this, metal–hydrogen exchange is less regiospecific in general than metal–halogen exchange. As mentioned above alkylbenzenes (with +I substituents) are metallated by butyllithium only in the presence of a strong base. Sodium

41

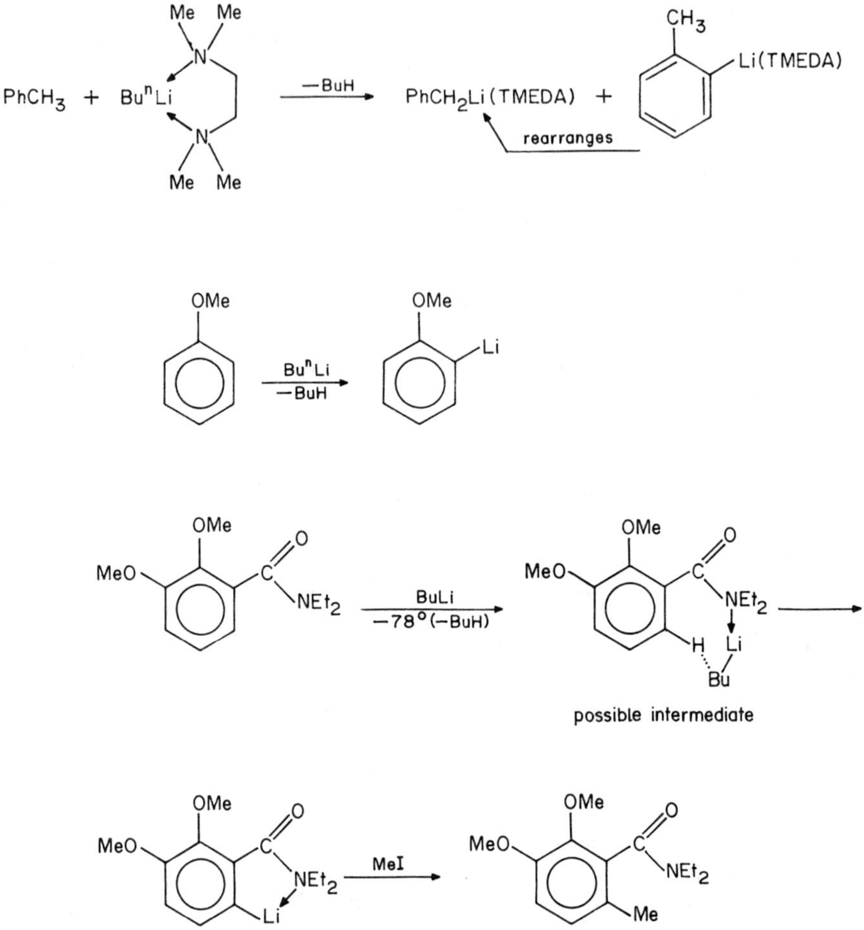

The first three periodic groups

Fig. 3.6 Metallation by organolithium reagents.

or potassium alkyls do attack them. Isopropylbenzene, for example, is metallated in the *meta* and *para* positions.

Aryl fluorides undergo metal–halogen exchange very slowly. The reaction between phenyllithium and fluorobenzene gives 2-lithiobiphenyl. This proceeds by metal–hydrogen exchange followed by loss of lithium fluoride to give a 'benzyne' intermediate (Fig. 3.7).

This mechanism has been confirmed using isotopically labelled fluorobenzene. Benzynes have been trapped by dienes in Diels-Alder reactions. They can also be generated by strong bases such as sodium amide in liquid ammonia; in this way bromobenzene can be used without the interference of metal–halogen exchange which is the preferred course with lithium alkyls.

42

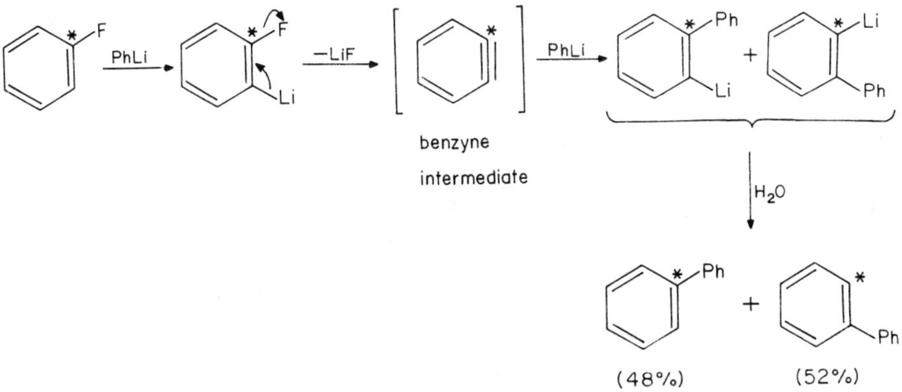

Fig. 3.7 Formation of a benzyne intermediate.

The presence of electron-withdrawing groups such as $-CO_2R$, $-COR$ or $-CN$ adjacent to a $\geqslant C-H$ bond enhances its acidity. Organolithium compounds are generally unsuitable for generating carbanions from ketones, esters or nitriles because they are strong nucleophiles (p. 45) as well as being strong bases. Bulky secondary amines such as diisopropylamine or 2, 2, 6, 6, -tetramethylpiperidine react with lithium alkyls as follows:

$$BuLi + Pr_2^i NH \longrightarrow Pr_2^i NLi + BuH$$

The resulting N-lithio derivatives are strong bases which can abstract protons from acidic carbon centres (if $pK_a < 30$), but which are too bulky to act as nucleophiles towards functional groups. They are therefore widely used to generate enolate ions.

$$Pr^n CH_2 COOMe \xrightarrow[\text{THF}, -78°C]{LiNPr_2^i} Pr^n \overset{\ominus}{C}HCOOMe$$

$$\xrightarrow{RBr} Pr^n CHRCOOMe$$

(d) METAL–METAL EXCHANGE. Metal–metal exchange reactions provide a convenient route to vinyl and allyllithium reagents. Phenyllithium exchanges rapidly with tetravinyltin in ether, when tetraphenyltin is precipitated nearly quantitatively.

$$4PhLi + (H_2C=CH)_4 Sn \longrightarrow 4H_2C=CHLi + Ph_4Sn \downarrow$$

Allyllithium can be made similarly. These exchange reactions take place with retention of configuration at carbon:

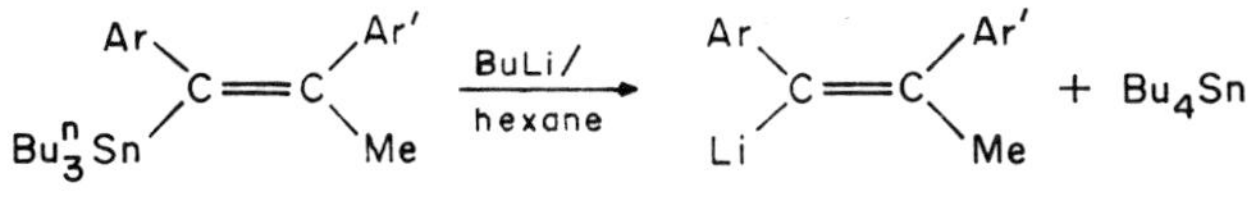

43

$$Et - \underset{SnBu_3}{\overset{H}{\underset{|}{C}}} -OR \xrightarrow[THF,\ -78°]{BuLi/} Et - \underset{Li}{\overset{H}{\underset{|}{C}}} -OR + Bu_4Sn$$

(e) DILITHIO DERIVATIVES. 1,1-Dilithiomethane has been obtained by pyrolysis of pure, salt free methyllithium (p. 35). The direct reaction between lithium metal and dibromomethane gives only about 6% yield. A two step synthesis through a dimercury derivative is more successful.

$$CH_2I_2 \xrightarrow{Hg} CH_2(HgI)_2$$

$$\xrightarrow{4Li} CH_2Li_2 + 2Hg + 2LiI$$

$$\xrightarrow{4Bu^tLi} CH_2Li_2 + Bu^t_2Hg + 2LiI$$

The preparation of 1,1-dilithioethane illustrates several general organometallic reactions, including hydroboration and the transfer of organic groups from one element to another. Dilithioethane is rather susceptible to decomposition into vinyllithium and lithium hydride through β-hydrogen transfer.

$$H_2C=CHBr \xrightarrow{Mg/THF} H_2C=CHMgBr \xrightarrow[(ii)Bu^nOH]{(i)B(OMe)_3} H_2C=CHB(OBu^n)_2 \xrightarrow[(ii)Bu^nOH]{(i)H_3B.THF}$$

$$H_3CCH[B(OBu^n)_2]_2 \xrightarrow[NaOH]{HgCl_2/} CH_3CH(HgCl)_2 \xrightarrow[\substack{Et_2O \\ 20\ °C}]{Li/} CH_3CHLi_2 \xrightarrow[\substack{D_2O \\ Me_3SiCl}]{} \begin{array}{l} CH_3CHD_2 \\ CH_3CH(SiMe_3)_2 \end{array}$$

$$H_2C=CHLi + LiH$$

3.3.2 Reactions of organolithium compounds and comparison with Grignard reagents

(a) GENERAL REACTIVITY. Some reactions of organolithium compounds which illustrate their great versatility in synthesis are summarized in Fig. 3.8. The reactions of the corresponding and less reactive Grignard reagents (p. 50) are often similar. Grignard reagents, for instance, do not metallate alkenes or arenes, although they do exchange with alkynes. Nucleophilic attack by Grignard and organolithium reagents on multiple bonds such as $>C=O$ or $-C\equiv N$ leads after hydrolysis to a wide range of organic products. It is generally most convenient to use the Grignard reagents to effect these classical transformations, as less rigorous precautions are required. In a few instances, however, organolithium reagents behave differently. Carboxylate salts, for example, react as follows:

$$R-\underset{O^-Li^+}{\overset{O}{\underset{\|}{C}}} + R'Li \longrightarrow \underset{\substack{R' \\ \text{stable to attack} \\ \text{by } R'Li}}{\overset{R}{\underset{O^-Li^+}{C}}}\overset{O^-Li^+}{} \xrightarrow{H_3O^+} \left[\underset{R'}{\overset{R}{\underset{OH}{C}}}\overset{OH}{} \right] \longrightarrow \underset{R'}{\overset{R}{C}}=O$$

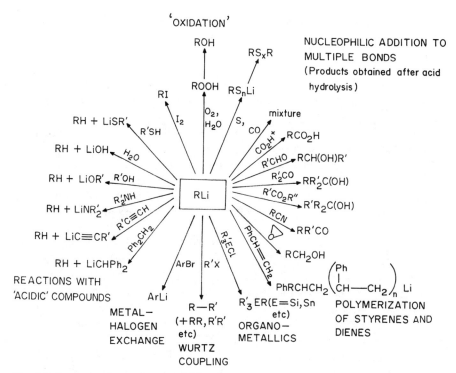

Fig. 3.8 Some reactions of organolithium reagents.

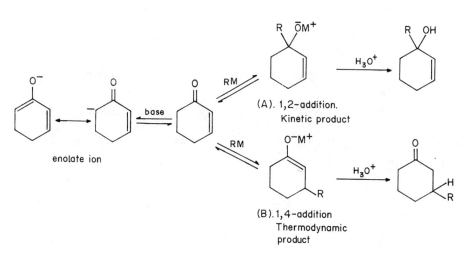

Fig. 3.9 1,2- and 1,4-addition of an organometallic compound to an $\alpha\beta$-unsaturated ketone.

The first three periodic groups

This provides a useful route to unsymmetrical ketones, or to aldehydes starting from lithium formate. (Grignard reagents do not react with carboxylate anions, apart from formates.) Aldehydes can also be obtained from formate esters or, better, orthoformate esters under mild conditions.

Organolithium reagents generally add in 1,2-fashion to $\alpha\beta$-unsaturated carbonyl compounds. This addition takes place rapidly at low temperatures and is kinetically controlled.

The lithium adduct (A) is formed irreversibly; lithium is strongly bound to oxygen. With the Grignard reagent the formation of (A) is reversible and the thermodynamically more stable product (B) gradually accumulates. Often a mixture of 1,2- and 1,4-addition products is obtained. Steric effects are important, favouring 1,4-addition. Essentially exclusive 1,4-addition results from the use of organocopper reagents or copper catalysts (p. 89) (Fig. 3.9).

(b) 'UMPOLUNG'. The carbonyl group in aldehydes and ketones is polarized $\overset{\delta+}{>}C=\overset{\delta-}{O}$. It is therefore susceptible to nucleophilic attack at carbon. Methods have been developed by which this polarization is effectively reversed so that the carbon atom itself becomes the nucleophilic centre. Such an inversion is known as 'umpolung'. An example is provided by the 1,3-dithiane system. An aldehyde may be converted into a cyclic dithioacetal by reaction with propane-1,3-dithiol in the presence of an acid. The two adjacent electronegative sulphur atoms make the $>$C—H bond of this acetal rather acidic. Treatment with butyllithium therefore affords a lithio derivative in which the carbon atom is susceptible to electrophilic attack. The 1,3-dithiane system is reconverted into a carbonyl group by acid hydrolysis in the presence of mercury(II) ions, which complex with the dithiol. The RCO group in the original aldehyde is thus 'equivalent' to $R—\bar{C}=O$ (Fig. 3.10).

(c) ADDITION TO ALKENES. The first observation of the addition of an organometallic compound to an alkene was made by Ziegler in 1927.

$$\underset{\text{red}}{Ph_2\bar{C}Me\cdot K^+} + PhCH{=}CHPh \longrightarrow \underset{\text{yellow}}{Ph_2C(Me)—CH(Ph)—\bar{C}HPh\cdot K^+}$$

Normally, however, alkyl sodium and potassium reagents metallate alkenes rather than add to them. Organolithium or Grignard reagents are also unreactive towards C—C multiple bonds. Vigorous conditions are needed to cause organolithiums to add to 1-alkenes unless the latter are activated e.g.

46

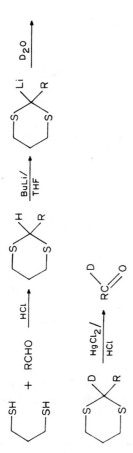

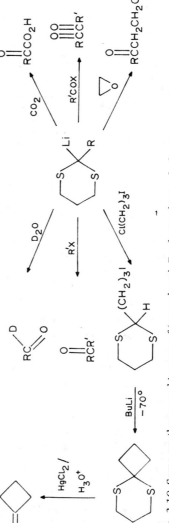

Fig. 3.10 Some syntheses making use of 'umpolung'. Products obtained after treatment with $HgCl_2/H_3O^+$.

The first three periodic groups

Here the driving force is provided by the formation of a carbanion adjacent to a stabilizing group (R_3Si— or dithioacetal). Organolithium reagents, however, add to conjugated dienes and hence initiate their polymerization. In hydrocarbon solvents n-butyl lithium and butadiene result in mostly 1,4-addition. The polymer contains a mixture of *cis* and *trans* stereochemistry about the double bonds.

$$BuLi + H_2C=CHCH=CH_2 \xrightarrow{\text{initiation}} BuCH_2CH=CHCH_2Li \xrightarrow{C_4H_6}$$

$$Bu(CH_2CH=CHCH_2)_2Li \xrightarrow[\text{propagation}]{C_4H_6} Bu(CH_2CH=CHCH_2)_nLi \longrightarrow \text{wwv}CH_2CH_2OH$$

living polymer

Me_2SiCl_2

H_3O^+ quenching

$CO_2,$ O_2, H_3O^+ wwv$SiMe_2Cl$

H_3O^+

$Bu(CH_2CH=CHCH_2)_nH$

wwvCO_2H wwvOH

35% *cis* 1,4-, **54%** *trans* 1,4- and **11%** 1,2- polybutadiene

Introduction of terminal functional groups

In the presence of polar ligands such as TMEDA 1,2-addition is favoured, giving the following type of polymer.

$$Bu\{CH_2CH(CH=CH_2)\}_nCH_2\bar{C}H(CH=CH_2)Li^+$$

The structures of the intermediates have been studied by n.m.r. but no clear detailed mechanism has yet emerged. These intermediates have 'living' characteristics. This means that they are capable of further polymerization on addition of more monomer at any time, provided of course that they are protected from oxygen and moisture. If the rate of initiation is fast relative to the rate of propagation, the resulting polymer has a narrow molecular weight range which is controlled by the mole ratio of initiator to monomer. The biggest use of alkyllithium catalysts is in the production of styrene-butadiene rubber for tyre treads. While simple olefins like ethene are not normally polymerized (BuLi·TMEDA in hydrocarbons does polymerize ethene), reactive ones like styrene (phenylethene) are. The 'living' character of the intermediates is particularly suitable for the formation of block copolymers by alternate separate

Table 3.4. World production figures for rubber in 1977

	Tonnes($\times 10^6$)
Emulsion polymer (styrene butadiene rubber) —free radical initiation.	4.8
Polyisoprene (natural rubber)	3.6
Polybutadiene (transition metal catalyst)	1.1
Polyisoprene (transition metal catalyst)	0.6
Polybutadiene and styrene butadiene rubber —alkyl-lithium initiation	0.8

addition of butadiene and styrene monomers. Furthermore lithium alkyls afford linear polymers, whereas the materials produced *via* free radical initiation are highly branched. (see Table 3.4).

3.3.3 *Organosodium and potassium compounds*

The organic group in these compounds has considerable carbanionic character. In general, paraffin hydrocarbons are the only suitable reaction media. Ethers are cleaved and aromatic hydrocarbons metallated. To prepare small quantities of organosodium or potassium compounds, the mercury alkyl method can be used.

$$R_2Hg + 2Na(\text{excess}) \xrightarrow{\text{petroleum}} 2RNa\downarrow + Na, Hg \text{ (amalgam)}$$

Exchange between sodium or potassium *tert*-butoxide and organolithium in paraffins is preferable, as the lithium alkoxide remains in solution.

$$RLi + Bu^tONa \longrightarrow RNa\downarrow + LiOBu^t \text{ (soluble)}$$

Direct reaction between sodium or potassium and alkyl and aryl halides is complicated by exchange and coupling reactions, which can lead to mixtures of products. These complications can be reduced by rapid stirring and the use of finely divided metal or amalgams. Phenylsodium can be made in this way: $PhCl + Na \rightarrow PhNa + NaCl$. Acidic hydrocarbons react with alkali metals in ether solvents. Cyclopentadiene, for example, affords sodium cyclopentadienide in tetrahydrofuran (p. 279). Triphenylmethylpotassium is obtained as a deep red solution from triphenylmethane and potassium in 1,2-dimethoxyethane. These carbanions, in which the negative charge is delocalized over several carbon atoms, do not attack ethers, in contrast to the simple alkyl or aryl carbanions present in methylsodium or phenylpotassium.

$$3C_5H_6 + 2Na \longrightarrow 2C_5H_5^-Na^+ + C_5H_8$$
$$Ph_3CH + K \longrightarrow Ph_3C^-K^+ + \tfrac{1}{2}H_2$$

Caesium reacts directly with toluene and even with alkanes.

(a) HYDROCARBON ANIONS.   Aromatic hydrocarbons containing two or more benzene rings joined (biphenyl, terphenyls), conjugated (1,4-diphenylbutadiene) or fused (naphthalene, anthracene, phenanthrene) react with alkali metals without loss or transfer of hydrogen. The formation of these addition compounds is facilitated by donor solvents such as tetrahydrofuran or 1,2-dimethoxyethene. The products are highly coloured; their formation is accompanied by the transfer of one or more electrons from the alkali metal to the lowest unoccupied molecular orbital (LUMO) of the hydrocarbon.

Solutions of sodium naphthalene in tetrahydrofuran are dark green, electrically conducting because the compound consists of ions $\bar{N}a(THF)_n^+ (C_{10}H_8)^{-\cdot}$ and paramagnetic because of the unpaired electron in a singly occupied π-orbital. The electron spin resonance spectrum consists of 25 lines. The interaction between the odd electron and the nuclear spins of four equivalent α protons gives rise to five lines of relative intensities $1:4:6:4:1$. Each of these lines is split into a further quintet by interaction with the four β protons.

Solutions of sodium naphthalene and similar compounds are very strong reducing agents, causing organic halides to form radicals which may react with the solvent, the hydrocarbon or its anion, or may dimerize.

$$RX + C_{10}H_8^{-\cdot} \longrightarrow R^\cdot + X^- + C_{10}H_8$$

Polytetrafluoroethene (Teflon) is etched, giving an active surface which can then be bonded strongly to an epoxy resin. With a source of acidic hydrogen such as water or an alcohol, sodium naphthalene affords 1,4-dihydronaphthalene. Polymers with functional groups at each end of the chain ('telechelic' polymers) can be made by polymerization of styrene with sodium naphthalene. The anionic radical which is produced in the initiation step dimerizes to a dianion containing two propagating centres.

$$PhCH{=}CH_2 + Na^+C_{10}H_8^{-\cdot} \longrightarrow PhCHCH_2^{-\cdot} Na^+ + C_{10}H_8 \quad \text{(Initiation)}$$
$$\downarrow$$
$$Ph\bar{C}HCH_2CH_2\bar{C}HPh \ (Na^+)_2 \quad \text{(Dimerization)}$$
$$\downarrow$$
$$Ph\bar{C}HCH_2(PhCHCH_2)_nCH_2\bar{C}HPh \ (Na^+)_2 \quad \text{(Propagation)}$$
$$\swarrow$$

(X = e.g. $\qquad$ $PhXCHCH_2(PhCHCH_2)_nCH_2XCHPh \qquad$ (Termination)
COOH, OH)

Reaction with further styrene yields a 'living' polymer which can be functionalized at both ends. Reaction with a different monomer such as butadiene at any stage gives a block copolymer.

3.4 Grignard reagents

3.4.1 Preparation

The organomagnesium halides, $RMgXL_n$ (L = base such as diethyl ether, n often 2), are much more extensively used than the dialkyls or diaryls R_2Mg. Barbier (1899) was the first to employ magnesium for organic synthesis. From a mixture of magnesium, methyl iodide and a ketone in diethyl ether he obtained a tertiary alcohol: It was his pupil, Victor Grignard (1871–1935), however, who isolated and analysed the organometallic reagents and applied them in many syntheses.

Grignard reagents are normally prepared by the slow addition of the organic halide (see Chapter 2) to a stirred suspension of magnesium turnings using dry apparatus and reagents. An atmosphere of dry oxygen-free nitrogen is also required as the products are readily hydrolysed and oxidized. Reaction cannot start until the reagents have penetrated in some places the thin oxide film which coats the metal. Activation by etching with a crystal of iodine (which also forms

MgI_2, a good drying agent for traces of moisture) or merely by stirring the metal under nitrogen is sometimes necessary. The formation of a Grignard reagent is strongly exothermic, so that once the induction period is over it commonly accelerates very markedly. Care is therefore necessary to avoid adding too much halide before the reaction is underway or the boiling of the solvent can get out of control. An exception is sometimes made for unreactive halides. Here the 'entrainment method' is useful. The inert halide is added together with 1,2-dibromoethane to an excess of the stoichiometric quantity of magnesium. 1,2-Dibromoethane (CARE–suspect carcinogen) reacts approximately according to the equation

$$BrCH_2CH_2Br + 2Mg \longrightarrow MgBr_2 + C_2H_4$$

exposing fresh metal which can attack the desired halide concurrently.*

While diethyl ether is usually the preferred solvent, the more basic tetrahydrofuran is useful for making arylmagnesium chlorides and vinyl Grignard reagents. Aryl chlorides, unlike the more expensive bromides and iodides, do not react with magnesium in diethyl ether, neither do vinyl chlorides. In contrast to organolithium compounds, Grignard reagents do not normally cleave ethers. Organomagnesium compounds can also be prepared in hydrocarbons, although only primary and aryl halides give good yields. The soluble products from chlorides are essentially MgR_2, as $MgCl_2$ precipitates from solution. Bromides and iodides afford solutions which contain unsolvated 'R_nMgX_{2-n}' $(n > 1)$.

Among side reactions which can arise in Grignard preparations, Wurtz coupling to form hydrocarbons R—R is the most significant. It occurs especially with bulky R groups and also when R is allyl or benzyl. Iodides are worse than chlorides or bromides. Excess magnesium and slow addition of the halide helps to reduce this problem. Coupling can be either through direct reaction between RX and RMgX or through radical coupling. The attack of an organic halide on magnesium is thought to occur by a free radical mechanism. It is those alkyl groups which form the most stable radicals which are most prone to these side reactions.

3.4.2 Composition and structures

Careful evaporation of solvent from solutions of Grignard reagents leaves colourless solids or viscous liquids. It is difficult to remove all traces of ether below 100°C. Prolonged pumping to constant weight at $90-120°/10^{-3}$mm leads to dismutation of 'RMgX' to $[R_2Mg]_n$ and MgX_2 (R = Me, Et; X = Cl, Br). Careful crystallization, however, has yielded materials suitable for X-ray analysis. Some

---

*Solutions of Grignard reagents can sometimes be prepared without the protection of an inert atmosphere, air being displaced by ether vapour during the exothermic reaction. Such a procedure is *not* recommended; if it is adopted boiling must not be allowed to stop before the formation *and* use of the reagent is completed.

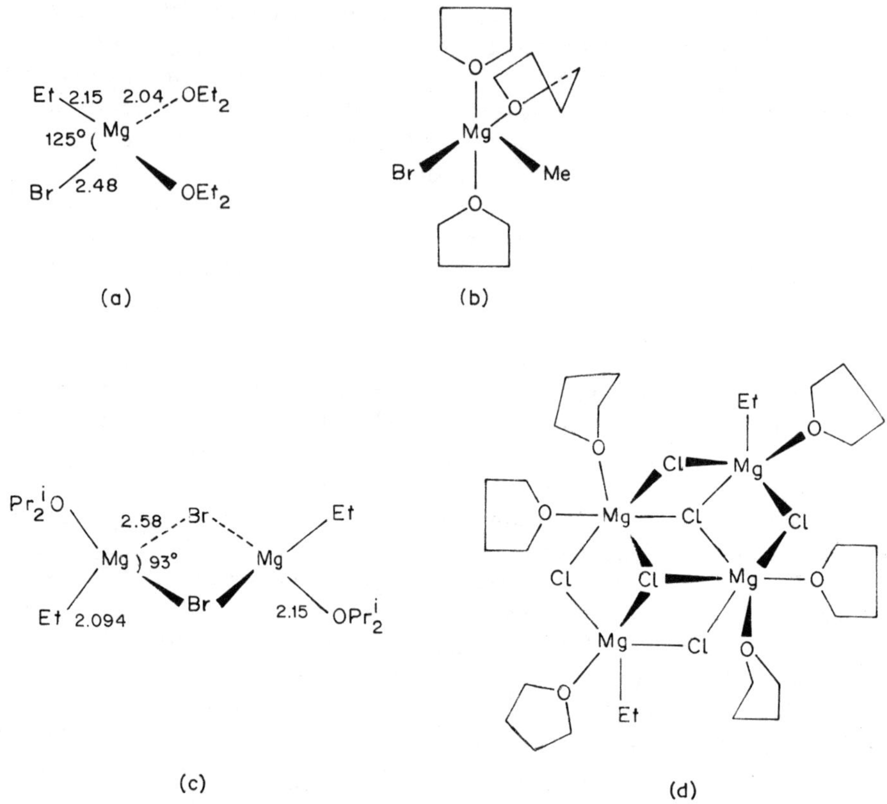

(a)

(b)

(c)

(d)

(a) EtMgBr.2Et$_2$O   (4-coordinate Mg)

(b) MeMgBr.3THF   (5-coordinate Mg)

(c) [EtMgBr.Pr$^i_2$O]$_2$ (4-coordinate Mg)

(d) [EtMg$_2$Cl$_3$(THF)$_3$]$_2$ (5- and 6-coordinate Mg).

Fig. 3.11 Structures of some solid Grignard reagents.

representative structures are shown in Fig. 3.11. In the simplest of these, EtMgBr · 2Et$_2$O, the magnesium atom is four-coordinate in a distorted tetrahedral environment. Pentacoordinate magnesium is present, however, in the adduct MeMgBr · 3THF. Associated species which contain bridging halogen atoms also occur and some of these have been studied by crystallography.

When dissolved in ethers, Grignard reagents have the composition 'RMgX'. On addition of dioxan, magnesium halide is precipitated as the dioxan complex MgX$_2$ · 2 dioxan, leaving R$_2$Mg in solution. An equimolar mixture of R$_2$Mg and

Table 3.5. Data on the Schlenk equilibrium MgX_2 $+ R_2Mg \rightleftharpoons 2RMgX$ at $25°C$

Grignard reagent	Solvent	K	$\Delta H^{\ominus}/$ ($kJ\,mol^{-1}$)	$\Delta S^{\ominus}/$ ($J\,mol^{-1}\,K^{-1}$)
EtMgCl	THF	5.52	15.9	67.3
EtMgBr	THF	5.09	25.5	99.1
EtMgBr	Et_2O	482	-15.6	-1.3
EtMgI	Et_2O	630	-20.6	-5.0

MgX_2 dissolved in an ether is identical to a Grignard solution, provided that sufficient time is allowed for equilibration. The rate of attainment of equilibrium falls $X = Cl > Br > I$; $R = $ primary $>$ secondary $>$ tertiary. This evidence points to the existence of an equilibrium, called the Schlenk equilibrium,

$$R_2Mg + MgX_2 \xrightarrow{ethers} 2RMgX; \quad K = [RMgX]^2/[R_2Mg][MgX_2]$$

This is consistent with the ready isotopic exchange which is observed when radioactive $^{25}MgBr_2$ or $^{28}MgBr_2$ is added to solutions of Et_2Mg or of EtMgBr. In ether RMgX predominates, whereas in THF there is a distribution close to random (Table 3.5). The difference between the two solvents arises from the greater degree of solvation by THF, especially of MgX_2.

Measurements of colligative properties (b.p. elevation, osmotic pressure etc.) show association of Grignard reagents in solution, particularly in less basic solvents (Fig. 3.12). Association increases at high concentration. In THF monomeric units predominate, whereas in diethyl ether dimerization is important. Association takes place through bridging halogen atoms. Such bridging groups are displaced from the metal by THF, but compete successfully with the weaker base diethyl ether, especially when $X = Cl$.

Solutions of Grignard reagents in ether conduct electricity, but the molar

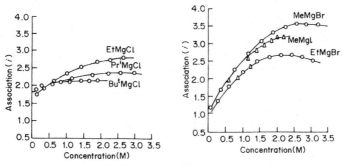

Fig. 3.12 Degree of association (i) as a function of concentration for Grignard reagents in diethyl ether. From E.C. Ashby, *Pure Appl. Chem.*, 1980, **52**, 545 (modified).

conductance is low. Magnesium migrates both to cathode and anode. Ion association must be important in solvents of low dielectric constant such as diethyl ether.

3.5 Beryllium

Much of the development of organoberyllium chemistry is due to Coates. The dialkyls are normally prepared from beryllium chloride and the Grignard reagent. After removal of most of the magnesium halide, dialkyls (ethyl to the butyls) are distilled under reduced pressure as etherates. Ether can be removed only by prolonged boiling under vacuum. Me_2Be is purified by repeated sublimation at low pressure. Small quantities of product free from ether can be made from beryllium and Me_2Hg. This method was applied to the preparation of $(CD_3)_2Be$ for spectroscopic measurements. The slow exchange of alkyl groups between triarylboranes and diethylberyllium over a period of days at room temperature has been used to make beryllium diaryls.

$$2Ph_3B + 3Et_2Be \longrightarrow 2Et_3B + 3Ph_2Be$$

Pyrolysis of dialkyls which possess β-hydrogen atoms proceeds mainly by loss of alkene. This provides a route to the polymeric beryllium hydride itself or to alkylberyllium hydrides. At higher temperatures Bu^t_2Be decomposes to the metal. It is possible to plate thin layers of beryllium on to surfaces at 280–300°C *in vacuo*.

$$(Me_3C)_2Be \xrightarrow[200°]{\Delta} BeH_2 \ (90\text{–}98\% \text{ purity}) + Me_2C{=}CH_2$$

$$(Me_2CHCH_2)_2Be \xrightarrow[200°]{\Delta} \frac{1}{n}[Me_2CHCH_2BeH]_n + Me_2C{=}CH_2$$

Alkylberyllium hydrides are oligomeric in benzene. They add rapidly to terminal alkenes (cf boron, p. 66).

Alkylberyllium hydrides can also be obtained in ether by the reaction,

$$RBeBr \xrightarrow[-LiCl]{LiH} RBeH \xrightarrow{Me_3N}$$

$$\Delta H = 13 \text{ kJ mol}^{-1}$$
$$\Delta S = 54 \text{ J K}^{-1} \text{ mol}^{-1}$$

54

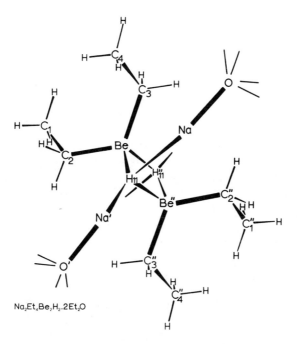

Na$_2$Et$_4$Be$_2$H$_2$.2Et$_2$O

Fig. 3.13

They can be isolated as ether or amine complexes which are dimeric in benzene. These hydrides contain a BeH$_2$Be bridge analogous to the BH$_2$B bridge in diborane, but differ sharply in that the BeH$_2$Be bridge is not split by bases such as trimethylamine. The proton n.m.r. spectrum of [(Me$_3$N)MeBeH]$_2$ is toluene-d$_8$ indicates the presence of *cis* and *trans* isomers. From a study of the spectrum over a range of temperature the enthalpy and entropy values for the isomerization have been obtained. The rather large entropy difference is due to the greater ordering of solvent molecules by the highly polar *cis* compared with the non-polar *trans* form.

Diethylberyllium forms an anionic complex, NaOEt$_2$[Et$_2$BeH], when stirred with sodium hydride in ether and part of its crystal structure is shown in Fig. 3.13. The presence of the Et$_4$Be$_2$H$_2$ unit should be noted, the [Et$_4$Be$_2$H$_2$]$^{2-}$ anion being isoelectronic with Et$_4$B$_2$H$_2$. It is likely that the bridging hydrogen atoms (located in the structure analysis) are electronically more closely associated with beryllium than with sodium. The four rather than two thin lines from oxygen indicate that the ethyl groups of the coordinated ether molecules can occupy two alternative positions.

There have been only limited studies on the alkylberyllium halides, 'RBeX', the analogues of Grignard reagents. Beryllium powder does react with alkyl halides on prolonged boiling. In ether the equilibrium R$_2$Be + BeCl$_2 \rightleftharpoons$ 2RBeCl lies mainly to the right. Acetone reacts with MeBeBr in the same way as MeMgBr. The initial product is a dimeric alkoxide which has the *trans* structure shown, with the alkoxide oxygen bonds all co-planar.

The second group alkyls form complexes with heterocyclic bases, many of which are brightly coloured. The occurrence of such coloured compounds is uncommon in main group chemistry. For example 2,2'-bipyridyl forms yellow Me_2Bebipy, red Et_2Bebipy and orange-red Et_2Znbipy. Similarly o-phenanthroline forms violet complexes with magnesium dialkyls. The colours may arise by transfer of an electron from the metal—carbon bonds as a group to the lowest vacant π-orbital of the heterocycle.

3.6 Calcium, strontium and barium

Organocalcium halides, RCaX (R usually I, in some cases Cl, Br) can be prepared from organic halides and metallic calcium in ethers, but to achieve good yields, great attention must be paid to experimental detail. In early reports such reactions were described as slow, erratic, difficult to initiate and complicated by Wurtz coupling giving R_2 and CaX_2. Impurities in the metal exert big effects; sodium, for example, is deleterious, whereas magnesium promotes formation of the organocalcium halide.

Calcium containing about 0.5% magnesium and only traces (0.002%) of sodium reacts smoothly at $-70°C$ with organic iodides in THF to give reasonable yields of RCaI (R = Me, 93%; Et, 68%; Pr^i, 57%; Ph (at $-30°C$), 97%). Aryl derivatives have also been obtained in diethyl ether. Addition of dioxan precipitates colourless ArCaI. dioxan, from which dioxan can be removed *in vacuo* at 110°C giving base-free ArCaI.

In general reactivity, organocalcium compounds behave more like organolithium derivatives than like Grignard reagents. Thus Bu^nCaI resembles Bu^nLi in forming Bu_2^nCO as well as Bu^nCO_2H (after hydrolysis) on reaction with carbon dioxide. Alkylcalcium halides also bring about the polymerization of butadiene or of styrene by an anionic mechanism.

Several dialkyls and diaryls of calcium have been prepared from the corresponding mercury derivatives R_2Hg and calcium or calcium amalgam in THF. Again experimental technique is crucial to success; the quality of the metal, its state of division and the cleanliness of its surface are especially important. Nevertheless, dimethylcalcium has been obtained in 96% yield as a white solid sparingly soluble in THF. Dialkyls and diaryls of Sr and Ba are even more difficult to prepare. Diethylstrontium attacks THF above $-30°C$. The reaction between Et_2Hg and Sr was initiated at $-10°C$ for 15 minutes and then completed during two hours at $-35°C$. The dibenzyls of Ca, Sr and Ba are yellow or yellow orange, as are Ph_2Sr and Ph_2Ba.

Much early work on organic compounds of Ca, Sr and Ba was carried out with solutions obtained by dissolving the metals in zinc dialkyls.

$$2ZnR_2 + M \xrightarrow[68°C]{benzene} MZnR_4 + Zn \ (M = Ca, Sr, Ba)$$

The complexes $MZnEt_4$ are monomeric in benzene. In cyclopentane as solvent, ^1H n.m.r. spectra over a range of temperatures indicate the presence of free dialkyls and a complex $CaZnEt_4$, which is probably present as a contact ion-pair. The exchange reaction presumably occurs *via* an alkyl-bridged species:

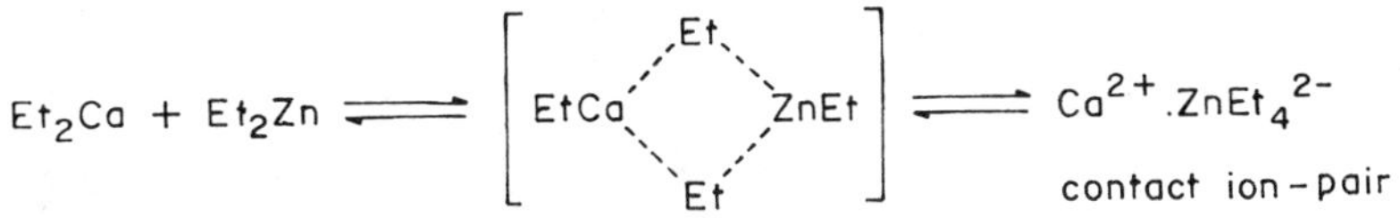

3.7 Zinc, cadmium and mercury

3.7.1 *Derivatives of the type RMX*

Ethylzinc iodide was first described by Frankland (1849) who made it from ethyl iodide and zinc, at a stage in the development of chemistry at which the ethyl group was represented C_4H_5 on account of uncertainties about atomic weights. Though diethylzinc, and other dialkyls R_2Zn, obtained by the thermal disproportionation of RZnI were valuable synthetic reagents until they were almost entirely superseded by Grignard reagents (from about 1900), the nature of RZnI long remained obscure. It was not until 1966 that its structure was unambiguously established. The alkylzinc iodides and bromides are formed from the alkyl halide and zinc which has been activated, for example, by copper. This may be done by alloying, formation of a zinc-copper couple by treating zinc powder with aqueous copper(II) acetate, or simply by mixing the two powdered metals. A particularly active form of metal can be obtained by reaction of zinc chloride with potassium in an ether solvent. The Schlenk equilibrium lies on the side of RZnX; this may be applied to the synthesis from the dialkyls.

$$R_2Zn + ZnX_2 \xrightarrow{ethers} 2RZnX \ (X = Cl, Br, I)$$

Ethylzinc chloride and bromide have tetrameric structures in the crystal, which are retained in solution in benzene. In contrast solid ethylzinc iodide forms polymeric chains. In all these structures zinc is four-coordinate. In ethers RZnX are monomeric through complex formation.

Organozinc halides are less reactive than Grignard reagents and generally have been replaced by the latter for organic syntheses. There are, however, a few

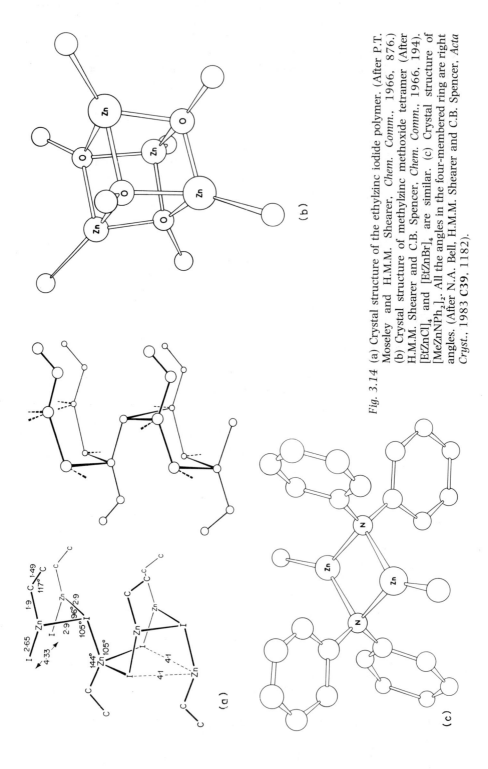

Fig. 3.14 (a) Crystal structure of the ethylzinc iodide polymer. (After P.T. Moseley and H.M.M. Shearer, *Chem. Comm.*, 1966, 876.) (b) Crystal structure of methylzinc methoxide tetramer (After H.M.M. Shearer and C.B. Spencer, *Chem. Comm.*, 1966, 194). [EtZnCl]₄ and [EtZnBr]₄ are similar. (c) Crystal structure of [MeZnNPh₂]₂. All the angles in the four-membered ring are right angles. (After N.A. Bell, H.M.M. Shearer and C.B. Spencer, *Acta Cryst.*, 1983 C39, 1182).

situations where they are especially useful. Allyl Grignard reagents are not easy to prepare in high yields. Good yields of allyl and benzylzinc bromides, however, result from activated zinc and the appropriate bromide in tetrahydrofuran. With aldehydes and ketones they afford, after hydrolysis, secondary or tertiary alcohols respectively. The allyl compounds react by allylic transposition, probably *via* a six-membered transition state.

Allyl and benzylzinc bromides are also useful for preparing hydrocarbons by crossed Wurtz coupling with reactive organic halides.

The Reformatsky reaction provides a method of converting α-bromoesters via the zinc reagent into β-hydroxy- or β-keto-esters. The reactions may be carried out either *in situ*, where zinc, the bromo ester and the carbonyl compound are all warmed together in a solvent (ether, benzene or 1, 2-dimethoxyethane), or in two stages by forming the zinc intermediate first. The actual zinc reagent is probably an enolate rather than an organozinc species:

59

The first three periodic groups

Cyclopropanes can be made stereoselecfively by the Simmons-Smith reaction. Diiodomethane, zinc and an alkene are heated together in ether. The intermediate is probably ICH_2ZnI.

Little is known of alkyl and arylcadmium halides. They cannot be obtained by reacting cadmium metal with organic halides, unless active cadmium is prepared by Rieke's method (p. 18). The dialkyls are usually prepared by the Grignard method. Their main use is to convert acyl halides into ketones. Pure Et_2Cd does not react with benzoyl chloride, but it does so in the presence of the magnesium salts which are by-products in the Grignard synthesis.

$$2EtMgX \xrightarrow[\text{ether}]{CdCl_2/} Et_2Cd + 2MgXCl \xrightarrow[\text{ii) } H_2O]{\text{i) PhCOCl}} 2PhCOEt\ (85\%)$$

Organomercury halides are accessible by a number of routes (Chapter 2). Generally the Grignard method is quite convenient. The polarizable $RHg—$ group ('soft' acid) prefers to bind the heavier halogens $(I > Br > Cl)$, so that if $X = Br$ or I, the product is mainly RHgBr or RHgI.

$$RMgX + HgCl_2 \longrightarrow RHgX + MgCl_2$$

The equilibrium $R_2Hg + HgCl_2 \rightleftharpoons 2RHgCl$ lies to the right, but it can be displaced by addition of ligands L which complex preferentially with $HgCl_2$ giving $HgCl_2L_2$ $(L = e.g.\ I^-,\ CN^-,\ Ph_3P)$. Organomercury halides are monomeric and, like the dialkyls, have linear structures. Their lack of association is consistent with the low Lewis acidity of many Hg(II) compounds.

The electrolysis of solutions of the lower alkylmercury halides in liquid ammonia results in the deposition on the cathode of an electrically conducting solid of metallic appearance, which decomposes at room temperature into equimolar proportions of R_2Hg and Hg. The methyl derivative is the easiest to obtain. It was the first example of an 'organic metal' to be described and appears to consist of MeHg cations and an equivalent number of free electrons. Its possible formulation as consisting of free radicals, or RHgHgR or amalgams of Hg + RHg·, has been excluded in various ways.

3.7.2 Organomercury compounds in organic synthesis

Organomercury compounds found application in organic synthesis from an early date. They are generally stable to attack by atmospheric oxygen although secondary and tertiary alkyls, allyls and benzyls do need some protection. They also resist attack by water and alcohols under neutral conditions and show little or no reactivity towards carbon electrophiles such as carbonyl compounds, oxiranes or alkyl halides. Mineral acids cleave both bonds in R_2Hg, but carboxylic acids only one; aryl-mercury bonds react more readily than alkyl-mercury, in spite of their greater strength.

Trihalomethylmercury derivatives have been developed, largely by D. Seyferth, as sources of dihalocarbenes. They are prepared by reactions of the type

$$PhHgBr + HCCl_3 \xrightarrow[\text{benzene}]{KOBu^t} PhHgCBrCl_2$$

A cyclopropane is formed in the presence of an alkene.

An iodomethyl derivative, prepared as follows, provides an alternative to the Simmons–Smith reagent.

The reaction between a mercury compound and chloroform in the presence of a base, mentioned above, is an example of a common type, in which a somewhat acidic C—H bond is mercurated.

With most metals, the reaction would proceed in the reverse direction (e.g. cleavage of organozinc derivatives by acid, HX). Carbon compounds which react in this way include alkynes, ketones and cyclopentadiene.

$$HC\equiv CH \xrightarrow[\text{alcoholic KOH}]{2RHgCl/} RHgC\equiv CHgR$$

$$H_3CCOCH_3 \xrightarrow[\text{ii) KI}]{\substack{\text{i)} Hg(NO_3)_2, HgO \\ CaSO_4}} H_3CCOCH_2HgI$$

$$Ph_2C=CH_2 \xrightarrow[\text{ii) NaBr}]{\text{i)} Hg(OCOCF_3)_2} Ph_2C=CHHgBr$$

3.7.3 Mercuration of aromatic compounds: comparison with thallation

$$ArH + HgX_2 \overset{K}{\rightleftharpoons} \underset{\pi-\text{complex}}{ArH.HgX_2} \overset{k_1}{\underset{k_{-1}}{\rightleftharpoons}} \left[\underset{\sigma-\text{complex}}{\overset{H \quad HgX}{\bigcirc}} + X^- \right] \overset{k_2, B}{\rightleftharpoons} ArHgX + BH^+ + X^-$$

Aromatic compounds undergo electrophilic mercuration. The old method involved heating the arene with mercury(II) acetate under reflux in acetic acid or in ethanol. Recently it has been found that mercury(II) trifluoroacetate in trifluoroacetic acid reacts at room temperature. The reactions are reversible; the isomer ratios depend on time, tending towards the statistical. The mechanism shown above has been proposed. The equilibrium constant K for π-complex formation has been estimated from changes in the UV spectra of arenes which occur on addition of $Hg(OCOCF_3)_2$. For benzene in CF_3CO_2H at 25°C, $K = 8.2 \, mol^{-1} \, dm^3$.

Multiple substitution of reactive aromatic compounds may occur. Thiophene gives a 2, 5-disubstituted and furan a 2, 3, 4, 5-tetrasubstituted derivative with mercury(II) acetate in boiling ethanol. The preferential mercuration of thiophene has been used for its removal from commercial benzene. Mercuration of ferrocene (p. 284) also occurs readily.

Thallation of aromatic compounds is similar. It is most efficiently carried out by using thallium(III) trifluoroacetate in trifluoroacetic acid. Reaction is often complete within a few minutes at room temperature; up to two days may be required for deactivated arenes like chlorobenzene. Thallation, like mercuration, is reversible. The reagent has a large steric requirement which normally hinders *ortho* substitution. With isopropylbenzene, for example the *para* isomer is formed preferentially at first but the *meta* isomer gradually builds up with increasing reaction time. This is a useful method of making *meta* substituted alkylbenzenes. Where a substituent is able to coordinate to the incoming thallium reagent, however, *ortho* substitution can result in spite of steric effects (Fig. 3.15).

The thallium group can be replaced by a wide range of substituents under mild conditions (Fig. 3.16). There is no need to isolate the arylthallium intermediate. Potassium iodide, for example, reacts with $ArTl(OCOCF_3)_2$ to give ArI and TlI (insoluble), presumably *via* unstable $ArTlI_2$.

3.7.4 Mercuration and thallation of alkenes and alkynes

(a) MERCURATION. These reactions are described by the general scheme

$$RCH{=}CH_2 + HgX_2 + HY \rightleftharpoons RCYHCH_2HgX + HX$$
$$(Y = e.g. \ OH, \ OR', \ OCOR', \ NR'R''; \ X = OCOCH_3 \ or \ OCOCF_3)$$

Mercury(II) acetate or trifluoroacetate is most often used. HY is often the solvent; where it is water ($Y = OH$), the reaction is termed oxymercuration or

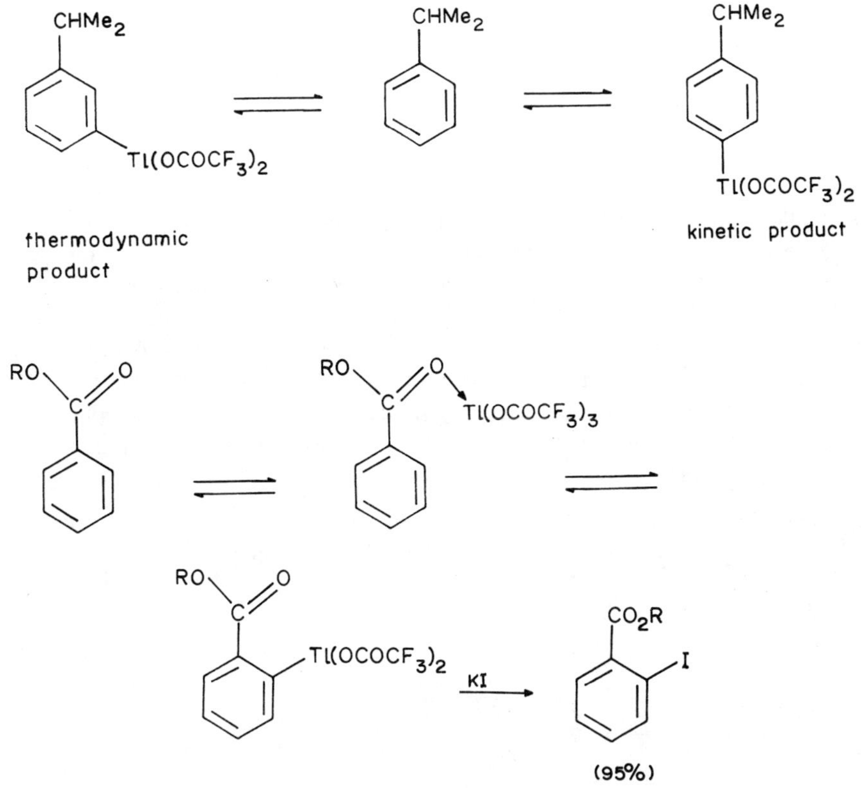

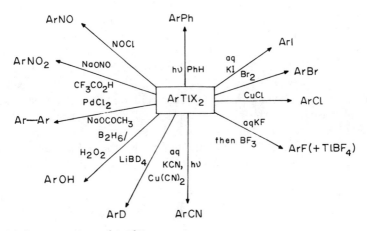

Fig. 3.15 Thallation of aromatic compounds.

(95%)

Fig. 3.16 Some reactions of ArTlX$_2$.

hydration. The addition obeys Markovnikov's rule, mercury becoming attached to the less substituted carbon atom of the alkene. The kinetics are consistent with rate determining attack of the nucleophile Y^- or HY on an intermediate bridged mercurinium ion. As with electrophilic bromination, the addition has *trans* stereochemistry. Mercury can readily be removed by reduction using sodium borohydride. The elements of water can be added across a double bond by treatment first with aqueous mercury(II) acetate followed by reduction. This Markovnikov hydration complements the anti-Markovnikov addition attainable by hydroboration (p. 70).

In alcohols as solvent, ethers are obtained. If more than one nucleophile is present, there is the possibility that a mixture will be formed. If neither the anion in HgX_2 nor the solvent is appreciably nucleophilic, however (e.g. $Hg(BF_4)_2$ in THF), vinylmercuration (p. 61) may be observed.

Acetaldehyde used to be made in Germany and in Japan by hydration of ethyne. Ethyne was passed into 20% aqueous sulphuric acid at about 70°C in the presence of mercury and iron salts. The mechanism may be related to the reactions discussed above. Serious consequences resulted in Minamata, Kyushu, Japan when Hg^{2+} from this process was discharged into the sea. It was converted anaerobically into very toxic methylmercury compounds, which were taken up by fish and crustacea as MeHgSMe. Several people died as a result of eating the fish. $MeHg^+$ is a very soft Lewis acid which complexes strongly with sulphur bases. Mercury is usually 2-coordinate, as in the L-cysteinato complex $MeHgSCH_2CH(\overset{+}{N}H_3)CO_2^- \cdot H_2O$. It also binds strongly to nucleosides, nucleotides and polynucleosides.

Organomercury compounds have been used medicinally as antisyphilitic drugs, general antiseptics and diuretics. They have been superseded by modern antibiotics in the first application, but they are still used in the others. One of the simplest organomercury antiseptics is phenylmercury acetate.

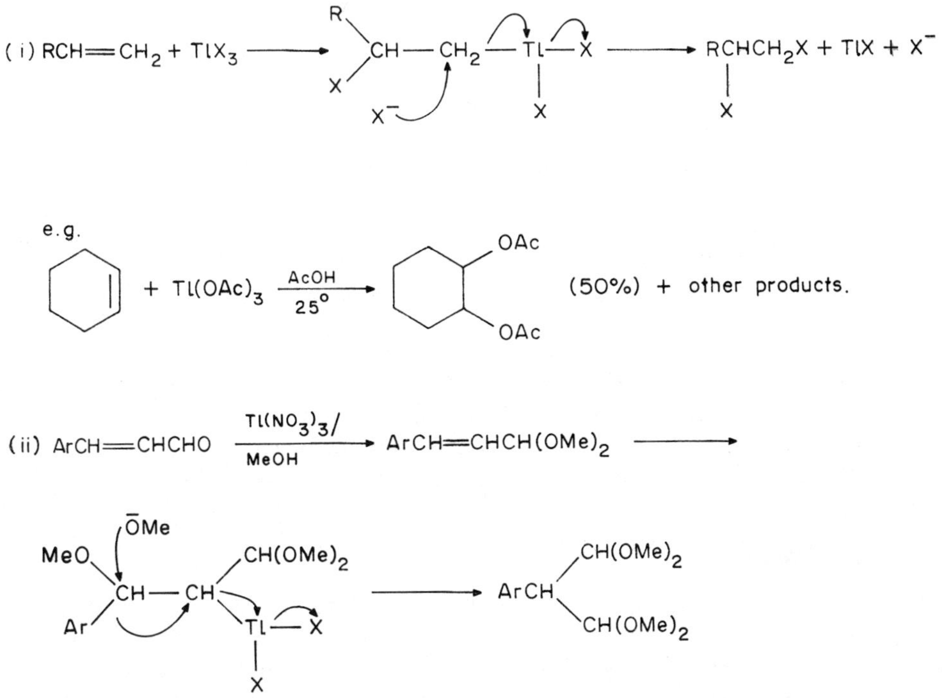

(i) $RCH=CH_2 + TlX_3 \longrightarrow$ [structure] $\longrightarrow RCHCH_2X + TlX + X^-$

e.g.

[cyclohexene] $+ Tl(OAc)_3 \xrightarrow[25°]{AcOH}$ [cyclohexane diacetate structure with OAc, OAc] (50%) + other products.

(ii) $ArCH=CHCHO \xrightarrow[MeOH]{Tl(NO_3)_3/} ArCH=CHCH(OMe)_2 \longrightarrow$

[structure] $\longrightarrow ArCH\begin{matrix}CH(OMe)_2\\CH(OMe)_2\end{matrix}$

Fig. 3.17 Two pathways for thallation of alkenes.

(b) THALLATION. Thallium(III) is a rather stronger oxidizing agent than mercury(II). $E^{\ominus}(Tl^{3+}/Tl^{+}) = 1.25\,V$; $E^{\ominus}(Hg^{2+}/Hg) = 0.85\,V$. Consequently thallation of alkenes is often followed by oxidation (loss of thallium(I)). Several pathways have been recognized, of which only two are mentioned here. Thallation may be accompanied by 1, 2-migration of substituents on elimination of TlX.

3.8 Organoboron compounds

Mononuclear organoboron compounds are of two types. First there are those which contain three-coordinate trigonal planar boron atoms. These include the trialkyl and triarylboranes, R_3B, and numerous derivatives R_nBX_{3-n} $(n = 1-3)$ in which R is formally replaced by a monovalent group X (X = halogen, OH, OR', SR', NR'$_2$). Generally these compounds are monomeric (contrast the Al analogues), but when X = H, dimerization through three-centre B—H—B bridges is normally observed. The boron atom has an empty $2p$ orbital, which may to a greater or lesser extent take part in π-bonding with substituents such as —NR'$_2$, OR' or F (p. 73).

Trialkylboranes can be prepared from the reaction of Grignard, organolithium

65

The first three periodic groups

or organoaluminium reagents on borate esters (trialkoxyboranes) or on the boron trifluoride ether complex.

$$B(OMe)_3 + 3RMgX \longrightarrow R_3B + 3Mg(OMe)X$$

The other boron halides BX_3 (X = Cl, Br, I) cleave ethers readily; this makes them particularly unsuitable when the Grignard method is adopted.

$$BCl_3 + EtOEt \longrightarrow EtOBCl_2 + EtCl$$

Partial substitution of chlorine by alkyl or aryl is conveniently achieved using organotin reagents.

$$Ph_4Sn + 2BCl_3 \longrightarrow 2PhBCl_2 + Ph_2SnCl_2$$

All trialkylboranes are readily oxidized in air. The lower members up to tri-n-butylborane are spontaneously inflammable, burning with the characteristic green boron flame. Peroxo intermediates are involved in the oxidations. Arylboranes are rather less sensitive to oxygen.

3.8.1 Hydroboration

By far the most widely used method for making boron–carbon bonds is by the addition of boranes to alkenes or alkynes. Much of the work in this area is due to H.C. Brown. This reaction is termed 'hydroboration'. Although it is reversible the equilibrium normally lies in favour of product rather than reactants in ether solvents at room temperature, so that the addition goes to completion. On heating above ca. 100°C organoboranes with a β-C—H function eliminate alkene, that is, the reverse reaction takes place. Isomerization of alkylboranes occurs by a series of these addition-elimination steps, leading eventually to terminal boranes. The alkene can then be displaced by a less volatile alkene. The direction of this isomerization is opposite to that predicted from relative stabilities of alkenes, for which the internal isomers are thermodynamically favoured. It is also opposite to that normally obtained using transition metal catalysts.

Although the reaction between diborane and an alkene proceeds only very slowly in the gas phase, in the presence of weak Lewis bases such as diethyl ether or dimethyl sulphide it occurs rapidly at room temperature. In early work it was customary to generate diborane *in situ* with the alkene substrate from Et_2OBF_3 and $NaBH_4$ dissolved in an ether such as diglyme. Diborane is not evolved until the second reaction is complete. Nowadays a solution of diborane in THF (approx.

$$3NaBH_4 + 4F_3B \cdot OEt_2 \longrightarrow 2B_2H_6 + 3NaBF_4 + 4Et_2O$$

$$7NaBH_4 + 4F_3B \cdot OEt_2 \longrightarrow NaB_2H_7 + 3NaBF_4 + 4Et_2O$$

$1 \, mol \, dm^{-3}$), which is obtained by passing the gas, generated as described, into the solvent, is often used. It is stable for several months at $0°C$ under a dry inert atmosphere and is commercially available. Another useful borane source is the complex $Me_2S \cdot BH_3$ which is supplied as a liquid m.p. $-40°C$ and which is stable at room temperature. Dimethyl sulphide (b.p. $38°C$) is volatile and is thus readily removed from products after reaction.

The addition of B—H takes place from the less hindered side of the double bond stereospecifically *cis*. The reaction is also generally quite regiospecific, boron entering at the less substituted position. Steric effects are important in determining the site of attack and the use of bulky boranes therefore leads to greater selectivity. The most generally accepted mechanism involves a four centre transition state.

The *cis* addition of 'R$_2$BH' to a substituted alkene can sometimes be complemented by the photochemical *cis* addition of a trialkylborane to the unsubstituted olefin, which affords the other geometrical isomer. Redistribution of organic groups about boron occurs very readily. Even though trialkylboranes exist as monomers, alkyl groups readily transfer between boron atoms. Unsymmetrical trialkylboranes are seldom stable above about $100°C$.

Table 3.6 Some hydroborating agents

Alkene	Product	Name	Properties
		Bis(3-methyl-2-butyl)borane Disisoamylborane 'Sia₂BH'	Can be stored for a few hours at 0°C in THF
		Dicyclohexylborane	m.p. 103–5°C. Keeps for 15 d at 0°C. Use as slurry in THF
		9-Borabicyclo[3.3.1]nonane '9-BBN'	m.p. 152–3°C, b.p. 195/12 mm. Stable. Relatively insensitive to air, moisture. Dimer in solid
 α = pinene		Diisopinocamphenylborane	Chiral. Useful for asymmetric syntheses *via* hydroboration
	(Me₂CHCMe₂)BH₂	t-hexylborane	Liquid, m.p. − 33°C. Isomerizes slowly at 25°C. Monomer in THF

All the species 'R₂BH' are dimers, R₂B $\overset{H}{\underset{H}{\cdots}}$ BR₂

Table 3.7. Selectivity in hydroboration using diborane and 9-BBN in THF

Alkene	$Bu^nCH{=}CH_2$		$MeEtC{=}CH_2$		$PhCH{=}CH_2$		$Pr^iCH{=}CHMe$		$Pr^iC{\equiv}CMe$	
	↑	↑	↑	↑	↑	↑	↑	↑	↑	↑
BH_3.THF	6	94	1	99	19	81	43	57	25	75
9-BBN	–	99	–	99	2	98	0.2	99.8	4	96

Hydroboration of alkenes with diborane (e.g. $BH_3 \cdot THF$ or $Me_2S \cdot BH_3$) gives mono-, di- or trialkylboranes, depending on the steric bulk of the alkene. Mono- and disubstituted alkenes generally yield R_3B, trisubstituted $(R_2BH)_2$ and tetrasubstituted $(RBH_2)_2$. Some useful bulky hydroborating agents can be prepared from substituted alkenes (Table 3.6). These show greater selectivity than diborane itself (Table 3.7). Their reactions with alkynes can stop at the alkenylboranes, whereas diborane itself gives dihydroboration.

The boron–alkyl linkage is not attacked by water or by mineral acids at ambient temperatures. Carboxylic acids, however, yield a hydrocarbon. Perhaps the most common use of hydroboration is to convert an alkene into an alcohol, by oxidative cleavage of the resulting borane by alkaline hydrogen peroxide. In this way formal anti-Markovnikov hydration of the alkene is achieved. Hydroboration is by no means confined to this; in fact the range of possible transformations is very large indeed (Fig. 3.18). The intermediate organoborane is often not isolated, the subsequent reactions being performed *in situ*.

Some of these reactions involve a 1,2-rearrangement which is typical of organoborate ions. In general terms this is represented by

69

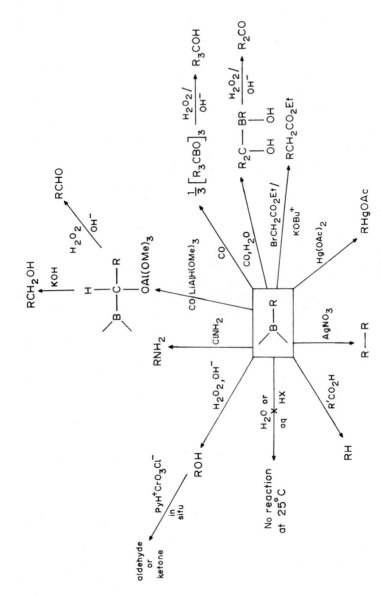

Fig. 3.18 Cleavage of $>$B—R bonds useful in organic synthesis.

The carbonylation of organoboranes presumably proceeds via a weak complex $R_3B \cdot CO$ which rearranges in a similar way to an acylborane. This can be converted as shown into a variety of organic products. In several of these reactions, however, only one or two of the three R groups on boron are used.

$$R_3B + \bar{C}\!\equiv\!\overset{+}{O} \longrightarrow \left[R_3\bar{B}\!-\!C\!\equiv\!\overset{+}{O}\right] \longrightarrow R_2B\!-\!\overset{O}{\underset{R}{C}}\overset{H^-}{\underset{OH^-}{\longrightarrow}}\ \cdots$$

$$\overset{O^-}{\underset{}{\underset{}{R_2\ BCHR}}}$$

$$\overset{OH^-}{\longrightarrow} RB\!-\!CR_2$$
$$\quad\quad \underset{OH\ \ O^-}{}$$

Similarly alkynyltrialkylborate ions, on protonation and oxidation yield ketones or tertiary alcohols according to the conditions.

$$R_3B + LiC\!\equiv\!CR^1 \longrightarrow R_3\bar{B}\!-\!C\!\equiv\!CR^1$$

$$\longrightarrow R_2B\!-\!\overset{R}{\underset{}{C}}\!=\!C\overset{E}{\underset{R^1}{}} \quad (E = \text{electrophile})$$

e.g. $R_3B + LiC\!\equiv\!CH \longrightarrow R_3\bar{B}C\!\equiv\!CH$

$$\xrightarrow{HCl} \overset{R}{\underset{R_2B}{}}\!C\!=\!CH_2 \xrightarrow[OH^-]{H_2O_2/} RCOCH_3 \quad (\text{ketone})$$

$$\downarrow aq.HCl$$

$$RB\!-\!CR_2CH_3 \xrightarrow[OH^-]{H_2O_2/} R_2CCH_3 \quad (\text{tertiary alcohol})$$
$$\underset{OH}{} \quad\quad\quad\quad \underset{OH}{}$$

As well as being hydroborating agents, boron hydrides can reduce certain functional groups. In Table 3.8 the behaviour of BH_3 and of 9-BBN in THF is compared with that of some other common hydridic reagents. It is therefore necessary to 'protect' groups such as $-CHO$, $-COR$ and $-CN$ if they are present in a molecule which it is intended to hydroborate.

The hydroboration of alkynes leads in the first instance to alkenylboranes. In the absence of sufficient steric effects, however, these react further. Using $H_3B \cdot THF$, for example, 1-alkynes give predominantly diboron derivatives.

$$BuC\!\equiv\!CH + H_3B \cdot THF \longrightarrow BuCH_2CH(B\!<)_2$$

71

Table 3.8. Summary of behaviour of various functional groups towards hydridic reagents

Functional/ Group	Reagent Solvent	$NaBH_4$ EtOH	BH_3 THF	9-BBN THF	$LiEt_3BH$ THF	$LiAlH_4$ THF
—CHO		+	+	+	+	+
—COR		+	+	+	+	+
—COCl		attacks solvent	—	+	+	+
—CO$_2$R		—	±	±	+	+
—CO$_2$H		—	+	±	—	+
—CN		—	+	±	+	+
—NO$_2$		—	—	—	+	+
C=C		—	+	+	—	—

+ reaction occurs;   — no reaction;   ± borderline, reaction often slow.

The problem is largely overcome by the use of bulky hydroborating agents, careful regulation of reaction conditions and if necessary, the use of excess of alkyne. The resulting alkenylboranes are useful synthetic intermediates.

3.8.2 *Lewis acidity of boranes*

Studies of the coordination compounds of boron, especially of its halides, alkyls and hydrides, have contributed substantially to our understanding of the factors which influence the stability of coordination compounds in general. The factors which particularly concern boron complexes are inductive and steric effects and reorganization energies, as explained below.

(a) INDUCTIVE AND STERIC EFFECTS. The formation of a coordination compound between, for example, a borane and an amine according to the equation

$$R_3B + NR_3' \rightleftharpoons R_3B \leftarrow NR_3'$$

involves transfer of charge from nitrogen to boron. This will be aided by the presence of electron attracting groups on boron and electron releasing groups on nitrogen. If such inductive effects were the only factors influencing the stability of the complexes, acceptor properties should decrease in the sequence $BF_3 > BCl_3 > BBr_3 > BMe_3 > BH_3$. In fact, towards a variety of donor molecules the order is $BBr_3 > BCl_3 > BF_3 \sim BH_3 > BMe_3$, showing that other factors predominate in the case of the halides. However acceptor strengths do decrease in the predicted order $BF_3 > MeBF_2 > Me_2BF > Me_3B$. The stabilities of the adducts formed between Me_3B and ammonia and the methylamines lie in the order $NH_3 < MeNH_2 < Me_2NH > Me_3N$. Considering inductive effects only, Me_3N should be the strongest donor. This prediction is supported by the ionization energies of the amines in the gas phase which fall smoothly from NH_3 (10.15 eV) to Me_3N (7.82 eV). The anomalously low free energy of formation of Me_3BNMe_3 is attributed to steric interference between methyl groups on nitrogen and those on boron. Towards the proton as acceptor, which is so small as to show no steric effects, the donor strengths of alkylamines measured in the gas phase or in inert solvents lie in the order $NH_3 < RNH_2 < R_2NH < R_3N$, predicted from inductive effects. By chance, the basic strengths in aqueous solution vary in the same way as their donor properties towards Me_3B. Here alkylammonium ions are stabilized by hydrogen bonding to the solvent which is least for Me_3NH^+ but greatest for NH_4^+. This offsets the differences which arise from inductive effects.

A striking demonstration of the effect of steric hindrance on donor strength is provided by the contrast between Me_3B adducts of triethylamine and quinuclidine. These amines should have essentially the same basicities through inductive effects, but the alkyl groups are pinned back away from the nitrogen in the latter, so that they do not interfere sterically with the trimethylborane. Whereas the adduct with Et_3N is hardly formed at all in the vapour phase at 100°C, the quinuclidine complex $HC(CH_2CH_2)_3N \rightarrow BH_3$ is only about 7% dissociated.

(b) REORGANIZATION ENERGIES. On coordination the shape of organo-borane acceptors R_3B changes from trigonal planar ($\angle\,'RBR = 120°$) to pyramidal ($\angle\,'RBR$ about 109°). In R_3B the p-orbital at right angles to the molecular plane is available for π-bonding with the attached group R. On complex formation this orbital becomes involved in the bond with the donor atom and the π-bonding with the groups R is essentially lost. π-Bonding between boron and the halogens is considered to decrease $BF_3 > BCl_3 > BBr_3$. It is this loss of π-bonding on formation of the adduct which is mainly responsible for the relative acceptor strengths of the boron trihalides $BF_3 < BCl_3 < BBr_3$. Even stronger π-interactions are present in trigonal boron-oxygen and boron-nitrogen compounds. Thus trimethylborate $B(OMe)_3$ is only a weak acceptor; it does form salts such as $Na[B(OMe)_4]$ and $Na[B(OMe)_3H]$. It seems that $B(NMe_2)_3$ forms no adducts which can be isolated, although it is very readily hydrolysed.

(c) ANIONIC COMPLEXES. Triphenylborane reacts with phenyllithium in ether to give lithium tetraphenylborate. This dissolves in water to form the ions Li^+ and BPh_4^-. The sodium salt is obtained by adding sodium chloride to the product of the reaction between $BF_3 \cdot OEt_2$ and a slight excess of phenylmagnesium bromide (after removal of magnesium as carbonate). Being soluble in water and some polar organic solvents it is a useful precipitating agent for large cations including K^+, Rb^+ and Cs^+, Tl^+, quaternary ammonium ions, Ph_4P^+, and $Co(C_5H_5)_2^+$. At one time potassium was often determined quantitatively by gravimetric analysis of $KBPh_4$, although nowadays this laborious technique has largely been superseded by instrumental methods. Ammonium tetraphenylborate decomposes on strong heating, providing a route to triphenylborane.

$$NH_4^+BPh_4^- \xrightarrow{240°} NH_3 + C_6H_6 + Ph_3B$$

Tetraalkylborates, e.g. $LiBMe_4$, are prepared similarly from LiR and R_3B, conveniently in benzene, which can be used for recrystallization (see p. 38 for structure). Trialkyl- and -aryl-boranes also react quantitatively with LiH or NaH to give trialkyl(aryl)hydridoborates.

$$LiH + Et_3B \xrightarrow[30\,min]{THF,\,65°C} LiBEt_3H$$

$$NaH + Bu_3^sB \xrightarrow[3\,h]{THF,\,65°C} NaBBu_3^sH$$

The products are useful reducing agents in organic chemistry. Those which contain bulky alkyl groups allow excellent steric control of the reductions, for example of cyclic ketones. Approach of the reagent, which adds H^- to carbon, takes place from the less sterically congested side of the molecule.

cis-2-methyl cyclohexanol
(>99%)

trans-2-methylcyclohexanol

3.8.3 Boron-oxygen compounds

Alkyl or aryl boron chlorides $RBCl_2$ and R_2BCl react readily with water to give boronic ($RB(OH)_2$) or borinic (R_2BOH) acids respectively. The alkylboronic acids are white crystalline solids which are slowly attacked by oxygen; the aryl analogues are essentially air-stable. Dehydration, carried out by azeotropic removal of water under reflux in benzene or toluene using a Dean-Stark apparatus, leads to the cyclic trimeric anhydrides or boroxines, $(RBO)_3$. The lower alkylboron chlorides are not very convenient to handle; they are spontaneously inflammable in air and very sensitive to moisture. Moreover their specific preparation from BCl_3 or R_3B is difficult; some R_2BCl is usually formed as well. Consequently other routes to boronic and borinic acids have been developed. Trialkylborates (trialkoxyboranes) react with Grignard reagents to give, after hydrolysis, alkylboronic acids, but yields often are rather low on account of the formation of some R_2BOH and R_3B as well.

$$B(OMe)_3 + RMgX \xrightarrow[\text{(ii) } H_3O^+]{\text{i) ether, } -78°C} RB(OH)_2 \ (+R_2BOH + R_3B)$$

Alkylboronic acids can be purified by conversion into diethanolamine chelates, which are crystalline air stable materials. Note that intramolecular coordination of nitrogen to boron gives the latter an octet configuration. Similarly dialkylborinic acids may be converted into chelate derivatives with 2-aminoethanol.

Monoalkylboron compounds may be obtained by hydroboration using catechol borane or $Me_2S.BHCl_2$.

The former is obtained from catechol (1, 2-dihydroxybenzene) and borane in

75

THF. It reacts cleanly with unhindered alkenes to yield, after hydrolysis, alkylboronic acids. Me_2SBHCl_2 is prepared by the redistribution reaction

$$2Me_2S.BCl_3 + Me_2S.BH_3 \rightleftharpoons 3Me_2S.BHCl_2 \quad (+ \text{ a little } Me_2S.BH_2Cl)$$

Hydroboration, followed by hydrolysis, alcoholysis etc. provides routes to many monoalkylboron derivatives.

$$Me_2S.BHCl_2 \xrightarrow[\text{THF}]{RCH=CH_2} Me_2S.BCl_2CH_2CH_2R$$

$$\xrightarrow{H_2O} RCH_2CH_2B(OH)_2$$
$$\xrightarrow{R'OH} RCH_2CH_2B(OR')_2$$
$$\xrightarrow{R'_2NH} RCH_2CH_2B(NR'_2)_2$$
$$\xrightarrow{R'NH_2} \tfrac{1}{3}(RCH_2CH_2BNR')_3$$

$Me_2S.BH_2Cl$, prepared similarly, is a precursor to a range of dialkylboron compounds.

$$Me_2S.BCl_3 + 2Me_2S.BH_3 \rightleftharpoons 3Me_2S.BH_2Cl$$

Trialkylboranes do not react at all readily with water or alcohols; temperatures of 150°C or more are needed to give R_2BOH or R_2BOR'. At these temperatures redistribution and elimination reactions pose a problem. Thiols, however, cleave one B—R bond at room temperature. The products R_2BSR' exchange the thiol group on treatment with amines, alcohols, water etc.

$$R_3B + R'SH \xrightarrow{-RH} R_2BSR'$$

$$\xrightarrow{H_2O} R_2BOH$$
$$\xrightarrow{R''OH} R_2BOR''$$
$$\xrightarrow{R''_2NH} R_2BNR''_2$$

Alkyl and arylboronic acids form cyclic esters with diols. The aryl derivatives are usually crystalline, air-stable materials which resist hydrolysis under neutral or acid conditions. They have been used to characterize compounds which contain the *cis* diol function and to 'protect' it during syntheses.

3.8.4 *Boron-nitrogen compounds*

The grouping B—N is isoelectronic with C—C. In 1926 Alfred Stock discovered borazine $(HBNH)_3$, which is isoelectronic with benzene and which has a similar planar ring structure. The analogy between B—N and C—C has been fruitfully extended, particularly by E. Wiberg and H. Nöth in Munich (Table 3.9). Commercial applications, apart from boron nitride itself, are very limited, on account of the ready hydrolysis of most of the compounds which contain 3-coordinate boron.

Although aminoboranes R_2N—BR_2 are isoelectronic with alkenes, the B—N bond has less double bond character than the C=C bond. Boron retains some Lewis acidity and nitrogen some basicity. This is shown by the tendency of the monomers to associate, usually giving cyclic dimers but sometimes trimers. Monomeric Me_2NBCl_2 is a colourless liquid which is rapidly hydrolysed in moist air. The molecule is planar in the gas phase. On standing the liquid gradually dimerizes to a white crystalline solid which is quite resistant to hydrolysis. Geometrical (*cis-trans*) isomerism has been observed using proton n.m.r. spectroscopy in compounds such as $(MeBClNMe_2)_2$.

$$Me_2NH + MeBCl_2 \longrightarrow Me_2NHBCl_2Me \xrightarrow[\substack{\text{heat or} \\ Me_3N(-HCl)}]{\text{ether}} Me_2NBClMe$$

Monomeric borazynes, the analogues of alkynes, are rare and are only curiosities. In C_6F_5BNAr (Ar = 2, 4, 6-trimethylphenyl) formation of oligomers is hindered by steric effects.

$$C_6F_5BCl_2 + ArNH_2 \longrightarrow C_6F_5B \equiv NAr + 2HCl$$

Normally the trimeric borazines, analogues of benzenes, are produced. Some methods of preparing alkylborazines are indicated. The borazine ring is retained on nucleophilic replacement of chlorine in chloroborazines by a variety of other groups.

$$3LiBH_4 + 3R'NH_2 \xrightarrow[-H_2]{\text{ether}} 3R'NH_2BH_3 \xrightarrow[-2H_2]{\text{heat}} (R'NBH)_3$$

$$3RBCl_2 + 3(Me_3Si)_2NH \longrightarrow (RBNH)_3 + 6Me_3SiCl$$

Planar regular rings are present in borazine and its derivatives. The B—N bond strength is high (about $500\,kJ\,mol^{-1}$) of the same order of magnitude as the C—C bonds in benzene. Hexamethylborazine may be heated to 450°C without significant decomposition. Compounds with B—H or B—Cl functions are less robust, being liable to undergo elimination of hydrogen or hydrogen halide.

The first three periodic groups

Table 3.9. Boron–nitrogen compounds and the isoelectronic carbon–carbon analogues.

Organic	Compounds	Boron-nitrogen	Compounds
Alkane	R_3C—CR_3	$R_3N \rightarrow BR_3$	Borane adduct
	$(R_3C)_2CR_2$	$[(R_3N)_2BR_2]^+X^-$	Boronium salt
Alkene	$R_2C{=}CR_2$	$R_2N{\rightleftharpoons}BR_2$	Monomeric aminoborane
Cycloalkane	$(R_2C)_n$	$[R_2N$—$BR_2]_n$	Dimeric, trimeric
		$\downarrow \quad \uparrow$	aminoborane
Alkyne	$RC{\equiv}CR$	$RN{\rightleftharpoons}BR$	Monomeric borazyne
Benzene	$(-RC{=}CR-)_3$	$(-RN$—$BR-)_3$	Borazine

Various borazine polymers have been prepared but found unsatisfactory for commercial application on account of their brittle nature and their susceptibility to hydrolysis and other chemical degradation.

Various heterocyclic compounds (borazarenes) have been made in which the B—N grouping forms part of an arene system. The following monocyclic species polymerized so rapidly that it could be identified only by its infrared and mass spectra. It behaved like a polarized butadiene. Other oligocyclic compounds,

$$Me_3NBH_3 + H_2C{=}CHCH_2CH_2NHMe \xrightarrow{\Delta} \quad \xrightarrow[heat]{Pd/C}$$

however, are much more robust. Many are stable in air, can be heated to fairly high temperatures and undergo some electrophilic aromatic substitution reactions.

$$\xrightarrow{H_3B.THF} \quad \xrightarrow{Pd/C \atop 300°}$$

9-aza-10-boranaphthalene

3.8.5 Spectroscopic studies of organoboron compounds

The characterization of organoboron compounds is greatly assisted by spectroscopic studies. This is especially true for boranes and carboranes which have complicated structures and for which the possibility of isomerism often exists. Data from as many techniques as possible are often required to assign a correct structure.

Boron–hydrogen stretching vibrations give rise to bands in the infrared spectrum in the region 2500–$2630\,cm^{-1}$ (B—$H_{terminal}$) and between 1500–$2200\,cm^{-1}$ (B—$H_{bridging}$). The latter are usually the less intense. A terminal BH_2 group is characterized by a strong doublet arising from symmetric and antisymmetric vibrations.

Boron has two naturally occurring isotopes, $^{11}_{5}B$ and $^{10}_{5}B$, relative abundance about 4:1. The mass spectrum therefore provides an excellent means of counting boron atoms in a molecule, provided that a molecular ion is observed. If n boron atoms are present, the relative intensities of the peaks at m/z = M ($^{11}B_n$), M-1 ($^{11}B_{n-1}{}^{10}B$), M-2 ($^{11}B_{n-2}{}^{10}B_2$) ··· are given by the coefficients of the expansion $(4x + 1)^n$—or, more exactly, $((81.17/18.83)x + 1)^n$. If $n = 4$, for example, the relative intensities are $345:320:111:17:1$.

Information about organoboron compounds can be gleaned from ^1H, ^{11}B and ^{13}C n.m.r. spectra. Both ^{10}B (I = 3) and ^{11}B (I = 3/2) nuclei possess spin and both are quadrupolar. In the proton spectra, coupling to both ^{10}B and to ^{11}B occurs, but the complex signals produced by the former are usually obscured by the more intense but simpler signals arising from the latter. A proton directly bonded to boron (^{11}B) gives rise to four lines of equal intensity, reflecting the four possible orientations of the ^{11}B nuclear spin ($3/2$, $\frac{1}{2}$, $-\frac{1}{2}$, $-3/2$). If attached to two equivalent boron atoms (e.g. the bridging hydrogens in diborane), seven lines of relative intensities $1:2:3:4:3:2:1$ are observed. Terminal protons resonate to low field of bridging protons. Coupling of ^{11}B to protons more remote from boron, for example in attached alkyl groups, can normally be ignored in interpreting the spectra.

^{11}B n.m.r. spectra are used almost exclusively in preference to ^{10}B spectra. The sensitivity of ^{11}B is about 10 times greater (Table 3.10). ^{10}B cannot interfere in ^{11}B n.m.r. spectroscopy as it has a completely different resonant frequency from ^{11}B. Bands are commonly broad on account of quadrupolar relaxation; this effect is more serious in trigonal than in tetrahedral derivatives. The range of chemical shifts in ^{11}B spectra is about 250 Hz. Electron density is important in determining the shift, but both inductive ($-I$) and perhaps most significantly mesomeric ($+M$) effects play their part. This is shown by the shift values for the series Me_nBX_{3-n} (X = F, OMe, NMe_2) for which the member BX_3 resonates to highest field. These are the compounds in which π-back-donation from X to boron is most important. ^{11}B nuclei in tetrahedral boron compounds such as BPh_4^- or $Me_3B.PMe_3$ are shielded, as expected from electron density considerations, relative to those in trigonal boron compounds.

As the natural abundance of ^{13}C is only about 1%, coupling to ^{13}C is not observed in ^{11}B spectra. ^{13}C signals may be broadened not only by coupling to ^{11}B but also on account of quadrupolar relaxation. Consequently a large number of

Table 3.10 Some ^{11}B chemical shifts (p.p.m. relative to $BF_3.OEt_2$ ($\delta = 0$))

Compound	$\delta(^{11}B)$	Compound	$\delta(^{11}B)$	Compound	$\delta(^{11}B)$
Me_3B	86.2	Me_3B	86.2	Et_3B	86.8
Me_2BF	59.0	$Me_2B(OMe)$	53.0	Ph_3B	68.0
$MeBF_2$	28.2	$MeB(OMe)_2$	29.5	$Me_3B.PMe_3$	-12.3
BF_3	11.5	$B(OMe)_3$	18.1	Me_4B^-	-20.7

pulses may be required (in FT n.m.r.) to obtain a reasonable signal, even with broad-band proton decoupling. The quadrupolar relaxation rate decreases with increasing temperature. At low temperatures therefore, relaxation is relatively fast, so that sharp resonance signals are observed, but with loss of ^{11}B—^{13}C coupling. As the temperature is raised, broadening occurs, followed by resolution into a 1:1:1:1 quartet from which the ^{11}B—^{13}C coupling constant can be measured.

3.9 Organometallic compounds of aluminium, gallium, indium and thallium

3.9.1 Alkylaluminium compounds

Alkylaluminium compounds are produced on a very large scale industrially. This is somewhat remarkable considering their extreme susceptibility to oxidation and hydrolysis, which makes handling hazardous. In 1975 about 20 000 tonnes were produced worldwide for sale and a further 90 000 tonnes were used as intermediates in the manufacture of linear alcohols and 1-alkenes. Their preparation depends on 'hydroalumination', or the addition of Al—H to an alkene. The key to the success of this method, which is much more suitable to industrial than to laboratory use, is the production of metal with an active surface. One way of achieving this is ball-milling the powder in contact with the metal alkyl.

The two main discoveries, both due to K. Ziegler, were first that Al—H bonds add to olefins, particularly in the absence of ethers and other bases, and second, that although aluminium metal does not form AlH_3 by direct reaction with hydrogen, it does take up hydrogen *in the presence of aluminium alkyl*.

With reactive alkenes such as ethene, the process is usually carried out in two stages, using preformed trialkylaluminium.

$$2Al + 3H_2 + Et_3Al \xrightarrow{150°C} 6Et_2AlH \xrightarrow[70°C]{C_2H_4} Et_3Al$$

With less reactive, bulky alkenes such as 2-methyl-1-propene, a single step suffices.

$$2Al + 3H_2 + 6Me_2C{=}CH_2 \xrightarrow[50-200\,atm]{110-160°C} 2(Me_2CHCH_2)_3Al$$

Some reactions of organoaluminium compounds are summarized in Fig. 3.18. By far the most important from a commercial standpoint are (1), the insertion of alkene into an aluminium—carbon bond

$${>}Al{-}R + H_2C{=}CH_2 \rightleftharpoons {>}Al{-}CH_2CH_2R \stackrel{C_2H_4}{\rightleftharpoons} {>}Al(CH_2CH_2)_2R \stackrel{C_2H_4}{\rightleftharpoons} {>}Al(CH_2CH_2)_nR$$

and (2) transalkylation, the displacement of one alkene by another.

$${>}AlCH_2CH_2R + H_2C{=}CHR' \longrightarrow {>}AlCH_2CH_2R' + RCH{=}CH_2$$

80

At low temperatures (90–120°C), using ethene, insertion is very much faster than transalkylation. Each of the alkyl chains grows by successive insertions of ethene. The rates of insertion are essentially independent of chain length. A Poisson type distribution* of chain lengths is produced in which the average length depends on the number of ethene molecules added per metal atom.

In the Alfol process, the mixture of aluminium alkyls is converted into linear primary alcohols by controlled oxidation followed by hydrolysis. The main use of C_6–C_{10} alcohols is to make plasticizers such as dialkylphthalates. C_{12}–C_{16} alcohols are precursors to surfactants and biodegradable detergents such as alkyl sulphates and alkyl sulphonates. Hydroformylation of alkenes provides an alternative route to these materials.

At about 180°C in the presence of excess ethene, alkene insertion and transalkylation take place at comparable rates. Under these conditions ethene is converted into a mixture of 1-alkenes, the aluminium alkyl acting as a catalyst. A geometrical distribution of oligomers is produced.† This tends to give too much butene. The aluminium alkyls are therefore sometimes mixed with this butene at 300°C/20 atm with very short contact times of about 0.5 s. This converts them into tri-i-butylaluminium (+ 1-alkenes) which then reacts with ethene in a separate reactor. The grown aluminium alkyls (now C_6–C_{10}) are then passed back to the transalkylation stage when they release the desired 1-alkenes.

$$\geq\!AlCH_2CH_2R + H_2C\!=\!CHC_2H_5 \longrightarrow \geq\!AlCH_2CH_2C_2H_5 + RCH\!=\!CH_2$$

$$\xrightarrow{C_2H_4} \geq\!Al(CH_2CH_2)_2C_2H_5 \quad \text{etc.}$$

In the Shell Higher Olefins Process (SHOP) a much more efficient catalyst based on transition metal compounds (probably nickel) is used to effect the oligomerization. Initially about 30% of the ethene is converted into the desired C_{10}–C_{18} olefins. The remainder (C_4–C_{10} and above C_{20}) are isomerized to internal alkenes and then converted into more useful fractions by metathesis.

The initial association between an alkene and an aluminium alkyl may involve interaction between the π-orbitals of the olefin and the Lewis acidic metal centre. Intramolecular association of this type is indicated by the lowering in the $\nu(C\!=\!C)$ stretching frequency in $Bu^i_2AlCH_2CH_2CH_2CH\!=\!CH_2$ compared with that in its ether complex or in the free alkene.

| (C=C)/cm⁻¹ | 1657 | 1635 | 1658 |

*The mole fraction X_p of $M(CH_2CH_2)_pCH_2CH_3$ is given by $X_p = (n^p e^{-n})/p!$ where n C_2H_4 are added to $\geq\!AlCH_2CH_3$.

†$X_p = \beta/(1 + \beta)^n$ where $\beta = k_2/k_1$; k_1 rate constant for insertion, k_2 for transalkylation.

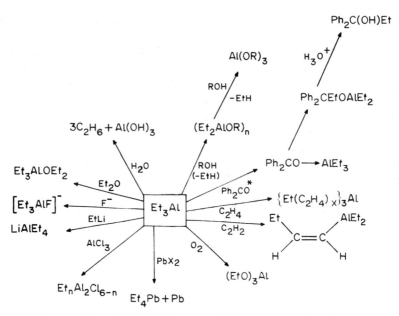

Fig. 3.19 Some reactions of triethylaluminium.* Side reactions when R has β-hydrogen e.g. ethyl.

e.g. $Et_3Al + Et_2CO \longrightarrow Et_2AlOCEt_3 \qquad\qquad \longrightarrow Et_3COH$ addition 55%

$\searrow Et_2AlOCHEt_2 + C_2H_4 \qquad \longrightarrow Et_2CHOH$ reduction 26%

$\searrow Et_2AlOCEt{=}CHMe + C_2H_6 \longrightarrow Et_2CO$ enolization 19%

The aluminiumalkyl-catalysed dimerization of propene, followed by thermal elimination of methane, provides an industrial route to isoprene, which itself is subsequently, polymerized by a transition metal aluminium alkyl catalyst:

$$H_3CCH_2CH_2C(CH_3){=}CH_2 \xrightarrow{\text{cracking}} H_2C{=}CHC(CH_3){=}CH_2 + CH_4$$

Polymerization of alkenes by Ziegler-Natta catalysts is discussed in Chapter 12.

3.9.2 The alkyls as Lewis acids

Towards amines, the acceptor strengths of the Group III methyls Me_3M, measured by the enthalpies of dissociation of the adducts, lie in the order $M = Al > Ga > In > B > Tl$ (Table 3.11). Apart from Me_3Tl possibly, for which there are few data, the trimethyls behave as hard acids, preferring nitrogen to phosphorus donors and oxygen to sulphur donors. In general the order of basicities seems to be $Me_3N > Me_3P > Me_3As > Me_3Sb > Me_3Bi$ and $Me_2O > Me_2S > Me_2Se \sim Me_2Te$.

Table 3.11. Enthalpies of dissociation (kJ mol^{-1}) of adducts Me$_3$ML

Base/acid	Me$_3$B	Me$_3$Al	Me$_3$Ga	Me$_3$In	Me$_3$Tl
Me$_3$N	74	Undisso-ciated at 135°C	88	83	Appreciably dissociated at 0°C
Me$_3$P	69	Undisso-ciated at 135°C	76	72	m.p. 27–28°C Probably more stable than Me$_3$N adduct
Me$_3$As	Appreciably dissociated at 0°C		42		
Me$_2$O		80	38		
Me$_2$S		80	33		
Me$_2$Se		68	42		

Measurements on trimethylaluminium are complicated by its dimerization. Formation of adducts with Me$_3$N, Me$_2$O and Me$_3$P goes to completion below at least 135°C, because the enthalpy of the reaction Me$_3$Al(*g*) + Me$_3$N(*g*) $\rightleftharpoons$ Me$_3$AlNMe$_3$(*g*) exceeds that of the dimerization 2Me$_3$Al(*g*) $\rightleftharpoons$ Me$_6$Al$_2$(*g*); $\Delta H^{\ominus} = -85$ kJ mol^{-1} (assuming that the entropy changes in both reactions are similar). Complicated equilibria result when the enthalpy of dissociation of the adduct is comparable with that of the dimer. This happens with Me$_2$S and Me$_2$Se. Steric factors are much less important for aluminium and the other Group III metals than for boron.

Not only are the organo derivatives of Al, Ga and In better acceptors than their boron analogues but they also are much more reactive towards cleavage by protonic acids, probably on account of the greater polarity of the metal–carbon bonds. This has already been noted in connection with their easy hydrolysis. Adducts with primary and secondary amines eliminate methane on heating.

e.g. 2Me$_3$Al $\longleftarrow$ NHMe$_2$ $\xrightarrow{110°C}$ Me$_2$Al$\langle$ ... $\rangle$AlMe$_2$ + 2CH$_4$

Similar dimers are formed by gallium and indium.

3.9.3 MOCVD

Mixed Group III and Group V compounds such as gallium arsenide (GaAs) and indium phosphide (InP) find important applications as semiconductors in optoelectronic devices such as solid state lasers, light-emitting diodes, solar cells and photodetectors. These devices require single crystal layers of the semiconductor material of exactly controlled thickness. The compounds MY (M = Al, Ga, In; Y = P, As) form solid solutions with each other over a wide range of compositions. By changing the composition of the mixed compounds in a layer, the semiconductor band gap can be adjusted to a desired value. The band gap determines the wavelength of a laser or the sensitivity range of a photodetector. To make these devices it is necessary to be able to deposit uniform layers over areas of at least $100\,cm^2$ to a thickness of between $20\,\text{Å}$ ($2 \times 10^{-9}\,m$) and $10^{-5}\,m$. The lattice dimensions of the layer must be the same as that of the substrate on which it is deposited, so that defects and dislocations are avoided. This is true, for example, of GaAs on AlAs. Semiconductor layers are usually grown from solution (liquid phase epitaxy) or from the gas phase (chemical vapour transport). A fairly recent development is to use the reactions of organometallic compounds as the source of the semiconductor material. This technique is termed metal-organic chemical vapour deposition (MOCVD). At high temperatures Me_3Ga and arsine react to give gallium arsenide in very high purity.

$$Me_3Ga + AsH_3 \xrightarrow{600°C} GaAs + 3CH_4$$

Similar reactions can be used to produce GaN, GaP, GaSb, the corresponding compounds of Al and In and also mixed solid solutions. The method can be extended to isoelectronic Group II/VI compounds such as ZnS, ZnSe or ZnTe ($Me_2Zn + H_2S$, H_2Se or Me_2Te) or Group IV/VI derivatives such as PbY (Y = S, Se, Te).

There are two disadvantages of using the free alkyls of the Group III elements. First, they are hazardous to handle. Secondly side reactions can sometimes cause problems; an involatile polymer $[InMePH]_n$ condenses in the inlet to the reactor when Me_3In and PH_3 are introduced. Both these difficulties are ameliorated by taking Lewis base adducts of the alkyls such as Me_3InPEt_3. They are generally crystalline and rather less sensitive to air and moisture than the free alkyls. They dissociate in the vapour phase giving a thermally stable Lewis base (e.g. Et_3P) which passes through the system unchanged.

The structures of some Lewis base adducts of Me_3Ga and Me_3In show some surprising features. In Me_3InPMe_3 the essentially planar Me_3In unit which is present in the free alkyl remains. The methyl groups attached to phosphorus and to indium eclipse each other, possibly on account of van der Waals attraction. Similar planar Me_3M units are present in the complexes with di- and tetramethylpiperidine. In spite of the long Ga—N bond ($2.20\,\text{Å}$, sum of covalent radii $2.01\,\text{Å}$) the former sublimes unchanged without elimination of methane. A linear polymeric structure involving trigonal bipyramidal indium coordination is found in the complex with the double ended ligand DABCO, 1,4-diazabicyclo-[2,2,2]octane.

3.9.4 *Organohalides*

While organoboron halides R_2BX and RBX_2 are monomeric, like R_3B or BX_3, the derivatives of Al, Ga and In associate through halogen bridges. Direct reaction between aluminium and methyl or ethyl halides gives the sesquihalides.

$$2Al + 3RX \longrightarrow R_3Al_2X_3 \quad (X = Cl, Br, I; \quad R = Me, Et, Ph).$$

The metal generally needs to be activated either with some preformed aluminium alkyl or with iodine or $AlCl_3$. Propyl and higher alkyl halides, however, give mainly $AlCl_3$ and hydrocarbons on reaction with aluminium. The sesquihalides disproportionate

$$2R_3Al_2X_3 \rightleftharpoons R_2AlX_4 + R_4Al_2X_2$$

$Me_3Al_2X_3$ reacts cleanly with salt, and the $Me_4Al_2Cl_2$ can be decanted from the complex halide and then distilled.

$$2Me_3Al_2Cl_3 + 2NaCl \longrightarrow 2Na[MeAlCl_3] + Me_4Al_2Cl_2$$

Another approach is by redistribution between R_3Al and $AlCl_3$. Halogen is a better bridging group than methyl to aluminium. The structure of $Me_4Al_2Cl_2$ (Fig. 3.1) is very similar to that of the Al_2Cl_6 dimer.

The fluoride $[Me_2AlF]_4$ is tetrameric in the crystal. It is isoelectronic with $[Me_2SiO]_4$ (p. 113) and has a very similar structure. The eight-membered ring is puckered and the bond angles $\angle$ AlFAl large (146°). Four coordinate aluminium is also present in alkoxides and amino- derivatives which result from partial reaction of R_3Al with alcohols or amines (p. 82). Cubane type tetramers e.g. $(PhAlNPh)_4$ are formed when triphenylaluminium is heated with an arylamine.

$$Ph_3Al + PhNH_2 \longrightarrow (PhAlNPh)_4 + 2PhH$$

Their structure should be contrasted with that of borazines. Other oligomers $(RAlNR')_n$ $(n = 6, 7, 8)$ have also been confirmed in other systems (Fig. 3.20).

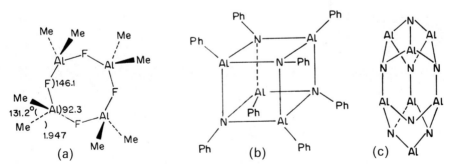

Fig. 3.20 Structures of some oligomeric organoaluminium compounds. (a) $[Me_2AlF]_4$, puckered ring; (b) $(PhAlNPh)_4$, cubane structure; (c) $(MeAlNMe)_7$, methyl groups omitted. See also Fig. 3.1.

The organogallium halides resemble the aluminium compounds closely. They can be made by redistribution between R_3Ga and $GaCl_3$. Indium reacts very slowly with alkyl bromides and iodides to give mixtures of $RInX_2$ and R_2InX which analyse as $R_3In_2X_3$. $MeInI_2$ can also be made from InI and MeI at room temperature. Solid Me_2InCl consists of units stacked one above the other, so that the indium atoms are five coordinate.

Electrophiles, such as protonic acids or halogens, readily cleave one methyl

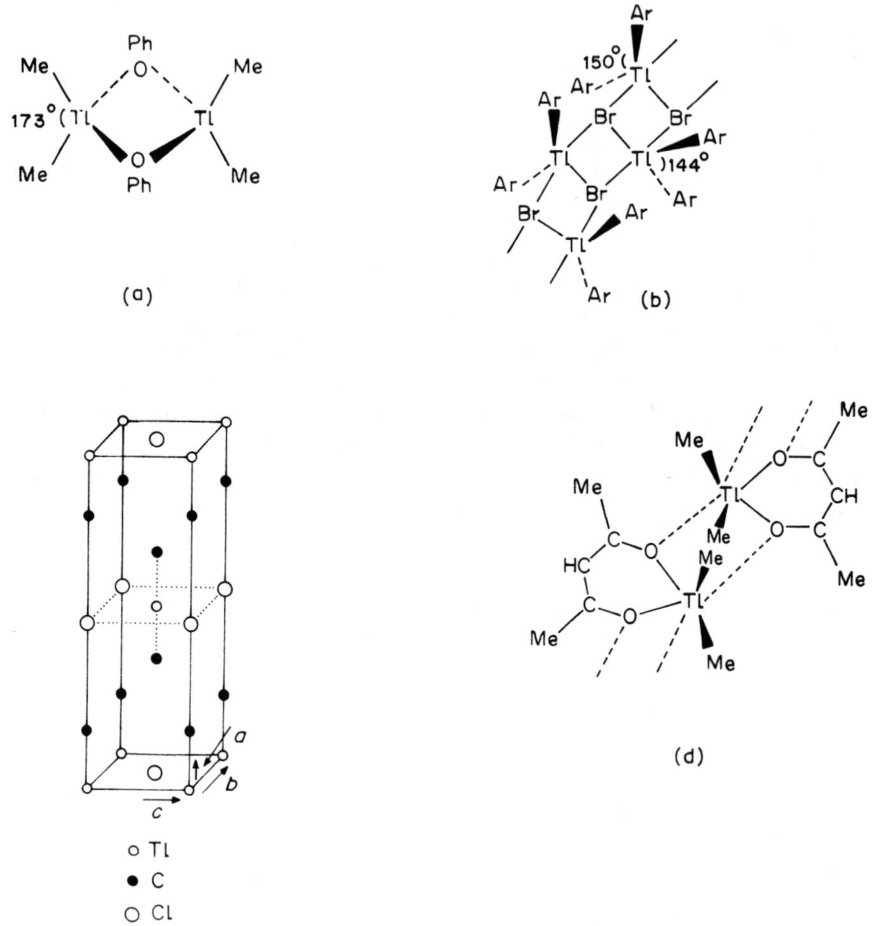

(a)

(b)

o Tl
• C
O Cl

Unit cell of Me_2 TlCl (From *Z. Naturforsch, Teil B*, 1974, **29**, 269).

(c)

(d)

Fig. 3.21 Different coordination environments in dimethylthallium compounds. (a) $[Me_2TlOPh]_2$; (b) $(2, 3, 5, 6-C_6F_4H)_2TlBr$; (c) Unit cell of Me_2TlCl; (d) Dimethylthallium acetylacetonate.

Table 3.12. Coordination environments in Me$_2$MX. (M = B, Al, Ga, In, Tl)

Compound	Coordination number of M	Environment of M	$\angle'CMC/°$
Me$_2$BF	3	Trigonal planar	120
(Me$_2$AlCl)$_2$	4	Tetrahedral	127
(Me$_2$InCl)$_n$	5	Trigonal bipyramidal	167
Me$_2$TlCl	6	Octahedral	180
(Me$_2$TlOPh)$_2$	4	Distorted tetrahedral	173
(Me$_2$TlSPh)$_2$	4	Distorted tetrahedral	163
(2,3,5,6-C$_6$F$_4$H)$_2$TlBr	5	Trigonal bipyramidal	144,150
Me$_2$Tl acac	6	Distorted pentagonal bipyramid (one equatorial site vacant)	170

group from Me$_3$Tl to form air-stable dimethylthallium derivatives.

$$Me_3Tl + HX \longrightarrow Me_2TlX + CH_4$$

$$Me_3Tl + X_2 \longrightarrow Me_2TlX + CH_3X$$

The halides Me$_2$TlX (X = C, Br, I) are high melting solids, which are sparingly soluble in all but strongly coordinating solvents. The fluoride Me$_2$TlF and hydroxide Me$_2$TlOH, which are prepared from the chloride and AgF or Ag$_2$O respectively in aqueous suspension, are very soluble in water. The hydroxide is apparently dissociated in solution into Me$_2$Tl$^+$ and OH$^-$ ions. The Me$_2$Tl$^+$ ion is isoelectronic with Me$_2$Hg and has a similar linear geometry.

The structures of several dimethylthallium derivatives have been determined in the solid state by X-ray crystallography. Some results are summarized in Table 3.12 and in Fig. 3.21. Thallium adopts coordination numbers of four, five and six. The trend to higher coordination numbers going down a Group is noted for Si to Pb (p. 102).

3.10 Organocopper and -silver compounds

3.10.1 *Organocopper compounds in organic synthesis*

Copper compounds have been used for many years in organic chemistry. The Ullmann synthesis of biphenyls in which an aryl iodide is heated with copper metal to temperatures over 200°C is considered to involve arylcopper intermediates rather than free radicals. The presence of 2- and 4-nitro groups aids coupling; 2-nitro-iodobenzene can be coupled by Cu$^+\bar{O}_3$SCF$_3$ in ammoniacal solution even at room temperature.

Copper(I) iodide reacts with monoalkyl and arylethynes in aqueous ammonia giving insoluble yellow, red or brown derivatives [Cu—C≡CR]$_n$. These reactions resemble those of mercury(II) (p. 61). These 'acetylides' couple with aryl iodides in pyridine under reflux. Similar species are intermediates in the oxidative coupling of alkynes which is catalysed by copper(I) salts, which provides a useful route to polyacetylenes.

$$ArI + Cu\text{—}C\equiv CPh \xrightarrow[\text{reflux}]{\text{pyridine}} ArC\equiv CPh \quad \text{(Castro reaction)}$$

$$RC\equiv CH \xrightarrow[\text{Cu}^+\text{ cat.}]{O_2\text{(air)}} RC\equiv C\text{—}C\equiv CR \quad \text{(Glaser reaction)}$$

$$RC\equiv CH + BrC\equiv CR' \xrightarrow[\substack{\text{base e.g}\\ \text{amine}}]{Cu^+/} RC\equiv C\text{—}C\equiv CR'$$

(Cadiot-Chodkiewicz reaction)

The most widely used reagents, however, which have been developed during the past 20 years, are the lithium organocuprates, R_2CuLi. They are prepared *in situ*, in ether or THF, by addition of two mol of alkyllithium to copper(I) halide below 0°C under a nitrogen atmosphere. It is convenient to use $Me_2S.CuBr$ as this is more resistant to oxidation in air than CuBr and can thus be introduced without contamination by copper(II).

R_2CuLi reagents couple readily with alkyl iodides or tosylates to give hydrocarbons.

$$R_2CuLi + R'I \xrightarrow{\text{ether}} R'\text{—}R + LiI + RCu$$

Vinyl halides, which are normally rather unreactive to nucleophiles are also attacked. Acyl chlorides afford ketones. This provides a useful alternative to organocadmium reagents (p. 60), although it is sometimes necessary to use a large excess of the copper complex to obtain satisfactory yields.

These reactions are examples of crossed Wurtz coupling, $RX + R'M \rightarrow R'\text{–}R + MX$. While organolithium reagents do react with alkyl halides to give hydrocarbons, metal–halogen exchange often also occurs, so that the desired product is contaminated with R—R and R'—R'. Elimination reactions giving alkenes also can cause problems. Moreover organolithium reagents attack many functional groups. This is less of a problem with less polar organometallics such as

88

those of zinc, mercury or tin, but except in rather special cases (e.g. allylzinc bromides + allyl halides), they are insufficiently reactive for coupling to occur. Such coupling, however, is catalysed by nickel and palladium complexes (p. 227).

Organocopper reagents add in a 1,4-fashion to $\alpha\beta$-unsaturated ketones. This contrasts with the behaviour of organolithiums, for which 1,2-addition predominates and of Grignard reagents, which are rather unpredictable (p. 46). Such additions have proved valuable in the synthesis, for example, of terpene and steroid derivatives. They can be used to introduce interannular substituents.

3.10.2 Complexes of copper(I) and silver(I) with alkenes and arenes

Copper(I) and silver(I) compounds form adducts with alkenes both in the solid state and in solution. With simple monoalkenes these are rather unstable, the bonding being much weaker than that present in palladium and platinum complexes (p. 196). Ethene and $AgBF_4$ are said to form four solid complexes of compositions 3:1, 2:1, 3:2 and 1:1 $C_2H_4 : AgBF_4$. Solid silver nitrate takes up ethene forming $C_2H_2 . 2AgNO_3$ as white needles which decompose above $-30°C$. Stability constants for complex formation in many of these systems have been obtained by, for example, measuring the distribution of alkene between an organic solvent such as carbon tetrachloride and aqueous silver nitrate.

The ability of aqueous copper(I) solutions to absorb alkenes and alkynes is used industrially. In the steam cracking of naphtha into alkenes, which is carried out at 770°C/1 atm, the main product is ethene. Propene and lesser amounts of alkynes, 1,3-butadiene and butenes are also produced. A chilled aqueous ammoniacal copper(I) acetate solution is used as the absorber. The more unsaturated the hydrocarbon, the more stable is the complex. In the first stage the alkynes combine with copper, the complex is withdrawn and the alkynes liberated by heating. The copper solution is then recycled at a lower temperature to remove the 1,3-butadiene, allowing butanes and butenes to pass on. The 1,3-butadiene is freed by heating the solution of complex in a desorber and is finally fractionated.

Gas chromatographic columns in which the stationary phase consists of silver nitrate dissolved in a polyethylene glycol are useful for separating alkenes. These are specifically retarded with respect to alkanes of the same carbon number.

The bonding of alkenes to copper and silver probably resembles that found in palladium or platinum complexes. The interactions are much weaker, probably on account of the filled d^{10} configuration of the former elements which lowers the participation of d-orbitals.

89

Silver perchlorate dissolves in benzene and a crystalline adduct $C_6H_6AgClO_4$ (dec. 145° in a sealed tube) can be isolated from the solution. A rather old X-ray diffraction experiment (1958) showed a chain structure.

3.11 Complexes of main group elements with unsaturated hydrocarbons

Copper(I) and silver(I) possess filled d^{10} shells of electrons. Nevertheless they form complexes, discussed above, with unsaturated organic ligands. Can main group elements which possess no filled d-orbitals (e.g. Al) or in which the (n-1) d-orbitals are very low in energy (e.g. Ga) form any similar complexes? It is now clear that they can although their extent and variety is unlikely to rival those of the d-block metals. Some evidence for weak π-interactions of this type is mentioned on p. 62 for mercury and on p. 33 for aluminium.

3.11.1 Cyclopentadienyls

The bonding in cyclopentadienylsodium, C_5H_5Na, is generally regarded as ionic in character. The $C_5H_5^-$ ion has six π-electrons and according to Hückel's $4n + 2$ rule is aromatic. Considerable covalent character, however, probably exists in cyclopentadienyl derivatives of the elements of Groups II, III and IV. The cyclopentadienyl group exhibits a variety of bonding modes which may be compared with those found in transition metal complexes (p. 278).

Bis-cyclopentadienylberyllium, obtained from $BeCl_2$ and sodium cyclopentadienide in THF, is a very reactive compound, unstable in air and hydrolysing with violence. It sublimes under reduced pressure and decomposes at about 70°C. It has a large dipole moment (2.46 D) in benzene suggesting an unsymmetrical structure. Its proton n.m.r. spectrum, however, shows only one sharp peak from $-100°$ to 50°C. A single crystal X-ray diffraction study, carried out at $-120°C$ revealed a 'slip sandwich', in which one ring is symmetrically bound (η^5), but the other is attached through only one carbon atom (η^1).

The molecules CpBeX (X = Cl, Br, Me) have C_{5v} symmetry in which the ring is bonded in a symmetrical fashion.

Bis(cyclopentadienyl)magnesium can be made in good yield by a reaction which is remarkable for an organometallic compound. Solid magnesium reacts with cyclopentadiene at 500–600°C, the metal then having an appreciable vapour pressure. The product m.p. 176°C, sublimes readily at 100°C. It has a symmetrical sandwich structure, similar to that of ferrocene (p. 279). It dissolves in hydrocarbons, suggesting that intermolecular interactions are low. The bis-cyclopentadienyls of calcium, strontium and barium are made from the metal and cyclopentadiene in THF or liquid ammonia. The bonding is considered to be largely ionic. The environment of the metal in solid Cp_2Ca shows that the organic ligands form weak bridges between the units, in a way reminiscent of the corresponding lanthanide compounds.

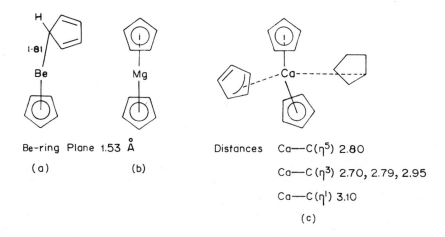

Be-ring Plane 1.53 Å   Distances   Ca—C(η^5) 2.80

     (a)   (b)   Ca—C(η^3) 2.70, 2.79, 2.95

                                            Ca—C(η^1) 3.10

                                                    (c)

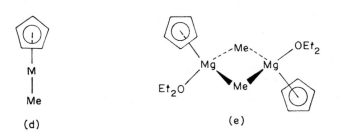

(d)   (e)

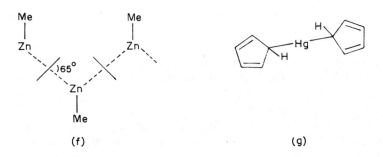

     (f)   (g)

Fig. 3.22 Structures of some cyclopentadienyl compounds of Group II elements. (a) Cp$_2$Be, a 'slip sandwich'; (b) Cp$_2$Mg, a symmetrical sandwich; (c) Environment of Ca in Cp$_2$Ca; (d) CpBeMe and CpZnMe$_{(g)}$; (e) Bridging methyl groups in [CpMg(OEt$_2$)]$_2$; (f) Bridging cyclic pentadienyl groups in MeZnCp$_{(s)}$; (g) Cp$_2$Hg, a fluxional η^1-Cp molecule.

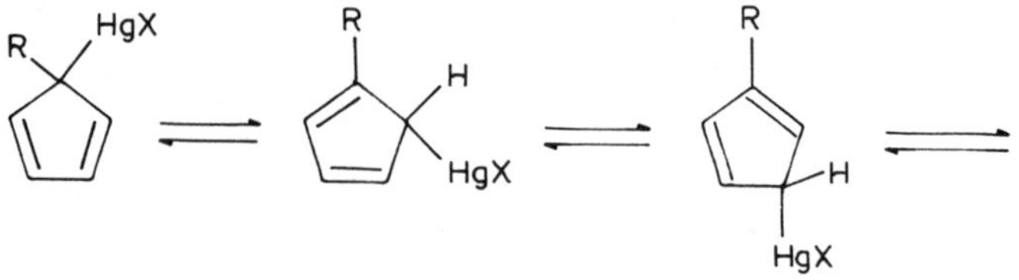

The first three periodic groups

Fig. 3.23 Fluxional behaviour of CpHgX (X = Cl, Br, Me, Cp).

The remarkable variety in structures is continued in the compounds of Zn, Cd and Hg. In the vapour CpZnMe has the same structure as CpBeMe. In the solid state the units associate into zigzag chains bridged by cyclopentadienyl ligands.

The ^1H n.m.r. spectra of Cp_2Hg, CpHgCl, CpHgBr and CpHgMe each show only a single line at room temperature. Of these examples, a frozen, limiting spectrum has been obtained only for CpHgCl, at $-113°C$. This shows conclusively that the cyclopentadienyl group is bonded η^1. Using ^{13}C n.m.r. the rearrangement has been shown to take place by a series of 1,2-shifts (cf. p. 300), with an activation energy of only $32\,kJ\,mol^{-1}$ (Fig. 3.23).

Treatment of $InCl_3$ with excess of sodium cyclopentadienide in THF, followed by sublimation affords CpIn rather than Cp_3In, which is formed in very low yield as a by-product. Cp_3In can be prepared using LiCp and $InCl_3$, but it is rather unstable. The indium atom in CpIn has a non-bonding pair of electrons and hence forms a 1:1 adduct with boron trifluoride. (cf. p. 292). In the vapour it is monomeric, and has C_{5v} symmetry. In the solid, however, it polymerizes into chains *via* bridging cyclopentadienyl groups.

Cyclopentadienylthallium is precipitated when cyclopentadiene is added to aqueous thallium(I) sulphate in the presence of sodium hydroxide. It is stable to air and moisture although like many organothallium compounds, it is slightly light sensitive. It is a useful reagent for exchanging the cyclopentadienyl group for chloride, especially in transition metal chemistry. The reaction is usually carried out in an inert polar medium such as dichloromethane. Insoluble TlCl is precipitated, leaving the desired cyclopentadienyl complex in solution. e.g. $2TlCp + HgCl_2 \rightarrow Cp_2Hg + 2TlCl$. CpTl is monomeric in the vapour, but, like CpIn, is associated in the solid state.

Cyclopentadienyl complexes are formed by most elements, including the very electropositive alkali metals and alkaline earths. While the bonding of the C_5H_5 unit to d-block transition elements is considered to be largely covalent in character, involving the d-orbitals of the metal (p. 205), compounds such as Cp_2Mg and CpIn have often been described as 'ionic'. It does seem plausible, however, that even here there is significant covalent interaction between the carbon $2p$ orbitals of the ligand and the valence orbitals (ns, np and possibly nd) of the main group element. This is most reasonable in cases such as CpTl in which the carbanionic nature of the cyclopentadienyl group is relatively low.

3.11.2 Arene complexes

Magnesium reacts with anthracene in tetrahydrofuran to give a brilliant orange complex Mg(anthracene) (THF)$_3$. This compound dissociates above room temperature to give a very active form of magnesium. Organic halides react with it to form Grignard reagents; allyl bromide affords a 98% yield, whereas rather low yields are obtained by the conventional method on account of coupling. While bulk magnesium reacts reluctantly with hydrogen, the complex rapidly forms MgH$_2$ in the presence of catalytic amounts of TiCl$_4$ or CrCl$_3$. The hydride has a high surface area and is very reactive, adding to alkenes to give magnesium alkyls and reducing organic halides to hydrocarbons.

On heating to 250°C hydrogen is liberated reversibly. The Mg/MgH$_2$ system could prove useful as a storage system for hydrogen; it can contain up to 7.65% hydrogen by weight. Hydrogen is a possible alternative fuel to petrol but there are problems in its transport and storage, particularly as a liquid. Metal hydride systems could provide a solution to these difficulties.

$$\text{Mg} + \text{anthracene} \xrightarrow{\text{THF}} \text{Mg(anthracene)} \xrightarrow{\text{H}_2} \text{MgH}_2 + \text{anthracene}$$

$$-\,\text{H}_2/250°\text{C}$$

Magnesium also reacts with 1,3-dienes in THF. Derivatives of composition Mg(C$_4$H$_6$), Mg(C$_4$H$_6$)$_2$ and Mg(C$_4$H$_6$)$_3$ have been isolated as solvates. They are almost certainly polymeric.

Salts of Ga(I), In(I) and Tl(I) are often readily soluble in aromatic hydrocarbons. Ga[GaCl$_4$], for instance, forms a 70% solution in benzene at room temperature. In spite of the similar ionic radii of Ga$^+$ (1.20 Å), In$^+$ (1.40 Å) and Tl$^+$ (1.50 Å) and K$^+$ (1.38 Å), KGaCl$_4$ is quite insoluble. Similarly complex halides of tin(II) and lead(II), such as M(AlCl$_4$)$_2$ also dissolve in arenes. Crystalline materials (η^6-C$_6$H$_6$)SnCl.AlCl$_4$ and (η^6-C$_6$H$_6$)M(AlCl$_4$)$_2$. C$_6$H$_6$ (M = Sn, Pb) have been isolated and their structures determined by X-ray diffraction.

If hot benzene is saturated with Ga[GaCl$_4$] and the solution cooled to 20°C, colourless crystals can be grown which have the composition [(C$_6$H$_6$)$_2$Ga$_2$Cl$_4$]$_2$.3C$_6$H$_6$. These crystals lose benzene readily under vacuum or on purging with nitrogen. The mesitylene complex, however, is somewhat more stable. In these compounds, bent sandwiches, (Arene)$_2$Ga, in which the arenes are symmetrically bonded to gallium, are linked together through bridging GaCl$_4$ units to form rings (arene = benzene) or chains (arene = mesitylene) (Fig. 3.24). Some related indium and thallium complexes with even more complicated structures have also been studied.

The ions Ga$^+$, In$^+$, Tl$^+$, Sn^{2+} and Pb^{2+} have the electronic configurations Core $(n-1)d^{10}ns^2$. The three p-orbitals are therefore vacant and can accept electrons from the filled π-orbitals of the arene. Transition elements are also capable of η^6-coordination with arenes (p. 204), but here there are important additional interactions involving the metal d-orbitals. Copper(I), silver(I) and gold(I) form

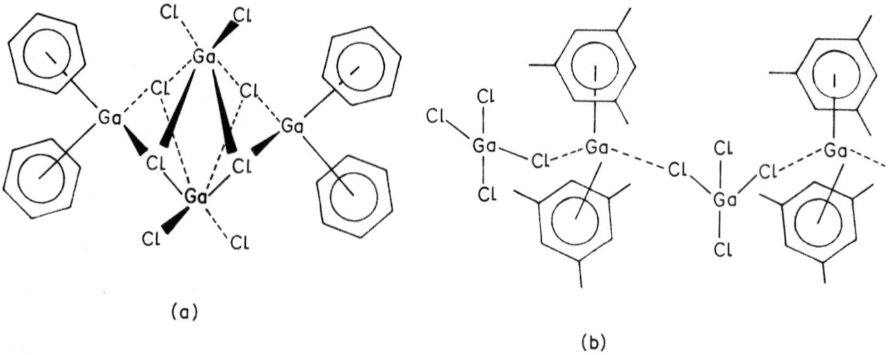

The first three periodic groups

(a)

(b)

Fig. 3.24 (a) Structure of $[(C_6H_6)_2Ga_2Cl_4]_2$ unit in $[(C_6H_6)_2Ga_2Cl_4]_2 \cdot 3C_6H_6$. (b) The chain structure of $[(C_6H_3Me_3)_2Ga_2Cl_4]$.

weak complexes with arenes. They differ from the Group III and IV compounds in that generally the arene is unsymmetrically bonded to the metal through only two or three of the carbon atoms of the ring.

3.12 References

Bogdanovic, B. (1985) Catalytic synthesis of organolithium and magnesium compounds and of lithium and magnesium hydrides. *Angew. Chem. (Int. Edn)*, **24**, 262.

Brown, H.C. (1975) *Organic Synthesis via Boranes*, Wiley-Interscience, New York.

Brown, H.C. and Singaram, B. (1985) An asymmetric synthesis that's really general. *Chem. Tech.*, 572.

Coates, G.E. and Wade, K. (1967) *Organometallic Compounds*, (*The Main Group Elements*, Vol. I), 3rd edn, Methuen, London.

Jutzi, P. (1986) π-Bonding to main group elements. *Adv. Organomet. Chem.*, **26**, 217.

Kauffmann, T. (1982) Heavy main group organometallics in organic synthesis. *Angew. Chem. (Int. Edn)*, **21**, 401.

Larock, R.C. (1986) *Organomercury Compounds in Organic Synthesis*, Springer Verlag, Berlin.

Miginiac, L. (1985) in *The Chemistry of the Metal–Carbon Bond*, (Vol. 3, Ch. 2), (Eds F.R. Hartley and S. Patai), Wiley, New York.

Moss, R. (1983) MOCVD. *Chem. Brit.*, **19**, 733.

Oliver, J.P. (1977) Structures of organic derivatives of Main Group II to III. *Adv. Organomet. Chem.*, **15**, 235.

Posner, G.H. (1980) *An Introduction to Synthesis using Organocopper Reagents*, Wiley, New York.

Schmidbaur, H. (1985) Arene complexes of univalent Ga, In and Tl. *Angew. Chem. (Int. Edn)*, **24**, 893.

Setzer, W.N. and Schleyer, P. von R. (1986) X-ray diffraction measurements on lithium compounds. *Adv. Organomet. Chem.*, **24**, 353.

Smith, K. (1982) Organolithium reagents. *Chem. Brit.*, **18**, 29.

Urbanski, T. (1976) History of the discovery of Grignard reagents. *Chem. Brit.*, **12**, 191.

Problems

1. Explain the following observation: The reaction of CH_3MgCl and $TlCl_3$ in diethyl ether gives a white solid, stable in air, which contains C, H, Cl and Tl (Compound A). If an excess of CH_3Li is used in place of CH_3MgCl, a volatile solid (Compound B), is obtained, which inflames in air. Compound B, when treated with dilute hydrochloric acid, produces Compound A, with evolution of a gas. (*University College, London*).

2. The proton n.m.r. spectrum of MeMgBr in THF consists of a single line at room temperature, but at $-80°C$ two lines of relative intensity 0.55:1 are observed, the resonance at high field being the more intense. If Me_2Mg is added to the solution at $-80°C$, the relative intensity of the high field resonance is increased. Comment on these results. (*University College, London*).

3. Diborane and dimethyl phosphine $((CH_3)_2PH)$ form an initial product *A* which contains 14.25%B, 40.90%P, and 31.66%C. The ^1H n.m.r. spectrum of *A* indicates the presence of three types of proton with relative abundances 6:3:1. When *A* is heated to 150°C, hydrogen is produced, together with a second product *B* which contains 14.63%B. Alkaline hydrolysis of 1 mole of *B* yields 6 moles H_2, and the ^1H n.m.r. spectrum of *B* shows two proton environments with relative abundances 3:1. Identify *A* and *B*. (*University of Southampton*).

4. A metal *A* reacts with dimethylmercury to give metallic mercury and the mercury free compound *B*, *B* contains 50.0% carbon and has the empirical formula C_3H_9A. The mass spectrum of *B* gives a molecular ion peak at $m/z = 144$, and the ^1H n.m.r. spectrum at 20°C consists of a sharp singlet at $\delta = -0.31$, which at $-65°C$ becomes two sharp singlets at $\delta = +0.07$ and $\delta = -0.50$, with relative intensities 1:2.

B reacts with methylamine to produce the complex *C* which has the molecular formula $C_4H_{14}NA$. On heating *C* at 70°C for 4 h methane is evolved to produce the compound *D* with the empirical formula $C_3H_{10}NA$, the mass spectrum of which gives a molecular ion peak at $m/z = 261$. Careful crystallization of *D* separates it into two geometrical isomers.

Upon heating *D* at 180°C for 24 h methane is lost to produce *E* which has the empirical formula C_2H_6NA. The mass spectrum of *E* shows a molecular ion peak at $m/e = 497$. The ^1H n.m.r. spectrum of *E* shows six different methyl group environments at $\delta = 2.83(1); 2.66(3); 2.49(3); -0.31(1); -0.33(3);$ and $-0.41(3)$; (the relative intensities of the absorptions are given in brackets).

Identify *A, B, C, D,* and *E* and derive the structures of *B, C, D,* and *E*. (*University of Exeter*).

The first three periodic groups

5. What organometallic intermediates are involved in the following transformations, and what will be the final organic products?

(a)

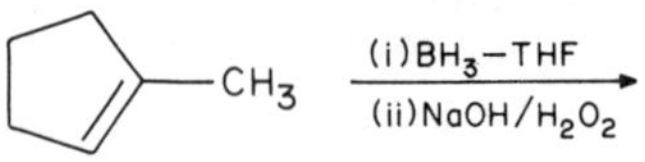

$$\text{(i) } BH_3-THF$$
$$\text{(ii) } NaOH/H_2O_2$$

(b)

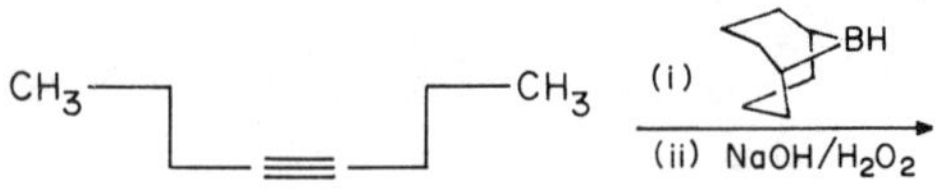

(i)

(ii) $NaOH/H_2O_2$

(c) 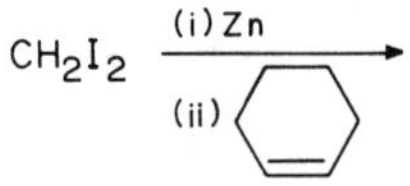 CH_2I_2

(i) Zn

(ii)

(d)

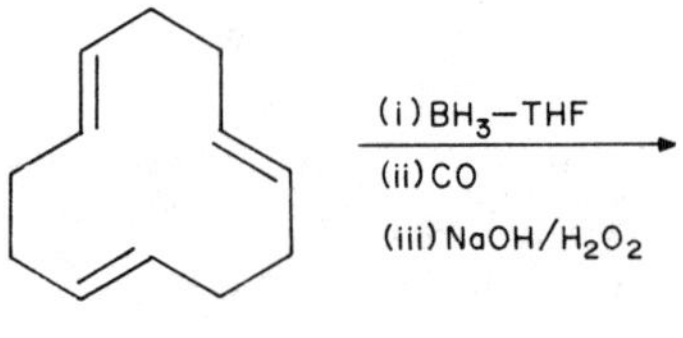

(i) BH_3-THF

(ii) CO

(iii) $NaOH/H_2O_2$

(e) 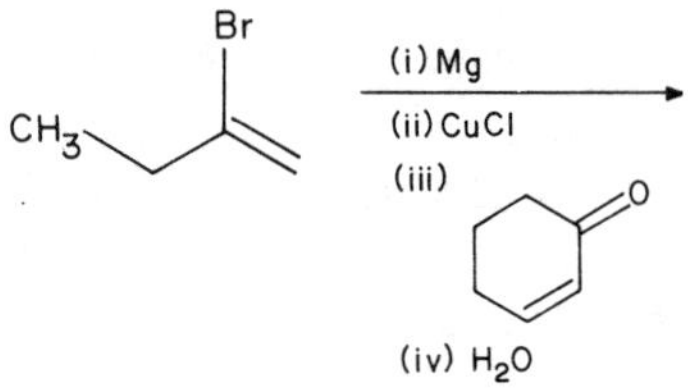 CH_3

(i) Mg

(ii) CuCl

(iii)

(iv) H_2O

(*University College, London*).

6. Outline a preparative route from non-organometallic starting materials to each of the following compounds.

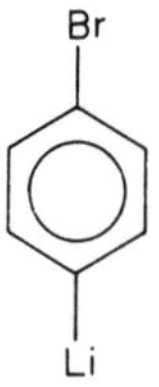

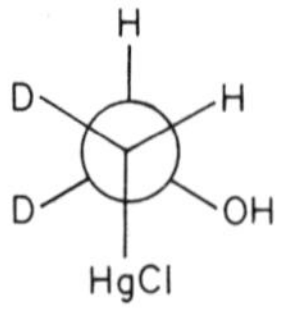

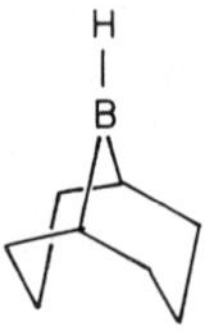

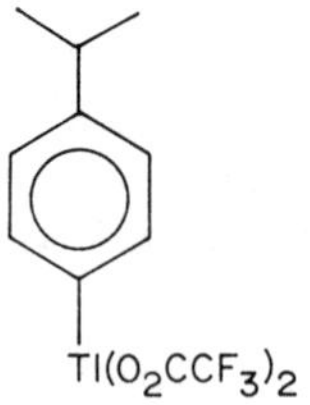

(*University College, London*).

96

7. Discuss the use of organoaluminium compounds in organic synthesis. Exemplify your account by reference to each of the following compounds.

(a)

(b) 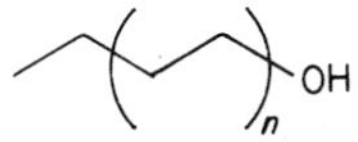 $(n = 4–6)$

(c)

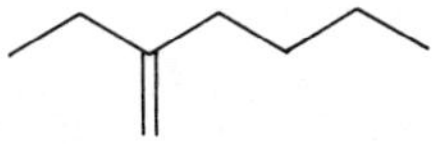

(d)

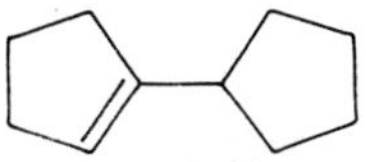

(e)   Isotactic polypropylene

(University College, London).

8. An optically active compound A, $C_{11}H_{18}O$, $(\sigma = 1685\ cm^{-1})$ on reaction with lithium dimethylcuprate gave B, $C_{12}H_{22}O$ $(\sigma = 1730\ cm^{-1})$ and an isomer C. Reaction of B with *meta*-chloroperbenzoic acid gave D, $C_{12}H_{22}O_2$, $(\sigma = 1745\ cm^{-1})$ converted by lithium aluminium hydride into E, $C_{12}H_{26}O_2$, $[\alpha]_D^{25\ \circ} = +83^\circ$. Similar treatment of C gave F, $C_{12}H_{26}O_2$, $[\alpha]_D^{25\ \circ} = +17^\circ$.

A compound identical with E in all respects except for its optical rotation, $[\alpha]_D^{25\ \circ} = -80^\circ$, was synthesized as outlined below.

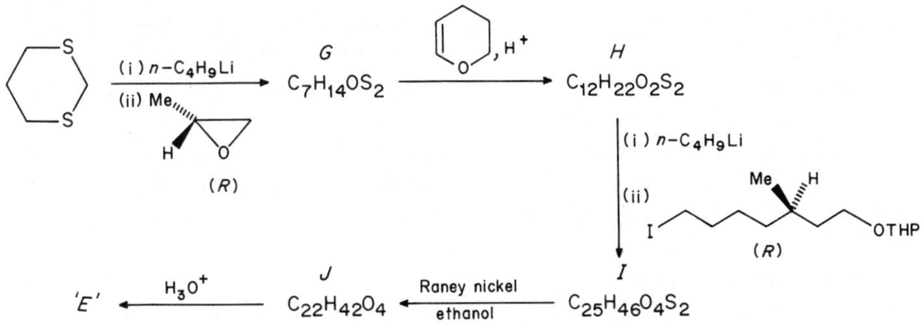

Identify A and delineate all the reactions described, with stereochemical detail. (THP = tetrahydropyranyl.)
(University of Exeter).

4 Organometallic compounds of elements of main groups IV and V

4.1 Introduction

In nearly all their organic derivatives, silicon, germanium, tin and lead are tetravalent. The tetraalkyls and -aryls are air stable materials which are not hydrolysed under neutral conditions. Their thermal stability decreases down the Group in parallel with the trend in metal–carbon bond energies (p. 5). Tetraphenylsilane can be distilled in air at 430°C/760mm. Tetramethylsilane starts to break up only above 600°C to a complicated mixture of products which probably arise through radical chain reactions. Liquid Me_4Pb, however, has been known to explode even at 90–100°C. Decomposition of tetramethyllead in the gas phase, begins at only about 250°C giving ethane and lead, together with small quantities of other hydrocarbons and hydrogen. The existence of free radicals was first demonstrated by Paneth in 1929 from pyrolysis of Me_4Pb in a flow system at low pressure. The radicals produced removed mirrors of lead, antimony or other elements forming their methyl derivatives.

It is perhaps in their propensity to oxidation and hydrolysis, however, that the greatest contrast between Group IV organometallics and those of neighbouring groups lies. Tetramethylsilane is so inert chemically that it is chosen as an internal standard for n.m.r. spectroscopy. (It absorbs at higher field than most organic compounds.) Trimethylaluminium, on the other hand, not only inflames in air but also reacts explosively with water. Me_3Al is coordinatively unsaturated; it is a six electron compound with an empty valence orbital and is hence very susceptible to nucleophilic attack. Tetramethylsilane behaves as a saturated compound. Group IV elements show little tendency to increase their coordination number above four unless they are attached to electronegative atoms or groups.

Although trimethylphosphine also has an octet configuration, it is spontaneously inflammable. The ease of oxidation of Group V alkyls must be associated with the presence of a non-bonding pair of electrons on the central atom. The less basic triaryls are air stable, although controlled oxidation to the oxides Ar_3MO is still readily achieved chemically. Bonds between carbon and Group V elements are normally stable to hydrolysis under neutral conditions. Cleavage of M—C bonds by acids, which involves electrophilic attack at carbon, becomes increasingly facile down both Group IV and Group V. This is influenced by the polarity as well as by the strength of the bond.

In addition to tetraorgano- derivatives R_4M, the Group IV elements form a large number of compounds R_nMX_{4-n}, where X is an electronegative substituent. The halides are very useful starting materials for the preparation of other species. The tin and lead halides, especially the fluorides, tend to associate through halogen bridges in the solid state. The metal atom can also increase its coordination number to five or six by complex formation with halide ions or with other Lewis bases.

Organic derivatives of Group IV find applications in diverse areas. Especially important are the silicone polymers, organotin compounds as stabilizers for plastics and as biocides and organolead antiknock agents for petrol.

4.2 Tetra-alkyls and -aryls of the elements of Group IV

4.2.1 Preparation

The first organosilicon compound to be prepared was tetraethylsilane. It was obtained by Friedel and Crafts in 1863 by reacting diethylzinc with silicon tetrachloride at 150°C. Nowadays in the laboratory, however, tetraorgano derivatives of Si, Ge and Sn are conveniently prepared by reacting the tetrachloride with excess Grignard or organolithium reagent. This is straightforward for silicon and germanium. Tin(IV) chloride forms a sparingly soluble adduct with diethyl ether, the usual solvent for Grignard reactions, much heat being evolved. It is advisable to add the $SnCl_4$ to excess Grignard solution to minimize the formation of partially substituted products R_2SnCl_2 and R_3SnCl. To aid the addition and to reduce the vigour of the reaction it is therefore expedient to dissolve the tin halide in benzene or toluene.

Organoaluminium compounds can also be used to transfer the organic groups. With tin halogen bridged complexes such as $R_2SnCl_2.AlCl_3$, $R_3SnCl.AlCl_3$ are formed which hinder further alkylation. This can be prevented by addition of donors such as NaCl or an amine which complex with the aluminium halide as it is formed.

$$4R_3Al + 4NaCl + 3SnCl_4 \longrightarrow 3R_4Sn + 4NaAlCl_4$$

By careful addition of a stoichiometric quantity of Grignard reagent to the halide, or better the alkoxide, stepwise attachment of organic groups can often be achieved e.g.

$$MeSi(OMe)_3 \xrightarrow{\text{PhMgBr}} PhMeSi(OMe)_2 \xrightarrow{\text{NpMgBr}}$$

$$(\pm)\text{-NpPhMeSi(OMe)} \xrightarrow{(-)\text{MenOH}} (\pm)NpPhMeSiO(-)Men$$

| racemic mixture | diastereoisomers |
| of enantiomers | |

$$(-)MenOH = (-)menthol; \quad Np = 1\text{-naphthyl}.$$

In this way a racemic mixture of the two enantiomers of NpPhMeSi(OMe) was obtained. This molecule is chiral on account of its having four different groups arranged tetrahedrally around the silicon atom. The diastereoisomeric menthoxides were separated by fractional crystallization at low temperature. Compounds of this type have been used to study mechanisms of substitution at silicon centres.

In industry, organosilicon compounds are made by the 'direct synthesis' (p. 109) or by hydrosilation (p. 24). Alloy methods (p. 19) are used in the manufacture of the antiknock compounds Me_4Pb and Et_4Pb. Antiknock compounds are blended into gasoline (petrol) to resist uneven combustion or detonation in spark ignition engines. Aromatic and branched aliphatic hydrocarbons are more resistant to 'knock' than straight chain paraffins. In the early part of this century it was impracticable to increase the content of knock resistant hydrocarbons in petrol by refining. In 1920 Midgley discovered that the addition of small amounts of Et_4Pb effected a similar improvement. It was soon found that it was also necessary to add dichloro- and dibromoethane as scavengers to convert lead oxides formed in combustion into the more volatile dihalides. Tetramethyllead, which has a higher vapour pressure than the tetraethyl, becomes more evenly distributed between cylinders in multi-cylinder engines. It is therefore more suitable for use with low grade petrols which contain components with a wide range of octane numbers.

In the manufacture of Me_4Pb or Et_4Pb, lead-sodium alloy reacts with the chloroalkane

$$4PbNa + 4RCl \longrightarrow R_4Pb + 3Pb + 4NaCl$$

The organolead product is separated by steam distillation and the metallic lead residue refined in a furnace for reuse. The alloy is prepared by mixing weighed quantities of molten lead and sodium under a blanket of nitrogen. The hot molten alloy is transferred to a flaking machine where it is cooled and stripped into solid flakes. The process provides a good example of industrial practice where as many components as possible are produced on one site, which is placed next to an oil refinery. Electrolysis of molten salt yields sodium metal and chlorine (Downs cell). Hydrogen and chlorine are produced by electrolysis of brine in a mercury cell and then burned together to yield hydrogen chloride. Caustic soda is sold as a by-product. Ethylene from the refinery is converted into ethyl chloride and dichloroethane by the action of HCl and chlorine respectively.

In recent years there has been growing awareness of the hazards from lead in the environment. The permitted amount of lead in petrol is gradually being reduced, although the UK is lagging behind some other countries such as Germany and Japan in this regard.

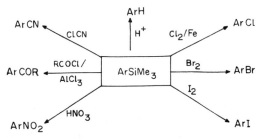

Fig. 4.1 Some reactions of arylsilanes.

4.2.2 *Reactions of tetra-organosilicon compounds*

The inertness of tetraalkylsilanes to thermal decomposition and to chemical attack was alluded to in the introduction. Electrophiles cleave alkyl–silicon bonds, but only in the presence of a Lewis acid:

$$Me_4Si + HCl \xrightarrow{AlCl_3} Me_3SiCl + CH_4$$

$$Et_4Si \xrightarrow[-EtI]{I_2/AlI_3} Et_3SiI \xrightarrow[-EtI]{I_2/AlI_3} Et_2SiI_2$$

Aryl—silicon bonds are more reactive. Acid cleavage of Et_3SiPh proceeds 10 000 times as fast as proton exchange in benzene. R_3Si behaves as an electron-releasing group which stabilizes the positively charged Wheland intermediate.

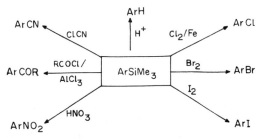

The relative rates of acid cleavage of R_3M—Ph increase dramatically down the Group as follows: M = Si, 1; Ge, 36; Sn, 3.5×10^5; Pb, 2×10^8. R_3Si substituents on an arene can similarly be replaced by a wide range of groups by reaction with suitable electrophiles (Fig. 4.1).

Vinyl—silicon bonds are also rather easily cleaved by electrophiles. It appears that the R_3Si group is able to stabilize β-carbonium ions. Allylsilanes are prone to attack even by fairly weak electrophiles. Acetals, aldehydes and ketones react in the presence of Lewis acids to form unsaturated alcohols. The reaction is accompanied by an allylic transposition:

Vinyl silane:

101

e.g. aliphatic Friedel Crafts.

$$H_2C=C\begin{smallmatrix}H\\SiMe_3\end{smallmatrix} \xrightarrow[AlCl_3]{H_3CCOCl/} H_2C=C\begin{smallmatrix}H\\COCH_3\end{smallmatrix} + Me_3SiCl$$

Allyl silane

$$Nu^- \begin{smallmatrix}Me\\|\\CH\\Me_3Si\end{smallmatrix} \begin{smallmatrix}CH_2\\CH\end{smallmatrix} E^+ \longrightarrow NuSiMe_3 + MeCH=CHCH_2E$$

e.g.

$$\begin{smallmatrix}Me\\|\\CH\\Me_3Si\end{smallmatrix}\begin{smallmatrix}CH_2\\CH\end{smallmatrix} + \begin{smallmatrix}R\\\delta+ \quad \delta-\\C=O\\R'\end{smallmatrix} \xrightarrow[-70°C]{TiCl_4} MeCH=CHCH_2C\begin{smallmatrix}OSiMe_3\\R\\R'\end{smallmatrix} \xrightarrow{H_2O}$$

$$MeCH=CHCH_2C\begin{smallmatrix}OH\\R\\R'\end{smallmatrix}$$

4.3 Organo-halides of the elements of Group IV

4.3.1 *Structures*

The full range of mixed methyl and phenyl halogen derivatives R_3MX, R_2MX_2 and RMX_3 ($X = F$, Cl, Br, I) is known for Si, Ge and Sn. The methyl silicon and germanium halides are volatile materials which can be distilled without decomposition. Like the tetrahalides themselves, they have essentially tetra-hedral, molecular structures, with little or no association between the molecules in the condensed states. The compounds of tin and lead, especially the fluorides, often show association in the solid state. This is consistent with the increase in polarity of the M—X bonds which occurs down the Group. This leads to a greater positive charge on the central atom which becomes more stongly acidic. Between Ge and Sn there is also a marked increase in the tendency to form structures in which the metal atom has coordination numbers of five and six. Organotin halides form a wide range of stable complexes with donor molecules. With the exception of the fluorides, however, such complexing by the silicon and germanium compounds is rare.

The range of organolead halides is restricted on account of the tendency, especially of the iodides, to decompose giving PbI_2. The trihalides $RPbX_3 (X = Cl$, Br, I) are extremely unstable and little is known about them. This parallels the lability of the tetrahalides; $PbCl_4$ decomposes above 0°C, $PbBr_4$ is even less stable and the existence of PbI_4 is doubtful.

The structures of the methyltin halides well illustrate the trend towards greater intermolecular association and towards increasing ionic character which is observed down the Group. This is most pronounced in the fluorides. Me$_3$SnF decomposes without melting at about $360°$ and is very sparingly soluble in most solvents. It is precipitated from aqueous solutions (p. 117) by addition of fluoride. X-ray diffraction shows a chain structure in which essentially planar Me$_3$Sn groups are held together by unsymmetrical fluorine bridges. There are only weak van der Waals interactions between the chains. The tin atoms are five coordinate in an essentially trigonal bipyramidal environment.

The other trihalides Me$_3$SnX (X = Cl, Br, I) contrast strongly with the fluoride. All are volatile materials which dissolve as monomers in non-polar organic solvents. Association through bridging halogen atoms occurs in the solid chloride, but this seems to be unimportant in the iodide.

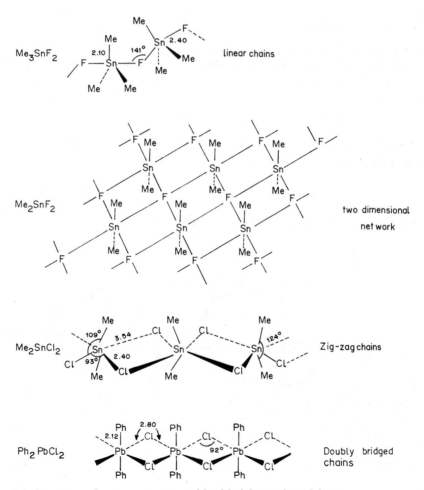

Fig. 4.2 Structures of some organotin and lead halides in the solid state.

Dimethyltin fluoride (m.p. $360°C$) crystallizes in an infinite two dimensional lattice resembling that of SnF_4. The tin atoms are octahedrally coordinated (Fig. 4.2). While Me_2SnCl_2 is unassociated in the vapour, in the solid association occurs through highly unsymmetrical chlorine bridges. The bromide and iodide are probably similar. In continuation of the trend to greater association, the lead fluorides R_3PbF and R_2PbF_2 are involatile, sparingly soluble solids which have high melting points or which decompose without melting. These properties are consistent with a high degree of ionic character. Ph_3PbCl, m.p. $205°C$ and the similar bromide have solid state structures similar to that of Me_3SnF, but in contrast to the molecular Ph_3SnCl. Ph_2SnCl_2 forms double chlorine-bridged chain polymers in which lead is octahedrally coordinated (cf. R_2SnF_2 but contrast Ph_2SnCl_2).

4.3.2 *Lewis acidity*

Intermolecular association in covalent halides occurs through Lewis acid–base interactions. This leads to the expectation that the halides R_nMX_{4-n} would also form complexes with donor molecules or ions and that this Lewis acidity would increase

$$Si \sim Ge < Sn < Pb; \ I < Br < Cl < F; \ R_3MX < R_2MX_2 < RMX_3$$

While Me_3SiF does not complex with neutral Lewis bases, it reacts with tetramethylammonium fluoride

$$Me_3SiF(g) + 2Me_4N^+F^-(s) \xrightarrow{5 \text{ atm}} (Me_4\overset{+}{N})_2Me_3SiF_3^{2-}$$

Fluoride ions also add to R_2SiF_2 and to $RSiF_3$.

$$RSiF_3 \underset{F^-}{\rightleftharpoons} RSiF_4^- \underset{F^-}{\rightleftharpoons} RSiF_5^{2-}$$

The ^{19}F n.m.r. spectrum of $Ph_2SiF_3^-$ in a non-polar solvent shows only one resonance at room temperature, but below $-60°$ this splits into two groups of resonances of relative intensity 2:1 ($F_{ax}:F_{eq}$). Rapid intramolecular scrambling between axial and equatorial positions is a common phenomenon in trigonal bipyramidal species (p. 130). This normally occurs through a square pyramidal transition state (Berry pseudorotation). The most stable trigonal bipyramidal structure is often the one in which the most electronegative substituents are in the axial positions.

Apart from the fluorides, organosilicon halides do not normally form isolable adducts with Lewis bases. They are, however, very susceptible to nucleophilic substitution. All are rapidly hydrolysed (p. 112). In contrast Me_3SiF is almost inert to hydrolysis and alcoholysis under neutral conditions. This is because of the high energy of the Si—F bond ($E(Si—F) > E(Si—O)$ but $E(Si—Cl) < E(Si—O)$) so that the equilibrium.

$$Me_3SiF + ROH \rightleftharpoons Me_3SiOR + HF$$

lies to the left. If HF is removed by base or by calcium ions, for example, the reaction proceeds to completion.

While tetraorganotin compounds themselves show no Lewis acidity, the halides R_nSnX_{4-n} form stable complexes with neutral and anionic donors. $RSnX_3$ and R_2SnX_2, like SnX_4, generally give 1:2 adducts $RSnX_3L_2$ or $R_2SnX_2L_2$ with monodentate ligands, in which the tin atom is octahedrally coordinated. R_3SnX, however, normally forms 1:1 complexes R_3SnXL which have trigonal bipyramidal structures. In polar, donor solvents, these complexes may ionize; R_3SnXL might, for example, give R_3SnL^+ and X^- ions.

The planar Me_3Sn arrangement which is adopted in solid Me_3SnF is also present in the pyridine complex $Me_3SnClpy$. Similar complexes are formed by Ph_3PO and by Me_2SO. If chlorine is replaced by more poorly coordinating anions such as BPh_4^-, cationic complexes, $[Me_3SnL_2]^+BPh_4^-$ ($L = $ e.g. H_2O, Me_2SO), can be prepared (Fig. 4.3).

Cis and trans geometrical isomers have been observed for the adducts $R_2SnX_2L_3$. The R_2Sn grouping is usually linear.

Similar complexes have been obtained from organolead halides, although so far in fewer numbers.

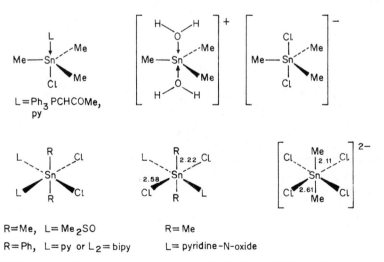

Fig. 4.3 Structures of some complexes of organotin halides with Lewis bases.

The main Groups IV and V

4.3.3 *Reactions*

Some reactions of trimethylchlorosilane are summarized in Fig. 4.4. Many of these involve nucleophilic attack at silicon in which chlorine is substituted by another group. Alkoxysilanes are obtained using metal alkoxides or alcohols in the presence of pyridine. If the latter reaction is carried out in a non-polar solvent such as light petroleum, pyridinium hydrochloride is precipitated and may be filtered off, leaving the alkoxysilane in solution. Hydrogen chloride cleaves carbon–oxygen bonds in alkoxysilanes (as it does in ethers also) to form an alkyl halide and a siloxane:

$$2Me_3SiOR' + 2\,HCl \longrightarrow (Me_3Si)_2O + 2R'Cl + H_2O$$

In order to avoid having to use an acceptor for the HCl in industry, the chlorosilane and alcohol are fed together into the top of a still. Hydrogen chloride is formed in the cool upper regions of the column and is taken off before it can cleave the products.

The Me_3Si— group is used for 'protecting' hydroxyl and other functions in organic chemistry. It is easily introduced by reaction with Me_3SiCl/pyridine or with hexamethyldisilazane:

$$(Me_3Si)_2NH + 2R'OH \longrightarrow 2Me_3SiOR' + NH_3$$

$$2Me_3SiOR' \xrightarrow{H_2O} (Me_3Si)_2O + R'OH$$

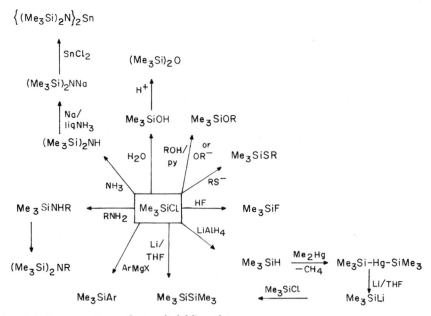

Fig. 4.4 Some reactions of trimethylchlorosilane.

106

It can be removed by hydrolysis under mildly acidic or basic conditions. Trimethylsilyl derivatives are thermally stable and are often more volatile than the precursor alcohols. Derivatives of monosaccharides, for example, can be analysed by gas–liquid chromatography whereas the parent molecules are far too involatile and would decompose.

Organotin halides $(X = Cl, Br)$ can be prepared by redistribution reactions. Good yields of R_3SnCl or of R_2SnCl_2 are obtained by heating together R_4Sn and $SnCl_4$ in the appropriate proportions. $RSnCl_3$ is made from R_4Sn and excess $SnCl_4$.

$$3R_4Sn + SnCl_4 \longrightarrow 4R_3SnCl$$

$$R_4Sn + SnCl_4 \longrightarrow 2R_2SnCl_2$$

$$R_4Sn + 3SnCl_4 \longrightarrow 4RSnCl_3 \quad (+ \text{ some } R_2SnCl_2 + \text{excess } SnCl_4)$$

No Lewis acid catalyst is required (contrast silicon p. 101). Methyltin halides also result from the 'direct synthesis':

$$2MeCl + Sn \xrightarrow[\text{315 C}]{Cu/} Me_2SnCl_2 \,(75\%) + \text{other halides}$$

Some reactions of alkyltin compounds are summarized in Fig. 4.5. This

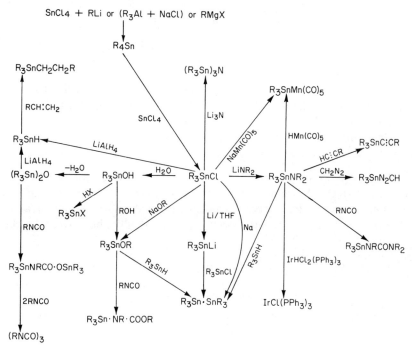

Fig. 4.5 Reactions of organotin compounds. (With kind permission of Professor K. Wade.)

107

illustrates the wide range of derivatives which can be prepared starting from the halides. Organotin hydrides R_3SnH, made from R_3SnCl by reduction with lithium aluminium hydride, are useful reagents in organic chemistry. The stability of the hydrides increases in the sequence $MH_4 < RMH_3 < R_2MH_2 < R_3MH$. While stannane is thermally unstable, decomposing to tin and hydrogen, R_3SnH can be stored indefinitely in clean glass apparatus in the absence of air. Organotin hydrides are characterized by sharp intense bands in their infrared spectra in the region $1800-1880\,\mathrm{cm^{-1}}$. Their proton n.m.r. spectra show Sn–H resonances at about $\delta 3.2$ to 4.0. These peaks are accompanied by satellites due to coupling with ^{117}Sn and ^{119}Sn (natural abundance 7.7% and 8.7% respectively). Both these nuclei have spin $I = \frac{1}{2}$, so the satellites appear as doublets to each side of the main resonance, with $^2J(Sn–H)$ of the order of $1500-2000\,\mathrm{Hz}$.

Tin hydrides add readily to double and triple bonds:

$$\equiv Sn—H + A = B \longrightarrow \equiv Sn—A—B—H$$

$$(A = B \text{ can be} > C = C <, -C \equiv C-, > C = O, -C \equiv N).$$

Sometimes no catalyst is required, but the reactions are accelerated by free radical initiators (cf. hydrosilation, p. 24). Both polar and radical mechanisms of addition have been found.

$$R_3SnH + H_2C = CHCO_2Me \longrightarrow R_3SnCH_2CH_2CO_2Me$$

$$HC \equiv CCH_2C \equiv CH + R_2SnH_2 \longrightarrow$$

Organotin hydrides reduce alkyl halides to hydrocarbons; this process can sometimes be made catalytic in tin by using a small amount of R_3SnCl and a stoichiometric quantity of $LiAlH_4$.

Amino- and alkoxy- derivatives $R_3SnNR'_2$ and R_3SnOR' are very versatile reagents. The former are the more reactive. Their reactions are of two main types:

(1), Attack by proton donors: $R_3SnNMe_2 + HX \rightarrow R_3SnX + Me_2NH$ ($X = NR'_2$, PR'_2, AsR'_2, OH, OR', SR', Halogen, transition metal complex, carbon 'acid'). This has an enormous range. Carbon 'acids' include hydrocarbons such as alkynes and cyclopentadiene. The reactions are aided by the loss of volatile dimethylamine.

(2), Addition to double bonds: e.g.

$$Me_3SnNMe_2 + O = C = O \longrightarrow Me_3SnO—\overset{\overset{\displaystyle O}{\|}}{C}NMe_2$$

The addition to isocyanates forms the basis of a commercial route to polyurethanes, catalysed by organotin alkoxides.

$$Bu_3SnOMe + R—N=C=O \longrightarrow Bu_3SnNRCO_2Me \xrightarrow{R'OH} Bu_3SnOR' + RNHCO_2Me$$

108

Industrial production of organotin compounds has increased more than 600 times since 1950. In 1978 world production reached 30 000 tonnes. The major application is as stabilizers for PVC. Typical materials are di-n-octyltin laurate or maleate, and for food packaging, the S,S'-bis(iso-octylmercaptoethanoate) $Oct^i_2Sn(SCH_2CO_2Oct^i)_2$. When about 2% of the tin compound is incorporated into the plastic it remains colourless, transparent and unaffected by sunlight. Its function is probably to act as a radical trap, a scavenger for hydrogen chloride and an antioxidant against breakdown of the polymer in the atmosphere.

Triorganotin compounds kill a variety of organisms including bacteria, fungi, moulds, algae, lichens, insects, molluscs and crustacea. This activity is greatest in trialkyltin compounds which have a total of nine or ten carbon atoms in the molecule. A typical preparation consists of aqueous bis(tri-n-butyltin) oxide, TBT, $(Bu_3Sn)_2O$, in combination with a quaternary ammonium compound. It is used, for example, in wood treatment and preservation, and in paper manufacture to prevent bacterial decay of ground wood pulp during storage. It eradicates moss, algae and lichens from masonry and prevents their regrowth for several years. Similar preparations protect water based paints and adhesives from fungal and bacterial attack. Organotin biocides were once thought to show low toxicity to mammals, but recent studies question this. Concentrated solutions cause eye and skin irritation. TBT is readily degraded in the environment to inorganic tin(II) compounds.

4.4 Silicones

4.4.1 The 'direct process': manufacture of alkylchlorosilanes and silicones

The discovery by E.G. Rochow (General Electric Co.) and by R. Muller (VEB Silikon Chemie) that methylchlorosilanes can be prepared by the direct reaction between methyl chloride and silicon opened up the way for commercial developments using silicon compounds. The reaction

$$2CH_3Cl(g) + Si(s) \longrightarrow (CH_3)_2SiCl_2(g) \quad \Delta H^\ominus = -302\ kJ\ mol^{-1}$$

is strongly exothermic but requires a catalyst. In industry ground silicon of metallurgical grade ($\geqslant 98\%$) is reacted with gaseous methyl chloride at 250–350°C and 1–5 atm in a fluid bed reactor. The best catalyst is copper, which is introduced by grinding, sintering or merely by mixing with the silicon. It is even possible just to add copper(I) chloride, which is reduced by silicon at the elevated temperatures used. An induction period occurs during which oxide coatings are removed and the catalyst is transported to active sites on the surface. Addition of electropositive elements such as Al, Ca or Mg reduces the induction period; Zn or Cd also helps to remove oxides and to mobilize CuCl by dissolution. A typical mixture consists of Si, 90–95%; Cu, 5–10%, Al, 0.1–1%; Zn, 0.1–1% and Sb, a 'promoter', 0.001–0.005%. The reaction has the characteristics of a heterogeneous gas–solid process. It has been suggested that methyl chloride dissociates on the surface to $CH_3^\cdot$ and $Cl^\cdot$ radicals. Radicals could also arise by decomposition of methylcopper formed as an intermediate. Alternatively chemisorbed CH_3Cl may react with activated silicon or silicide phases such as Cu_3Si. A typical product mixture consists of (%, b.p.) Me_2SiCl_2 (75, 70°C), $MeSiCl_3$ (10, 66°C), Me_3SiCl (4, 58°C), $MeSiHCl_2$ (6, 41°C) and smaller quantities of $SiCl_4$, Me_4Si, $HSiCl_3$ and disilanes. On account of the close boiling points of these materials careful fractionation is necessary to separate them.

The direct synthesis also works with methyl bromide, but methyl fluoride and methyl iodide do not react satisfactorily. Phenylchlorosilanes, however, can be prepared starting from chlorobenzene. Other related reactions include:

$$3MeOH + Si \longrightarrow (MeO)_3SiH + H_2 \text{ (catalysed by copper, no migration of catalyst occurs)}$$

$$3HCl + Si \longrightarrow HSiCl_3 + H_2 \text{ (no catalyst required)}.$$

Redistribution of groups between silicon occurs approximately statistically, because the $\Delta G°$ values for the reactions are very small (ca. $+4$ to -4 kJ mol^{-1}).

$$SiX_4 + SiY_4 \rightleftharpoons SiX_{4-n}Y_n$$

Exchange between 'labile' groups (X, Y = halogen, OR, SR, NR_2) occurs rapidly in the presence of acid or base catalysts, so that the most volatile component can be distilled out preferentially. For redistributions involving Si–C bonds, strong Lewis acid catalysts such as $AlCl_3$ or Al_2O_3 are required. These reactions involve electrophilic substitution at carbon and do not proceed through organoaluminium intermediates. Methylchlorosilanes obtained by the direct process are interconverted commercially in this way e.g.

$$MeSiCl_3 + Me_3SiCl \overset{AlCl_3}{\rightleftharpoons} Me_2SiCl_2$$

4.4.2 Hydrolysis of chlorosilanes–silanols

Hydrolysis of alkyl or arylchlorosilanes initially yields silanols $R_3Si(OH)$, $R_2Si(OH)_2$ or $RSi(OH)_3$. Except where R is a bulky group such as tert-butyl or cyclohexyl, these compounds are very prone to self-condensation, giving siloxanes. Consideration is brought about by traces of acids or bases, so it is essential to ensure carefully buffered neutral conditions in their preparation.

$$Me_2SiCl_2 \xrightarrow[MgCO_3]{H_2O, Al_2(SO_4)_3} Me_2Si(OH)_2 \longrightarrow HOSiMe_2(OSiMe_2)_nOSiMe_2OH$$
$$\text{m.p. } 100°C$$

Silanols are associated through intermolecular hydrogen bonding. Solid diethylsilane diol, m.p. 96° has a structure in which hydrogen bonded chains are further linked to neighbouring chains, also through hydrogen bonds, to form layers.

Me_3SiOH and Ph_3SiOH are more strongly acidic than the corresponding alcohols Me_3COH and Ph_3COH. Ph_3SiOH is about as strong an acid as phenol. This is surprising, as silicon is less electronegative than carbon; inductive effects should make the silanols less acidic. It is suggested that $(d\pi–p\pi)$ interactions remove electron density from oxygen and hence from hydrogen. Alkali metal silanolates are formed when concentrated aqueous alkali reacts with trialkylsilanols. A better method of preparing them is to treat the silanol with alkali metal in ether.

$$2Me_3SiOH + 2K \xrightarrow{\text{Et}_2O} 2Me_3SiOK + H_2.$$

X-ray diffraction studies have shown that crystalline Me_3SiOK is tetrameric with a cubane-like structure.

4.4.3 Structure of siloxanes and related compounds

Hydrolysis of Me_3SiX (X = halogen, OR, NR_2, SR etc.) leads, *via* Me_3SiOH to hexamethyldisiloxane, b.p. 99°C. Electron diffraction has shown that the Si—O—Si bond angle is 148°, very much wider than the C—O—C angle in the corresponding ether $(Me_3C)_2O$ (111°). At room temperature there is considerable bending motion; the linear transition state for inversion is only about 1 kJ mol^{-1} above the shallow minimum which corresponds to the bent, ground state structure. Because of the very low energies involved, this 'ball and socket' motion, which is found in all siloxanes, persists even at low temperatures.

The wide bond angle and the short Si—O bond distance (sum of covalent radii = 1.76Å) are suggestive of significant $(p \rightarrow d)\pi$ bonding between oxygen and silicon. $(Ph_3Si)_2O$ actually has a linear skeleton. The existence of such π-bonding between silicon and electronegative atoms such as nitrogen, oxygen or fluorine, which possess non-bonding electron pairs, has been inferred from a variety of chemical and structural studies. Thus trisilylamine $(H_3Si)_3N$ has a planar structure, in contrast to trimethylamine, which is pyramidal. The formation of the three coplanar Si—N σ-bonds leaves a lone pair of electrons in the remaining p orbital of nitrogen which can be delocalized over the molecule by donation into $3d$ silicon orbitals. Maximum overlap occurs for the planar structure. (This is the conventional explanation. Recent work suggests that antibonding σ^* orbitals may also act as acceptors, see p. 160.)

Trigermylamine also has a planar skeleton; from vibrational spectra, however, it seems that $N(SnMe_3)_3$ is pyramidal. When the Group IV element (X = Si, Ge, Sn) is bonded to a second row (Y = P, S) or to a third row (Y = As, Se) element, the bond angles in $(Me_3X)_3Y$ or $(Me_3X)_2Y$ are narrow. This indicates less $(p \rightarrow d)\pi$ bonding than in the nitrogen or oxygen compounds but does not necessarily mean that it is absent.

Table 4.1. Structural data for some compounds $(H_3X)_3Y$ and $(Me_3X)_3Y$ $(Y = N, P)$ and $(H_3X)_2Y$ and $(Me_3X)_2Y$ $(Y = O, S)$

Compound	Bond length X—Y (Å)	∠'XYX (°)	Structure
$(H_3C)_3N$	1.47	108	Pyramidal
$(H_3Si)_3N$	1.74	120	Trigonal planar
$(Me_3Si)_2NH$	1.74	125	Planar
$(H_3Ge)_3N$	1.84	120	Trigonal planar
$(H_3Si)_3P$	2.25	95	Pyramidal
$(Me_3Si)_3P$	–	90–100	Pyramidal
$(H_3C)_2O$	1.42	111	Tetrahedral angle
$(Me_3Si)_2O$	1.63	148	Wide angle
$(Me_3Ge)_2O$	1.77	140	Wide angle
$(Me_3Sn)_2O$	1.94	140	Wide angle
$(Me_3Si)_2S$	–	100	Narrow angle
$(Ph_3Si)_2O$	1.62	180	Linear

4.4.4 Silicone polymers

The hydrolysis of difunctional silicon derivatives Me_2SiX_2 $(X = halogen, OR, NR_2)$ can lead to ring or chain siloxanes which are made up of Me_2SiO units. Note that the monomeric species $Me_2Si{=}O$, corresponding to a ketone $Me_2C{=}O$ is not obtained.

$$Me_2SiCl_2 \xrightarrow{H_2O} Me_2Si(OH)_2 \longrightarrow (Me_2SiO)_n + HO(SiMe_2O)_nH$$

Consider the relative energies of single and double bonds between oxygen and carbon and silicon respectively $(kJ\ mol^{-1})$:

$$E(C{-}O)\ 358;\ E(C{=}O)\ 803;\ E(Si{-}O)\ 464;\ E(Si{=}O)\ 640.$$

The bond energy $E(C{=}O)$ is more than twice that of a single bond; $2E(C{-}O) = 716\ kJ\ mol^{-1}$. Thus a monomeric ketone is thermodynamically preferred over the polymer. These and related data suggest that whereas $(2p–2p)\pi$ overlap is strong, $(3p–2p)\pi$ or $(3p–3p)\pi$ overlap is much weaker. The poorer overlap of $3p$ orbitals may arise on account of the greater size and diffuseness of $3p$ relative to $2p$ orbitals. Moreover the lateral extension of the p orbitals may be sufficient to give good overlap only at the short bond distances which are observed in bonds involving elements of the First Short Period. In addition, as discussed above, formal single bonds between elements of the Second Short Period (Si, P) and atoms which bear lone pairs of electrons (N, O and F) can be strengthened by $(2p–3d)\pi$ interaction (see also p. 128).

When excess water is used to hydrolyse Me_2SiCl_2, approximately equal quantities of cyclic and linear products are formed. The volatile cyclic material consists of about 80% tetramer ($n = 4$) (b.p. 176°C), together with some trimer ($n = 3$) (b.p. 134°C) and larger rings ($n = 5$–10). These materials can be separated by fractional distillation or by gas liquid or high pressure liquid chromatography. The cyclic trimer has a planar six-membered ring both in the vapour (electron diffraction) and in the solid state (X-ray diffraction). The ring in the cyclic tetramer however is puckered and adopts a saddle shaped conformation. Puckered rings are also present in all the higher oligomers.

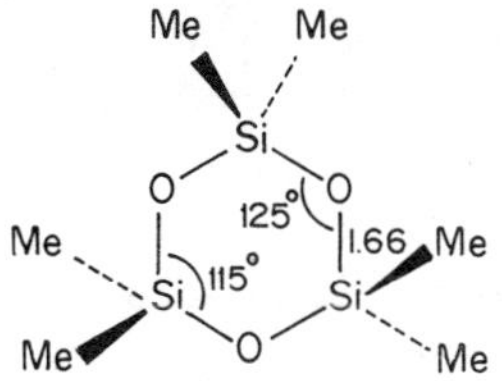

The cyclic trimer is unstable with respect to the tetramer by 16–20 kJ mol^{-1}, on account of ring strain in the former. Its presence in hydrolysates is due to its rapid formation in the early stages of the reaction under kinetic control. Cyclic and linear siloxanes can be interconverted using either acidic or basic catalysts. If long chain linear materials, for example, are heated with potassium hydroxide, $(Me_2SiO)_n$ ($n = 3,4$) can be distilled away. Conversely cyclic siloxanes are polymerized by OH$^-$, OR$^-$ as follows:

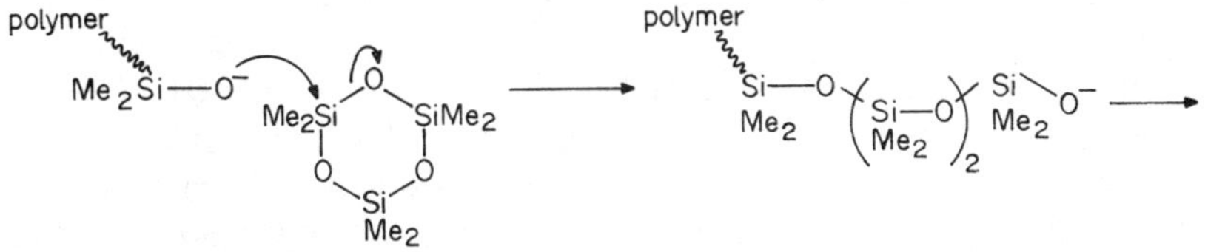

Heating to 80°C with 1–2% of sulphuric acid which contains up to 8% oleum has a similar effect. Polymethylsiloxanes produced in this way still carry end groups which are susceptible to further condensation. To stabilize the linear polymers, their ends can be blocked by Me$_3$Si groups. $(Me_3Si)_2O$ is added to the polymerizing mixture; the relative proportion of Me$_3$Si— and Me$_2$Si= units determine the average chain length of the resulting siloxane polymer. The polymers thus obtained do not all have the same chain length, but there is a Gaussian distribution about the mean.

Silicones are manufactured on a very large scale; in 1975 about 24 000 tonnes of silicon were converted into chlorosilanes for this purpose in the USA alone. About half the production is of methylsilicone fluids $Me_3SiO(SiMe_2O)_nSiMe_3$ (n up to 3000). The rest consists of resins, rubbers and some composites which include copolymers, blends and laminates.

Methyl and phenyl silicones exhibit very good thermal and oxidative stability in comparison with hydrocarbons. Methylsilicone oils (linear polydimethylsiloxanes) are stable up to at least 150°C in air for unlimited periods. In closed or vacuum systems they can be used up to 300°C. Antioxidants are often added to increase their resistance to degradation in the atmosphere. The ignition temperatures of the high molecular weight polymers are above 400°C.

Over the range 0–200°C the temperature coefficient of viscosity of silicone fluids is only about one tenth of that of mineral oils. In contrast the isoelectronic polymers $(Me_2SiCH_2)_n$ have normal viscosity characteristics. This means that silicones can be used over very much wider ranges of temperature. Some can still be poured well below $-50°C$. Polysiloxane chains are very flexible; as noted above for hexamethyldisiloxane, the bond angles in the chains are readily deformed. Moreover there are two mutually perpendicular $(2p-3d)\pi$ systems which together have approximately cylindrical symmetry about the Si—O bonds. This means that there is little resistance to torsional motion within the molecule. There is also essentially free rotation of methyl groups about the carbon—silicon bonds. (Barriers to rotation $(kJ\,mol^{-1})$: About Me—Si, 6.7; Me—C, 15.1: About Si—O, 0.8; C—O, 11.3.)

The viscosity of silicone fluids is also little affected by pressure. A pressure of 2000 atm causes the viscosity of mineral oil to increase between 50 and 5000 times, whereas the viscosity of a typical silicone fluid increases only about 14 times under the same conditions. Even when subjected to extreme pressures, which would make mineral oils become solid, silicone fluids remain in the liquid state.

As a result of these properties, silicone fluids find application for example as heat transfer media, hydraulic fluids, brake fluids, lubricants and vacuum pump oils. One of their earliest uses was as dielectrics; the dielectric constant changes very little with temperature and is also essentially independent of frequency over the range 10^2–10^7 Hz. On thermal breakdown organic polymers have the disadvantage that they commonly produce carbon, which conducts electricity. Organosiloxanes are more robust and also tend to decompose or oxidize to products which are themselves insulators, such as silica or volatiles of low molecular weight.

The surface effects of silicones also lead to useful applications. They derive from the presence of both polar Si—O and non-polar hydrocarbon groups in the same material. Textiles or papers treated with silicone oils become water repellent because the polar groups orient themselves close to the fibres, presenting a hydrophobic hydrocarbon exterior. Water penetration into bricks, concrete and other building materials is also greatly reduced by impregnation with silicones. Glass vessels drain more efficiently after coating. Dimethyl and phenylmethylpolysiloxanes are used as paint additives as they improve the water repellent properties of the painted surface as well as reducing the surface tension of the paint.

In order to make the silicone adhere more strongly to the surface which is being treated cross links are often introduced which anchor the two together. This can be achieved by including some MeSiH groups in place of Me_2Si in the polymer chain (MeSiHCl$_2$ is one product from the 'direct synthesis'). Linkage can occur by reaction with OH groups on the surface, for example:

$$\equiv Si-H + M-OH \longrightarrow \equiv Si-O-M + H_2$$

Silicones confer anti-stick properties on surfaces. Moulds for rubbers, as in tyre manufacture, or for plastics are usually silicone treated to give easy release. Silicones are also useful surfactants and anti-foam agents.

Rubbers in general are polymers with long chain lengths in which there are occasional cross links. They can thus be deformed under stress without flow. Increase in the proportion of cross links results in increasing rigidity which leads ultimately to hard resinous materials.

Elastomers designed for fabrication under high pressure consist of linear polymers R$_3$Si(OSiR$_2$)$_n$OSiR$_3$ with n about 4000, cross linked to form a network with about 300 monomer units between the links. A filler such as finely divided silica is also included. Cross links are introduced by heating after the mixture has been formed into the desired shape by rolling, extrusion or moulding. Radical sources such as benzoyl peroxide are often used to initiate cross linking (Fig. 4.6). Better control over the structure is achieved by introducing a few vinyl substituents, which are more susceptible to attack by free radicals. Rubbers which cure at room temperature are also available. They are produced by mixing two components. In one type the first component is a polysiloxane with some vinyl side chains plus a trace of hydrosilation catalyst (H$_2$PtCl$_6$). The second component consists of a polysiloxane which has a few Si—H units. Cross linking occurs through hydrosilation of Si—CH=CH$_2$ by Si—H. The vinyl precursors are also made in the same way. Addition of Si—H bonds to alkenes or alkynes is catalysed at ambient temperatures by platinum complexes present in very low

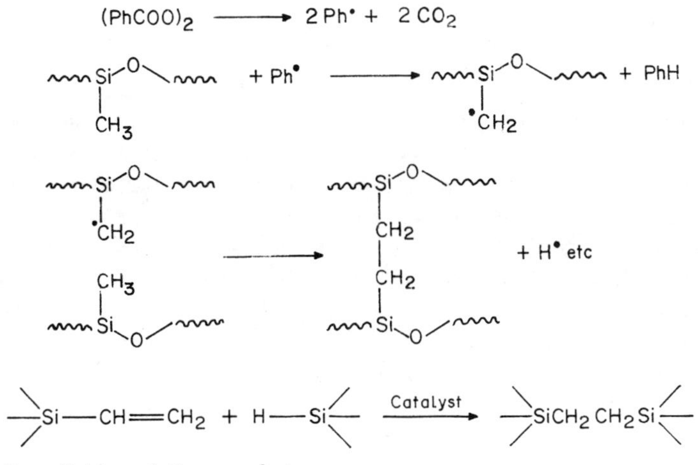

Fig. 4.6 Cross-linking of siloxane chains.

concentrations.

$$Cl_3SiH + H_2C{=}CH_2 \xrightarrow{H_2PtCl_6} Cl_3SiCH_2CH_3$$

$$Cl_3SiH + HC{\equiv}CH \xrightarrow{H_2PtCl_6} Cl_3SiCH{=}CH_2$$

Silicone resins are rigid polymers produced by hydrolysis of R_2SiCl_2 mixed with an appreciable quantity of $RSiCl_3$. More control over their structure is obtained by treating hydroxyl ended polysiloxanes $HO(SiMe_2O)_nH$ with organosilicon esters $RSi(OR')_3$. They are extensively cross-linked and are used for making electrically insulating material as a glass-cloth laminate. Hydrolysis of trifunctional organo-silicon derivatives $RSiX_3$ on their own normally leads to amorphous highly cross-linked polymers. In a few cases crystalline oligomers have been isolated. Some of these are cage molecules; the octamer $(MeSiO_{1.5})_8$ is based on a cube of silicon atoms with Si—O—Si bond angles of about $145°$.

Silicones are apparently completely non-toxic. Silicone rubbers have been used to make surgical implants for the human body.

4.5 Hydrolysis of organohalides of germanium, tin and lead

Hydrolysis of alkylchlorogermanes proceeds readily, although much less is known about the resulting products than about the corresponding siloxanes. Me_3GeCl affords $(Me_3Ge)_2O$, and Me_2GeCl_2 first yields $Me_2Ge(OH)_2$ which condenses to $(Me_2GeO)_n$. The latter has been isolated in three different forms, a cyclic trimer, a cyclic tetramer and a polymer.

The metal atom in organotin- and organolead-oxygen compounds commonly assumes corrdination numbers of five or six, in contrast to silicon or germanium derivatives in which the coordination number is four. This is illustrated by the structure of trimethyltin hydroxide, which is obtained as a crystalline sublimate, m.p. $118°C$, when trimethyltin bromide is heated with 55% aqueous sodium hydroxide. In the solid there are essentially planar Me_3Sn groups symmetrically bridged by OH units. Each five-coordinate tin atom is in a trigonal bipyramidal environment. The hydroxide is converted into the monomeric oxide $(Me_3Sn)_2O$

b.p. 86°C/24 mm by treatment with sodium in benzene, but the oxide readily picks up water again in the air.

$$2Me_3SnOH \underset{H_2O}{\overset{Na/-H_2}{\rightleftarrows}} (Me_3Sn)_2O$$

In contrast to Me_3SiCl, which is vigorously and completely hydrolysed, Me_3SnCl dissolves in water to give the hydrated ion $[Me_3Sn(OH_2)_2]^+$. This behaves as a weak acid (pK_a about 6.5)

$$Me_3Sn(OH_2)_2^+ + H_2O \rightleftharpoons Me_3Sn(OH)(OH_2) + H_3O^+$$

Me_2SnCl_2 also gives acid solutions which contain $[Me_2Sn(OH_2)_4]^{2+}$ and $[Me_2Sn(OH)(OH_2)_3]^+$ ions. As the pH is raised polynuclear μ-hydroxo complexes are formed and eventually $[Me_2Sn(OH)_3]^-$ and $[Me_2Sn(OH)_4]^{2-}$. Alkaline hydrolysis eventually gives dimethyltin oxide, $(Me_2SnO)_n$, which is probably a cross-linked polymer, quite unlike the dimethylsiloxanes. Mössbauer measurements suggest that the tin atoms are five coordinate. The stannonic or plumbonic acids $[MeMO(OH)]_n$ result from the Meyer reactions (p. 131) of methyl iodide with alkaline aqueous potassium stannate(II) or an alkaline suspension of lead(II) oxide respectively, followed by neutralization with carbon dioxide. They show amphoteric behaviour, dissolving in hydrohalic acids to form trihalides $MeMX_3$ and in alkali hydroxides to give anions $[MeM(OH)_4]^-$.

4.6 Catenated organic derivatives of the elements of Group IV

4.6.1 Bond energies

Catenation, the property of forming compounds which contain bonds between like atoms, is most widely found in the chemistry of carbon. Most organic compounds rely for their framework on carbon–carbon bonds. These can be single bonds only as in aliphatic or alicyclic hydrocarbons, or multiple bonds as in alkenes, arenes or alkynes.

Estimated strengths of homonuclear bonds between elements of Main Groups IV, V and VI are listed in Table 4.2. While the bonds become progressively weaker down Group IV, the maximum strength in Groups V and VI comes at phosphorus and sulphur. It is in the chemistry of those elements which form the strongest homonuclear bonds, i.e. C, Si, P and S, that the most robust catenated derivatives and the biggest variety of such compounds are found.

117

Table 4.2. Bond energy terms for homonuclear bonds (kJ mol^{-1})

E(C—C)	358	E(N—N)	159	E(O—O)	142
E(Si—Si)	220	E(P—P)	214	E(S—S)	268
E(Ge—Ge)	189	E(As—As)	167	E(Se—Se)	159
E(Sn—Sn)	151	E(Sb—Sb)	142		

Values quoted vary from source to source. Those given above serve only as a guide to trends and should not be accepted as accurately valid in organometallic compounds.

4.6.2 *Organopolysilanes*

Many textbooks give the impression that catenation among Group IV elements, apart from carbon, is rather limited. This is certainly not so as far as organopolysilanes are concerned. Polymeric compounds $Me_3Si(SiMe_2)_nSiMe_3$ with chain lengths of at least 3000 units have been prepared and cyclic compounds $(Me_2Si)_n$ which have up to 35 silicon atoms in the ring have been identified. A wide range of aryl compounds is also known. These materials are not only thermally stable but in contrast to silicon hydrides and halides, are resistant to hydrolysis and to oxidation. Polysilanes such as $Si_{12}Me_{26}$ can be distilled in air at 300°C. The difference between these organic derivatives and the hydrides is probably associated with the lower mobility and reactivity of alkyl and aryl substituents compared with hydrogen.

The most general method of preparation employs the Wurtz reaction, in which alkyl or arylchlorosilanes react with alkali metals or magnesium, often in tetrahydrofuran.

$$2Me_3SiCl \xrightarrow[\text{THF}]{\text{Na/K}} Me_3Si\text{–}SiMe_3 \text{ (b.p. } 113°C)$$

$$2Me_2SiCl_2 \xrightarrow[\text{THF}]{\text{Na/K}} (Me_2Si)_n + (Me_2Si)_6 + \text{other rings}$$
$$\text{polymer} \quad \text{m.p. } 254°C$$

In the early stages of the reaction with Me_2SiCl_2, while some chloro compound still remains, a polymeric species is produced. The Si—Si bonds in this are cleaved by alkali metals and the resulting metal derivatives react with Me_2SiCl_2 to produce cyclic materials. The major product is $(Me_2Si)_6$. Other cyclic compounds $(Me_2Si)_n$ are also formed, especially under non-equilibrium conditions. Rings from $n = 5$ up to $n = 24$ have been isolated and others up to $n = 35$ identified using high pressure liquid chromatography.

When neat Me_2SiCl_2 reacts with sodium metal in an autoclave, polymeric $(Me_2Si)_n$ is formed. When heated to 320°C under argon it rearranges to a polymer which has a backbone of alternating carbon and silicon atoms. This can be drawn into strands which on heat treatment at 1300°C in vacuo produce fibres of β-SiC of very high tensile strength.

$$Me_2SiCl_2 \xrightarrow[\text{autoclave}]{Na} (Me_2Si)_n \xrightarrow[320°C]{\Delta} \left(\begin{matrix} Mc \\ -Si-CH_2 \\ H \end{matrix} \begin{matrix} Me \\ -Si-CH_2 \\ H \end{matrix} \right)_n \xrightarrow[\text{vac.}]{1300°C} \beta\text{-}SiC$$

Planned syntheses have been devised for the preparation of specific linear polysilanes. This is illustrated by the following example starting from the cyclic hexamer.

$$(Me_2Si)_6 \xrightarrow{Cl_2} Cl(SiMe_2)_6Cl \xrightarrow{MeMgX} Me_3Si(SiMe_2)_6Cl$$

$$\xrightarrow[Cl(SiMe_2)_6Cl]{Na/} \underset{(n=12,18,24\cdots)}{Me_3Si(SiMe_2)_{n-2}SiMe_3}$$

Linear polysilanes show strong absorption in the ultraviolet. The wavelength at which maximum absorption occurs and the molar absorbance both increase with increasing chain length in a way resembling conjugated polyenes.

The Si—Si bonds in aryldisilanes and in polysilanes are cleaved by lithium in tetrahydrofuran to give lithio derivatives. Triphenylsilyllithium can be obtained in solution either from Ph_3SiCl or from $Ph_3SiSiPh_3$. The Ph_3Si^- anion is isoelectronic with triphenylphosphine and has a similar pyramidal structure which undergoes slow inversion (barrier $> 100\,kJ\,mol^{-1}$) (p. 127). It is a strong nucleophile; some of its reactions are shown below (Fig. 4.7).

Alkali metals do not cleave hexaalkyldisilanes, although alkoxides in the presence of crown ethers do. The usual route to Me_3SiLi is via the mercury derivative.

$$2Me_3SiH + HgMe_2 \xrightarrow{-CH_4} (Me_3Si)_2Hg \xrightarrow{2Li} 2Me_3SiLi + Hg$$

In the solid state trimethylsilyllithium exists as hexameric units $(Me_3SiLi)_6$. There are six lithium atoms in a highly folded chair arrangement and each trimethylsilyl group bridges three lithium atoms. The structure is very similar to that of cyclohexyllithium (p. 37).

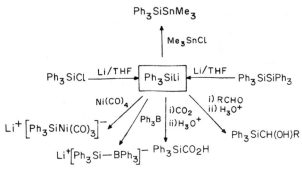

Fig. 4.7 Reactions of triphenylsilyllithium.

119

4.6.3 *Catenated organo-germanium, -tin and -lead compounds*

The chemistry of alkyl and arylpolygermanes closely resembles that of the silanes, although it has not yet been explored quite so fully. Ge—Ge bonds are somewhat more readily cleaved by alkali metals or by halogens. Cyclic oligomers, similar to the silicon compounds, $(Me_2Ge)_n$ $(n = 5, 6, 7)$ and $(Ph_2Ge)_n$ $(n = 4, 5, 6)$ have been isolated from the reactions of R_2GeCl_2 $(R = Me, Ph)$ with alkali metals. Germanium(II) iodide also provides a starting material for polygermanes. On treatment of GeI_2 with trimethylaluminium a mixture of air stable methylpolygermanes results.

$$GeI_2 + AlMe_3 \longrightarrow Me_3Ge(GeMe_2)_nGeMe_3 \quad (n = 2-22)$$

Apart from organic derivatives, compounds containing tin—tin bonds are generally very unstable. Distannane, Sn_2H_6, is formed in low yield, together with stannane, SnH_4, by reduction of aqueous alkaline stannate(II) solutions with sodium borohydride. The preparation of Sn_2Cl_6 has been claimed, but it disproportionates to $SnCl_4$ and $SnCl_2$ even at $-65°C$.

Organopolystannanes result from reactions of alkali metals with organotin halides. These are often carried out in liquid ammonia.

$$Me_3SnBr + Na \xrightarrow[NH_3]{liq.} Me_3SnNa \xrightarrow{Me_3SnBr} Me_3SnSnMe_3$$

$$Me_2SnBr_2 + 2Na \longrightarrow Me_2SnNa_2 \xrightarrow{Me_2SnBr_2} NaSnMe_2SnMe_2SnMe_2Na$$

$$\xrightarrow{MeI} Me_3SnSnMe_2SnMe_3.$$

In this way fully methylated tetra and pentastannanes have also been prepared. A good way of forming tin–tin bonds under mild conditions is reaction of organotin hydrides with aminotin compounds (p. 107).

$$Et_3SnH + Et_3SnNMe_2 \longrightarrow Et_3Sn—SnEt_3 + Me_2NH$$

$$3Et_2SnH_2 + 3Et_2Sn(NMe_2)_2 \longrightarrow (Et_2Sn)_6 + 6Me_2NH$$

The decomposition of organotin hydrides is catalysed by amines. The dihydrides yield cyclic compounds e.g.

$$Ph_2SnH_2 \xrightarrow{pyridine} (Ph_2Sn)_6$$

$$\xrightarrow[formamide]{dimethyl-} (Ph_2Sn)_5$$

An X-ray diffraction study of the crystalline hexamer has shown that the six membered ring is in the chair form. Rings of varying sizes are obtained depending on the reaction conditions and on the attached groups. In the case of the ethyl compounds, for example, cyclic materials $(Et_2Sn)_n$ with $n = 5, 6, 7$ and 9 have been isolated.

Tin(II) halides and Grignard or organolithium reagents afford air-sensitive oils or amorphous solids, usually yellow or red in colour, which have the approximate empirical composition R_2Sn. They are similar to the materials obtained from alkali metals and organotin halides. These products do not contain divalent tin but are mixtures of linear and cyclic polystannanes. The linear compounds are coloured, the more branching which is present the deeper the colour. The extent of branching can be determined by degradation by chlorine at $-70°C$ which gives a mixture of R_3SnCl, R_2SnCl_2 and $RSnCl_3$.

$$(R_3Sn)_2SnR(SnR_2)_2SnR_3 \xrightarrow{Cl_2} 3R_3SnCl + 2R_2SnCl_2 + RSnCl_3$$

$$(R_2Sn)_6 \xrightarrow{6Cl_2} 6R_2SnCl_2$$

These polystannanes presumably arise through the intermediacy of trialkyl/aryltin lithium or Grignard reagents. One way of preparing R_3SnLi is to treat tin(II) chloride with three molar equivalents of RLi in ether. Further reaction with $SnCl_2$ yields $(Ph_3Sn)_3SnLi$ ($R = Ph$) which may be trapped with Ph_3SnCl.

$$2RLi + SnCl_2 \rightleftharpoons 'R_2Sn' + 2LiCl$$

$$'R_2Sn' + RLi \rightleftharpoons R_3SnLi$$

$$3Ph_3SnLi + SnCl_2 \longrightarrow (Ph_3Sn)_3SnLi$$

$$\xrightarrow{Ph_3SnCl} (Ph_3Sn)_4Sn$$

$$\text{m.p. } 280° \text{ (dec.)}$$

The strengths of M—M bonds decrease down Group IV, so it is not surprising that organoderivatives which contain lead–lead bonds are rather unstable. When lead(II) chloride is treated with methylmagnesium iodide, the first product is probably polymeric dimethyllead, but this disproportionates first into lead and hexamethyldilead and finally on warming into lead and $PbMe_4$.

$$PbCl_2 \xrightarrow{MeMgI} 'Me_2Pb'$$

$$6'Me_2Pb' \longrightarrow 2Pb + 2Me_3Pb—PbMe_3 \longrightarrow 3Me_4Pb + 3Pb$$

4.7 Divalent organic compounds of the elements of Group IV

The elements of Group IV form oxides and halides in two formal oxidation states, $+2$ and $+4$. The $+2$ compounds of carbon and silicon (except for CO) are only short lived species which are unstable to dimerization (CX_2) or to disproportionation or polymerization (SiO, SiX_2). For all the other elements representatives of both oxidation states can be isolated. The $+2$ state becomes increasingly favoured relative to the $+4$ state down the group.

121

The main Groups IV and V

Treatment of dihalides of Ge, Sn or Pb with Grignard reagents does not afford monomeric divalent compounds R_2M, but leads rather to polymers in which the metal is tetravalent because of the presence of metal–metal bonds. The challenge to synthesise truly divalent organic compounds of these elements has been tackled in several ways.

Dimethylsilylene Me_2Si has been generated by photolysis of polysilanes such as $(Me_2Si)_6$.

$$(Me_2Si)_6 \xrightarrow{h\nu} (Me_2Si)_5 + Me_2Si$$

It is a short lived species which can be trapped as it is formed by, for example, 2-butyne or triethylsilane. If the irradiation is performed in a solid argon matrix at 10 K, spectra of Me_2Si can be obtained, as it is stable under these conditions. Me_2Si is a bright yellow species absorbing at 450 nm.

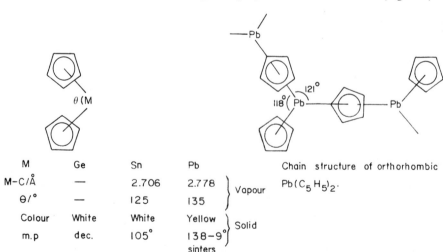

Bis(cyclopentadienyl) complexes Cp_2M (M = Ge, Sn, Pb) have been prepared by reaction of cyclopentadienylsodium and an appropriate dihalide in tetrahydrofuran. Electron diffraction shows that in the vapour these molecules have angular structures on account of the electron pair on the metal. In the solid state the germanium compound polymerizes within 3 h at ambient temperature, the tin compound within five days. The cyclopentadienyl groups are coordinated in a symmetrical pentahapto fashion (p. 191). Association through the rings occurs in the crystalline lead compound to give a polymeric chain structure (Fig. 4.8).

M	Ge	Sn	Pb	
M–C/Å	—	2.706	2.778	Vapour
θ/°	—	125	135	
Colour	White	White	Yellow	Solid
m.p	dec.	105°	138–9° sinters	

Chain structure of orthorhombic $Pb(C_5H_5)_2$.

Fig. 4.8 Structures of $Cp_2Ge_{(g)}$, $Cp_2Sn_{(g)}$ and $Cp_2Pb_{(c)}$.

122

In crystalline decaphenylstannocene, $(\eta^5\text{-}C_5Ph_5)_2Sn$, the C_5 rings are parallel and adopt a staggered conformation. The phenyl groups are arranged like a paddle wheel about each ring. In this structure an S_{10} inversion rotation axis is present passing through the centre of each ring and the metal atom.

It has been suggested that the dialkyls MR_2 are unstable kinetically to disproportionation or polymerization because the monomers are coordinatively unsaturated, possessing both filled non-bonding and low lying empty orbitals. If association between monomer units could be prevented, it might be possible to isolate dialkyls. The use of bulky groups such as Me_3SiCH_2- or Me_3CCH_2- has proved most successful in stabilizing transition metal alkyls (p. 219) of low coordination number. Me_3SiCH_2MgCl and $SnCl_4$, however, gave $Sn(CH_2SiMe_3)_4$ and $SnCl_2$ afforded the distannane $Sn_2(CH_2CMe_3)_6$ with neopentylmagnesium chloride. Success was achieved, however, with the even bulkier group $(Me_3Si)_2CH-$.

$$(Me_3Si)_2CHCl \xrightarrow[Et_2O]{Li} (Me_3Si)_2CHLi \xrightarrow[\substack{Et_2O \\ -20°C}]{MCl_2} M\{CH(SiMe_3)_2\}_2 + 2LiCl$$

$(M = Sn, Pb)$

Isoelectronic amino derivatives of divalent Ge, Sn and Pb were similarly prepared, starting from hexamethyldisilazane.

$$(Me_3Si)_2NH \xrightarrow{Na} (Me_3Si)_2NNa \xrightarrow{MX_2} \{(Me_3Si)_2N\}_2M$$

$$\xrightarrow{(Me_3Si)_2CHLi} \{(Me_3Si)_2CH\}_2M. \quad (M = Ge, Sn, Pb)$$

These were in turn converted into the alkyl compounds as shown.

The yellow germanium (m.p. $180°C$) and brick-red tin complexes (m.p. $136°C$) are thermally stable but air-sensitive solids which can be sublimed in vacuo, but the purple lead analogue is rather heat and photosensitive. They are monomeric in the vapour and in their solutions in benzene. X-ray diffraction of the tin compound shows that it dimerizes in the solid state. The monomer has both a lone pair and vacant p orbital. It acts as a Lewis base towards transition metal complexes and as a Lewis acid towards pyridine. Dimerization may therefore occur through mutual acid–base interaction to form a bent tin—tin double bond.

$$Cr(CO)_6 + SnR_2 \xrightarrow[hexane]{hv} (OC)_5Cr \leftarrow SnR_2 + CO$$

$$py + SnR_2 \xrightarrow[temp.]{low} py \leftarrow SnR_2 \quad \{R = -CH(SiMe_3)_2\}$$

Trivalent radicals $Mr_3^{\cdot}$ ($M = Si, Ge, Sn$) and $M^{\cdot}\{N(SiMe_3)_2\}_3$ ($M = Ge, Sn$) are obtained similarly from R_3MCl. Apart from the silicon species $t_{1/2} = 10$ min, they persist for very long periods in hexane in the absence of air.

The main Groups IV and V

4.8 *p*π–*p*π Bonding in compounds of Groups IV and V

Multiple bond formation between the elements of the First Short Period C, N and O is a common phenomenon. Relative to single bonds, multiple bonds between these elements are much stronger than those which involve one or more elements from later Periods. While $(2p–2p)\pi$ overlap is rather strong, similar overlap involving p orbitals of higher principal quantum number is less effective. Thus CO_2 and Me_2CO monomeric, whereas SiO_2 and $(Me_2SiO)_n$ polymeric (p. 112). It is instructive to note that $2E(C—C) = 712 \text{ kJ mol}^{-1}$, whereas $E(C{=}C)$ is only 602 kJ mol^{-1}. Ethene is thermodynamically unstable with respect to polyethene. While ethene does not polymerize under normal conditions, it does so very rapidly when catalysed (p. 371). Ethenes $RCH{=}CH_2$ with electron withdrawing substituents $(R = CN, CO_2Et \text{ or } Ph)$ are especially susceptible to polymerization initiated by free radicals or by traces of acids or bases. Can species such as $R_2Si{=}SiR_2$ or $R_2Si{=}CR'_2$ be obtained if pathways to polymerization can be blocked? If $2E(Si—Si) = 440 \text{ kJ mol}^{-1}$ and $E(Si{=}Si) = 314 \text{ kJ mol}^{-1}$, the prospect from thermodynamics does not look very hopeful. Nevertheless many attempts have been made to prepare such compounds and recently these have met with some success.

Pyrolysis of dimethylsilacyclobutane at 600°C gives a mixture of ethene and dimethylsilaethene. The latter species has been detected in the gas phase by mass spectrometry. It has also been trapped in a solid argon matrix at 10 K and its infrared spectrum measured. It is very reactive, dimerizing readily and adding to dienes in a Diels-Alder reaction:

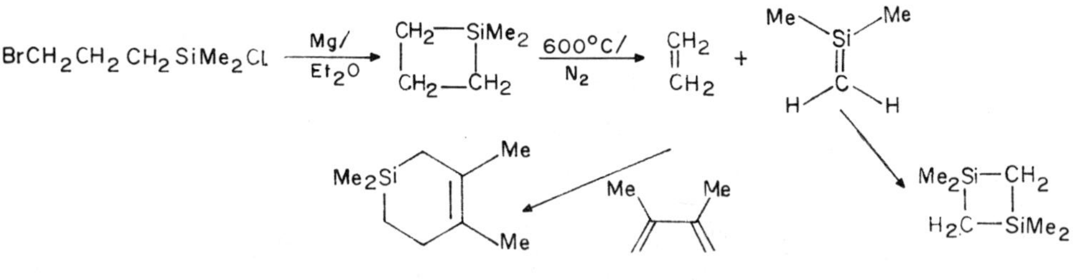

The introduction of very bulky substituents hinders dimerization and in the following example also prevents intermolecular rearrangement of small Si_3 rings to larger oligomers (Ar = 2, 4, 6-trimethyl-phenyl).

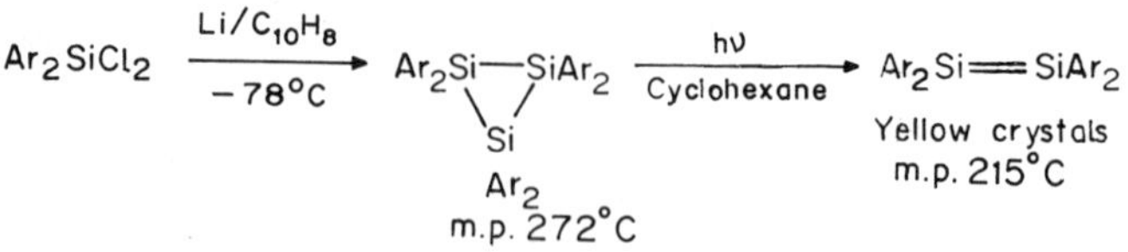

The bulkier the substituents the more stable the silene becomes. While $Me_2Si{=}SiMe_2$ has been observed only in a frozen matrix at low temperatures, $Bu_2^t Si{=}SiBu_2^t$ persists for a few hours under ambient conditions. $Ar_2Si{=}SiAr_2$ and

124

$Bu^tArSi\!=\!SiArBu^t$ are crystalline materials whose structures have been determined by X-ray diffraction (Ar = 2, 4, 6-trimethylphenyl). The latter has a planar $C_2Si\!=\!SiC_2$ skeleton with a $Si\!=\!Si$ bond distance of 2.14 Å. This bond length may be compared with a $C\!=\!C$ distance of 1.33 Å in *trans*-1,2-diphenylethene and a $Si\!=\!C$ distance of 1.70 Å in $Me_2Si\!=\!C(SiMe_3)SiMeBu_2^t$.

Silabenzene and silatoluene have been obtained as highly reactive intermediates from dehydrochlorination of silacyclohexadienes. They can be trapped as Diels-Alder adducts.

4.9 Organic derivatives of the elements of Group V

4.9.1 *Types of compound*

Phosphorus, arsenic, antimony and bismuth all form a large number of trialkyls and triaryls, R_3M, which have pyramidal structures. There are also numerous mixed derivatives of general formula R_nMX_{3-n}, where X is a univalent atom or group such as H, NR_2', OR', SR' or halogen. The pentavalent state is also important for P, As and Sb, although it is not favoured by bismuth. While the elements all form pentaphenyls, Ph_5M (Ph_5Bi is very unstable) (p. 135) it is the mixed derivatives R_5MX_{5-n} (X = halogen) and the related oxygen compounds formally derived from them by hydrolysis that are most extensive. Phosphorus in particular forms very strong formal double bonds to oxygen (as in $R_3P\!=\!O$) and this property tends to dominate much of its chemistry.

4.9.2 *Organophosphorus chemistry*

A series of seven volumes devoted to the subject of organophosphorus chemistry, completed in 1976, lists over 100 000 compounds.* It is therefore impossible to do more than touch upon some aspects which are relevant to the behaviour of the heavier elements of Group V and to organometallic chemistry as a whole.

Conventionally organophosphorus compounds have not been included in the area defined as "Organometallic", but rather as part of organic chemistry.

---

*Kosolapoff, G.M. and Maier, L. 1976 *Organic Phosphorus Compounds*, Wiley–Interscience, New York.

A convenient method of preparing trialkyl and triaryl phosphines in the laboratory is to use organolithium or Grignard reagents. Mixed phosphines $R'PR_2$ and R'_2PR can also be made in this way, starting from $R'PCl_2$ or R'_2PCl instead of from PCl_3.

$$PCl_3 + 3RMgCl \longrightarrow R_3P + 3MgCl_2$$

Another approach, much used before the introduction of Grignard reagents which still finds application in industry on account of convenience and cost (p. 138) is the mixed Wurtz reaction.

$$PCl_3 + 3ArCl \xrightarrow[\text{reflux/benzene}]{6Na} PAr_3 + 6NaCl$$

If triphenylphosphite $P(OPh)_3$ is used instead of PCl_3, excellent yields of both trialkyl and triarylphosphines result by this method using petroleum as a solvent.

Phosphines R_nPH_{3-n} $(n = 0-2)$ are converted into phosphides by alkali metals in liquid ammonia or in ethers. Treatment of these with alkyl halides provides an efficient method of forming P—C bonds. Unsymmetrical phosphines, which are useful ligands in transition metal chemistry, can be made in this way.

$$RPCl_2 \xrightarrow{\text{LiAlH}_4} RPH_2 \xrightarrow[\text{liq. NH}_3]{Na} RPNa_2 \xrightarrow{2R'X} RPR'_2$$

$$PH_3 \xrightarrow[\text{liq. NH}_3]{2Na} PHNa_2 \xrightarrow{2MeI} PHMe_2 \xrightarrow[\text{liq. NH}_3]{Na} Me_2PNa \xrightarrow{ClCH_2CH_2Cl} Me_2PCH_2CH_2PMe_2$$

Partial substitution of chlorine by alkyl is more cleanly carried out using less reactive organometallics than Grignards. Cadmium, mercury, tin and lead compounds have all been used e.g.

$$R_2Cd + 2PCl_3 \xrightarrow[\text{ether}]{-20°C} 2RPCl_2 + CdCl_2$$

The Friedel–Crafts reaction using PCl_3 and a stoichiometric quantity of $AlCl_3$ affords aryldichlorophosphines. The product forms a complex with the aluminium chloride from which it has to be displaced by a stronger Lewis base such as $POCl_3$ or pyridine.

$$C_6H_6 + PCl_3 \xrightarrow[\text{2-8 h, reflux}]{AlCl_3} PhPCl_2.AlCl_3 + HCl \xrightarrow{POCl_3} PhPCl_2 + Cl_3Al.OPCl_3$$

Trialkyl- and triarylphosphines have pyramidal structures, in which the CPC angles are about $100°$. By virtue of the lone pair of electrons on phosphorus they can act as proton acceptors (Brønsted bases) and as electron pair donors (Lewis bases). Some pK_a values for phosphonium ions and the corresponding ammonium ions are given in Table 4.3. A strong regular variation in base strength is found for the series $PH_3 < MePH_2 < Me_2PH < Me_3P$. This contrasts with the irregular variation found for the ammonium ions (p. 73). While phosphine is an extremely

Table 4.3. pK$_a$ values for ammonium and phosphonium ions

Ammonium ion	pK$_a$	Phosphonium ion	pK$_a$
NH$_4^+$	9.21	PH$_4^+$	-14
MeNH$_3^+$	10.62	MePH$_3^+$	-3.2
Me$_2$NH$_2^+$	10.64	Me$_2$PH$_2^+$	3.9
Me$_3$NH$^+$	9.76	Me$_3$PH$^+$	8.65
Ph$_2$NH$_2^+$	0.79	Ph$_3$PH$^+$	2.73

$$R_3NH^+ + H_2O \rightleftharpoons R_3N + H_3O^+; \; pK_a = -\log_{10}K_a;$$
$$K_a = [H_3O^+][R_3N]/[R_3NH^+]$$

weak Brønsted base, trimethylphosphine is comparable with trimethylamine.

Measurements of gas phase dissociation constants have shown that the Lewis basicity towards hard reference acids such as BMe$_3$, BF$_3$ or BCl$_3$ falls down Group V in the order Me$_3$N > Me$_3$P > Me$_3$As > Me$_3$Sb ($\gg$ Me$_3$Bi). Towards the softer acid BH$_3$, however, the order is Me$_3$P > Me$_3$N > Me$_3$As. Orders of basicity are strongly influenced by the nature of the acceptor. This is especially evident from the chemistry of transition elements, which often form stable complexes with phosphines and arsines for which there are no amine analogues (Chapter 5). Phosphines and arsines are, in the classification of Pearson, typical soft bases which form their most stable complexes with soft acids such as polarizable transition metal and heavy post transition metal acceptors.

Ammonia and amines are rapidly undergoing inversion through a planar transition state from one pyramidal structure to another. This motion resembles an umbrella turning inside out and back again. It occurs about 10^5 times every second. The potential energy barriers to inversion for phosphines, arsines and stibines, however, are so much greater than for amines that the rates are almost immeasurably slow at room temperature. Pyramidal compounds with three different substituents MRR'R'' are chiral and can be resolved into enantiomers. One route by which this has been achieved is through quaternary phosphonium or arsonium cations. These form diastereoisomeric salts with an optically active anion which can be separated by fractional crystallization. The resolved quaternary salt is then converted into the chiral phosphine or arsine by reduction. The optically active phosphine (+)MePr^{n}PhP is racemized on heating under reflux in toluene (110°C) for 3 h. The corresponding arsine is configurationally even more stable. It has a half life of about 1 month at 218°C.

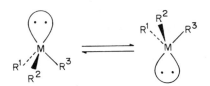

$(+)\left[\text{MePrPhAsCH}_2\text{Ph}\right]^{+}(+)\text{A}^{-}$

$(\pm)\left[\text{MePrPhAsCH}_2\text{Ph}\right]^{+}\text{Br}^{-}+ (+)\text{A}^{-}$

$(-)\left[\text{MePrPhAsCH}_2\text{Ph}\right]^{+}(+)\text{A}^{-}$

$(+)[\text{MePrPhAsCH}_2\text{Ph}]^{+}(+)\text{A}^{-} \xrightarrow{\text{Br}^-} (+)[\text{MePrPhAsCH}_2\text{Ph}]^{+}\text{Br}^{-}$

$\xrightarrow{\text{LiAlH}_4} (+)[\text{MePrPhAs}]$

Optically active phosphines have been used as ligands in transition metal complexes to induce asymmetric hydrogenation of suitable organic compounds (p. 184).

The alkyl derivatives R_3M ($M = P$, As, Sb, Bi) are all extremely susceptible to oxidation. Me_3P inflames in the air and even under more controlled conditions gives a complex mixture of products with oxygen *via* radical chain reactions. The triaryls, however, are air stable. Careful oxidation with aqueous hydrogen peroxide or a suspension of mercury(II) oxide affords the oxides R_3MO (not $M = Bi$).

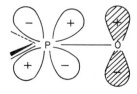

The phosphorus—oxygen bond in $R_3P{=}O$ is conventionally written as a double bond, implying pentavalent phosphorus (contrast $R_3N \rightarrow O$ in amine oxides; octet expansion in nitrogen is forbidden). This implies that $3d$ orbitals of phosphorus accept electron density from the oxygen lone pairs. This $(2p–3d)\pi$ bonding can involve both of the oxygen lone pairs and two orthogonal $3d$ orbitals of phosphorus, leading to a cylindrically symmetrical π cloud about the $P{=}O$ bond.

The bond energy $E(P{=}O)$ in phosphine oxides in about $580\,\text{kJ mol}^{-1}$, compared with $E(P{-}O)$ in phosphites of $385\,\text{kJ mol}^{-1}$. The formation of this very strong bond provides the driving force for many reactions of trivalent phosphorus compounds. The tendency to form formal double bonds $M{=}O$ decreases from P to As to Sb. While arsine oxides are monomeric and may be written $R_3As{=}O$, stibine oxides such as Ph_3SbO are polymeric. Their structures thus involve formal single bonds $(Ph_3Sb\!-\!O\!-\!)_n$. Other typical reactions of R_3M ($M = P$, As, Sb) include the addition of sulphur, selenium, halogens ($X = Cl$, Br, I) and alkyl halides (to give quaternary salts).

Phosphine oxides are extremely inert compounds. Me_3PO is stable up to $700°C$; Ph_3PO can be nitrated with a mixture of concentrated nitric and sulphuric acids to give $(3\text{-}O_2NC_6H_4)_3PO$. Reduction to the phosphine is difficult, but can be achieved with lithium aluminium hydride. R_3PO and R_3AsO form adducts through oxygen with a wide range of Lewis acids. Typical complexes include $HgCl_2L_2$, $SnCl_4L_2$ and $NiBr_2L_2$ ($L = Ph_3PO$ or Ph_3AsO). Phosphine oxides are used to complex actinides such as uranium and plutonium in their separation by solvent extraction.

4.9.3 Organo-halides

Halides R_2MX and RMX_2 are formed by all the elements P to Bi. In general they have pyramidal molecular structures. The rather high melting points of $MeBiCl_2$ ($242°C$) and Me_2BiCl ($116°C$) suggest that these compounds may be associated in the solid state. Phosphorus forms a complete set of halides Me_nPX_{5-n} (Table 4.4). The fluorides are monomeric, volatile species, which have trigonal bipyramidal structures. Electron diffraction studies show that the fluorine atoms preferentially occupy the axial positions. The axial $P\!-\!F_{ax}$ bonds are longer than the equatorial $P\!-\!F_{eq}$. ^{19}F n.m.r. studies confirm these findings for Me_3PF_2

Table 4.4. Organohalides of Group V elements

				cf. PF_5	Molecular structures
$Me_4\overset{+}{P}.X^-$	$Me_3\overset{+}{P}X.X^-$ Me_3AsF_2 molecular	$Me_2\overset{+}{P}X_2.X^-$ Me_2AsF_3 probably associated	$Me\overset{+}{P}X_3.X^-$	cf. $PCl_4^+PCl_6^-$ AsF_5	Auto-ionized.
$Me_4As^+X^-$	Me_3AsCl_2 molecular, weak Cl bridges	Me_2AsCl_3 dec. 40°C	$MeAsCl_4$ dec. 0°C	$AsCl_5$ dec. $-50°C$	
$Me_4Sb^+X^-$	Me_3SbX_2 molecular ($X = F$, Cl, Br, I)	Me_2SbCl_3 dec. $>120°C$ Me_2SbBr_3 dec. $-15°C$	$MeSbCl_4$ dec. 25°C	$SbCl_5$ dec. 140°C	
	no alkyls Ph_3BiX_2 molecular ($X = F$, Cl, Br)	–	–		

and Me_2PF_3, but $MePF_4$ gives only one signal even at low temperatures (as does PF_5 itself). This could indicate a square pyramidal (C_{4v}) structure for $MePF_4$, but a more likely explanation is that fast Berry pseudorotation is occurring in which the equatorial and axial fluorines are interchanged. The spectrum of $MeNHPF_4$ does reveal two signals at low temperature which coalesce as the temperature is raised.

The fluoro compounds act as Lewis acids ($PF_5 > RPF_4 \gg R_2PF_3 \sim R_3PF_2$). Fluoride ion, for example, complexes with $MePF_4$ and Me_2PF_3 (cf. $MeSiF_3$, p. 104) yielding $MePF_5^-$ and $Me_2PF_4^{2-}$ respectively.

In the crystalline state the other halides Me_nPX_{5-n} ($X = Cl$, Br, I) adopt autoionized structures $[Me_nPX_{4-n}]^+X^-$. In this they resemble PCl_5 ($PCl_4^+PCl_6^-$ in solid) and PBr_5 ($PBr_4^+Br^-$ in solid). These halides may be stabilized by complexing the anion with a Lewis acid such as PCl_5 or $AlCl_3$. This property is exploited in their preparation.

$$PCl_3 + R'Cl + AlCl_3 \longrightarrow R'PCl_3{}^+AlCl_4^-$$

$$RPCl_2 + R'Cl + AlCl_3 \longrightarrow RR'PCl_2{}^+AlCl_4^-$$

$$R_2PCl + R'Cl + AlCl_3 \longrightarrow R_2R'PCl^+AlCl_4^-$$

Some reactions of $RPCl_2$ ($R = $ e.g. Me, Ph) are summarized in Fig. 4.9. Nucleophilic substitution of chloride leads to derivatives RPX_2 ($X = H$, OR', SR', NR$_2'$, etc). Hydrolysis affords an alkyl or arylphosphinic acid. This is tautomeric and exists essentially in the 'keto' form (cf. $H_2P(O)OH$ from PCl_3).

Halide		'Keto'		'Enol'	Esters
PCl_3	$\xrightarrow{H_2O}$			$P(OH)_3$	$P(OR')_3$
$RPCl_2$	$\xrightarrow{H_2O}$			$RP(OH)_2$	$RP(OR')_2$
R_2PCl	$\xrightarrow{H_2O}$			$R_2P(OH)$	$R_2P(OR')$

This 'keto' structure is favoured on account of the formation of a strong P=O bond in each case. The corresponding arsenic compounds react differently; the precursor arsonous acid $RAs(OH)_2$ dehydrates to form $(RAsO)_n$ which consists mainly of cyclic trimers together with some other oligomers. The antimony compounds are polymeric. Hydrolysis of R_2AsCl affords the arsine oxides $(R_2As)_2O$. As in $(R_3Si)_2O$, the bond angle at oxygen is rather wide (137° when $R = Me$).

The reactions in Fig. 4.9 also emphasize the ease of addition to three-coordinate phosphorus compound to give four- and five-coordinate species. The procedures used to convert $\equiv PCl_2 \xrightarrow{SO_2} \equiv P=O$ and $\equiv PCl_2 \xrightarrow{H_2S} \equiv P=S$ are generally applicable in phosphorus chemistry. Sulphur dioxide, however, reduces

130

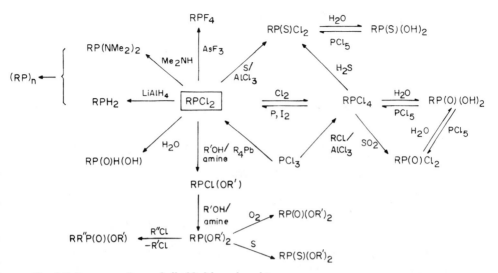

Fig. 4.9 Some reactions of alkyldichlorophosphines.

Table 4.5. Reactions of R_nPCl_{5-n}

Reagent	$R\overset{+}{P}Cl_3.Cl^-$	$R_2\overset{+}{P}Cl_2.Cl^-$	$R_3\overset{+}{P}Cl.Cl^-$	Other product(s)
SO_2	$RP(O)Cl_2$	$R_2P(O)Cl$	R_3PO	$SOCl_2$
H_2S	$RP(S)Cl_2$	$R_2P(S)Cl$	R_3PS	HCl
$R'OH$	$RP(O)(OR')_2$	$R_2P(O)(OR')$	R_3PO	$R'Cl + HCl$
H_2O	$RP(O)(OH)_2$	$R_2P(O)(OH)$	R_3PO	HCl

arsenic(V) to arsenic(III). Hydrolysis and alcoholysis of the halides R_nPX_{5-n} proceeds as shown in Table 4.5, and leads to the acids $RPO(OH)_2$ or $R_2PO(OH)$ or their esters. Hydrolysis takes place very readily. As the free chlorides are not very stable, their complexes with $AlCl_3$ or PCl_5 are often used instead.

4.9.4 *Organo-arsenic and -antimony compounds*

Reactions of organoarsenic compounds are summarized in Fig. 4.10. Alkyl-arsonic and dialkylarsinic acids can be prepared by the Meyer reaction. An alkyl halide or sulphate is treated with an alkali metal arsenite in aqueous solution.

$$As(ONa)_3 + RX \longrightarrow \; \hat{}\,NaX + RAsO(ONa)_2 \xrightarrow{H_3O^+} RAsO(OH)_2$$

This reaction also is apparently driven by the formation of the As=O bond; this involves formal oxidation of arsenic from the $+3$ to the $+5$ state. Arylarsonic acids are often obtained from aryldiazonium salts by the Bart reaction (p. 26) or

131

Table 4.6. Acid dissociation constants of some organo-oxo-acids

Acid	pK_1	pK_2	pK_3
$H_2PO(OH)$	1.1	–	–
$Me_2PO(OH)$	3.1	–	–
$Me_2AsO(OH)$	6.3	–	–
$HPO(OH)_2$	1.3	6.7	–
$MePO(OH)_2$	2.3	7.9	–
$MeAsO(OH)_2$	3.6	8.3	–
$PhAsO(OH)_2$	3.5	8.5	–
$PhSbO(OH)_2$	Very weak		
$PO(OH)_3$	2.2	7.2	12.4
$AsO(OH)_3$	2.2	6.9	11.5
Sb_2O_5, aq.	Very weak		

one of its modifications. Arylstibonic acids $ArSbO(OH)_2$ may be prepared similarly. The latter are amphoteric as they dissolve in concentrated hydrochloric acid forming the anions $ArSbCl_5^-$, which may be precipitated as pyridinium salts. Some acid dissociation constants of oxo-acids of P, As and Sb are listed in Table 4.6. While the phosphorus and arsenic compounds are of comparable strength, the antimony derivatives are very weak acids. Phenylstibonic acid, nominally $PhSbO(OH)_2$, is polymeric and insoluble in water, although it dissolves as a monomer in acetic acid. It probably forms the octahedral anion $PhSb(OH)_5^-$ in alkaline solution. In this it resembles not only Sb_2O_5 which forms $Sb(OH)_6^-$ and telluric acid, which in $Te(OH)_6$ rather than H_2TeO_4, but also related tin complexes which form ions $Me_2Sn(OH)_4^{2-}$ and $MeSn(OH)_5^-$ in strongly alkaline solutions (p. 117). There is a general trend for oxoanions to adopt octahedral rather than tetrahedral structures on moving from the First to the Second Long Period.

Dimethylarsinic acid crystallizes as hydrogen bonded dimers, similar to those formed by carboxylic acids. In solid propylarsonic acid, $C_3H_7AsO(OH)_2$, the molecules are linked together in chains through hydrogen bonds. Arsonic and arsinic acids are readily reduced, in contrast to the phosphorus analogues (Fig. 4.10). By suitable choice of reagent $MeAsO(OH)_2$ yields $MeAsX_2$, $(MeAs)_n$ or $MeAsH_2$.

Apart from their use as pharmaceuticals (p. 142) organoarsenicals such as alkyl- or arylarsonic acids have found some application as herbicides, fungicides and bactericides. They are being superseded by the less toxic tin compounds. All arsenic compounds should be treated as poisonous. The halides cause severe burns on the skin which heal only with difficulty. This property was put to evil use

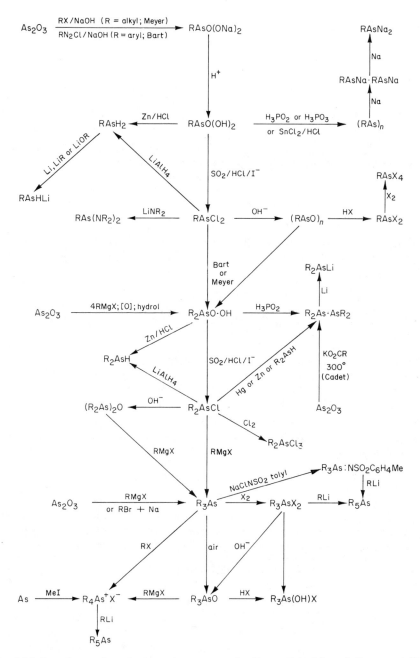

Fig. 4.10 Reactions of organoarsenic compounds. From *Principles of Organometallic Chemistry* 1968 1st Edn, Chapman and Hall, London. Reproduced by kind permission of Professor K. Wade who devised this scheme.

in World War I, when some of these compounds were developed as chemical warfare agents. One, called Lewisite, was a mixture of chlorovinyl derivatives obtained from $AsCl_3$ and ethyne in the presence of aluminium chloride.

$$AsCl_3 + HC \equiv CH \xrightarrow{AlCl_3} ClCH = CHAsCl_2 (+ (ClCH = CH)_2 AsCl \text{ etc}).$$

Arsenic and antimony show a fairly strong affinity for sulphur. Halides $RMCl_2$ (M = As, Sb) react with H_2S or NaSH to give oligomeric sulphides $(RMS)_n$ and with thiols to yield $RM(SR')_2$. Dithiols afford stable and relatively harmless products. This property was used to combat the notorious Lewisite.

4.9.5 Organobismuth compounds

The range of organobismuth compounds is rather restricted. The trialkyls and triaryls can be prepared by the Grignard method. Trimethylbismuth is spontaneously inflammable in air. It reacts with halogens by substitution, rather than by addition.

$$Me_3Bi + Br_2 \longrightarrow Me_2BiBr + MeBr$$

It does not form quaternary salts. Instead methyl iodide affords $MeBiI_2$ and ethane.

$$Me_3Bi + 2MeI \longrightarrow MeBiI_2 + C_2H_6$$

These reactions are in marked contrast to those of its congeners earlier in the Group, especially Me_3Sb. The latter is an extremely strong reducing agent; as well as adding halogens and sulphur to give Me_3SbX_2 and Me_3SbS respectively, it reacts even with concentrated hydrochloric acid on heating in a sealed tube with liberation of hydrogen.

$$Me_3Sb + 2HCl \longrightarrow Me_3SbCl_2 + H_2$$

It also reduces PCl_3 to elementary phosphorus.

Triphenylbismuth forms white, air stable crystals, m.p. 78°C. The molecule is pyramidal with the three phenyl groups arranged like a propeller. The phenyl groups are very easily cleaved, for example by acids, thiols and iodine.

$$Ph_3Bi + 3HCl \longrightarrow BiCl_3 + 3PhH$$

$$Ph_3Bi \xrightarrow[-PhH]{PhSH} Ph_2BiSPh \xrightarrow[-2PhH]{2PhSH} Bi(SPh)_3$$

Chlorine and bromine do give dihalides Ph_3BiX_2 which have covalent trigonal bipyramidal structures, but they decompose fairly readily on heating to Ph_2BiX and PhX. Even Ph_3BiF_2, from Ph_3BiCl_2 and potassium fluoride in aqueous alcohol, decomposes at 190°C to Ph_2BiF and fluorobenzene.

Triphenylbismuth is a very weak Lewis base. In contrast to Ph_3M (M = P, As, Sb) only a few complexes with transition metals have been prepared.

4.9.6 *Pentaalkyl and pentaaryl derivatives*

The search for pentaalkyl and -aryl derivatives of the elements of Group V provides a fascinating story, especially because some rather important discoveries have been made along the way. It began in 1862 when Cahours claimed to have prepared pentamethylarsenic from tetramethylarsonium iodide and dimethylzinc. In 1896 Lachman attempted to make pentamethylnitrogen by a similar route; at that time chemists' imaginations were not restricted by such ideas as the octet rule. Schlenk and Holz did isolate a red compound from the reaction

$$Me_4N^+I^- + Ph_3C^-Na^+ \longrightarrow Me_4N^+CPh_3^- + NaI$$

which has since been shown to be ionic. In general, however, only amines and hydrocarbons were obtained.

Undaunted, G. Wittig tried again during the 1930s and '40s to make R_5N, taking the strongly nucleophilic reagent phenyllithium instead of the resonance stabilized species Ph_3C^-. An intermediate ylide $Me_3\overset{+}{N}-\overset{-}{C}H_2$ was formed which was trapped as a zwitterion with benzophenone.

$$[Me_3\overset{+}{N}-CH_3]Cl^- \xrightarrow[-PhH, -LiCl]{PhLi} Me_3\overset{+}{N}-\overset{-}{C}H_2 \xrightarrow{Ph_2CO}$$

ylide

$$Me_3\overset{+}{N}CH_2C(O^-)Ph_2 \xrightarrow{HX} [Me_3\overset{+}{N}-CH_2C(OH)Ph_2]X^-$$

Zwitterion

Even if nitrogen is unlikely to form pentacoordinate derivatives, phosphorus is a much more promising candidate. 'Octet expansion' is common; PF_5, for example, has a covalent trigonal bipyramidal structure in both the gaseous and condensed states. $Me_4P^+I^-$, however, reacts with phenyllithium exactly analogously to $Me_4N^+I^-$. This is because the carbon—hydrogen bonds in the methyl groups are too acidic to let the nucleophile attack phosphorus. By taking $Ph_4P^+I^-$, which has significantly less acidic C—H functions, pentaphenylphosphorus, m.p. 124°C dec., was isolated as a colourless crystalline solid, which is stable in air at room temperature. Subsequently pentaphenyl derivatives of all the heavier Group VB elements were prepared by similar routes. While Ph_5Sb (m.p. 170°C) is somewhat more stable thermally than the phosphorus compound, possibly because the molecule is less congested sterically, Ph_5Bi can be kept for only a few days under nitrogen at room temperature. It forms a violet crystalline powder. Some routes to the pentaphenyl compounds are summarized in Fig. 4.11.

An X-ray diffraction study of pentaphenylphosphorus reveals a slightly distorted trigonal bipyramidal structure; $P-C_{axial}$, 1.987 Å; $P-C_{eq}$, 1.850 Å, $C_{ax}-P-C_{ax}$, 176.9 Å. Pentaphenylarsenic is isomorphous, indicating a similar structure. The antimony compound, however, is one of the very few Main Group derivatives to have square pyramidal geometry, both in the solid state and in solution in dichloromethane. The *p*-tolyl derivative, $(4-MeC_6H_4)_5Sb$, is trigonal

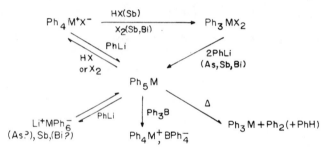

Fig. 4.11 Pentaphenyls of Group VB.

bipyramidal in the solid. Proton n.m.r. spectra of its solution show equivalence of methyl groups even at low temperatures which indicates rapid interchange of axial and equatorial substituents, presumably by pseudorotation. There is better evidence for such dynamic processes from n.m.r. studies of spirocycles. Here the barriers to rearrangement can be measured; they lie between $50-80\,\text{kJ mol}^{-1}$.

The only simple pentaalkyls of phosphorus which have yet been prepared contain CF_3 groups, which assume axial positions in the most stable conformer. Me_5As, m.p. $-6°C$, is thermally unstable, decomposing at $100°C$ to Me_3As, methane and ethene as the major volatile products. The antimony analogue, b.p. $126-7°C$, obtained from methyllithium and Me_3SbBr_2 or Me_4SbBr, is much more robust. Infrared and Raman spectra support a trigonal bipyramidal structure (D_{3h}). Acids cleave one methyl group to give a wide range of products.

$$Me_5Sb + HX \longrightarrow Me_4SbX + CH_4 \quad (X = \text{halogen, CN, } RCO_2\text{—, OR})$$

4.9.7 *Ylides*

(a) PHOSPHORUS YLIDES–THE WITTING REACTION: Triphenylphosphonium salts $Ph_3P^+CHRR'.X^-$, which possess α—C—H functionality react with strong bases such as phenyl or n-butyllithium to form orange solutions which contain phosphorus ylides e.g.

$$Ph_4PMe \xleftarrow[X]{PhLi} \left[Ph_3\overset{+}{P}CH_3\right] I^- \xrightarrow{PhLi} Ph_3P=CH_2 \xrightarrow{Ph_2C=O}$$

$$\downarrow$$

$$(Ph_3\overset{+}{P}-\overset{-}{C}H_2)$$

$$Ph_3\overset{+}{P}-CH_2 \longrightarrow Ph_3\overset{+}{P} \quad CH_2$$
$$\underset{-O-CPh_2}{\diagdown\!|} \qquad \underset{-O}{|} \quad \underset{CPh_2}{||}$$

The dipolar character of phosphorus ylides is shown by their reactions with nucleophiles (e.g. aldehydes and ketones) or with electrophiles (e.g. proton donors). Ylides $Ph_3P=CHR$ (R = alkyl) are readily hydrolysed, even by cold water, so it is usual to carry out reactions involving them under anhydrous conditions. An inert atmosphere (usually nitrogen) is also necessary as oxidation

$$Ph_3P=CHR+H_2O \;\rightleftharpoons\; Ph_3\overset{+}{P}-CH_2R + OH^- \longrightarrow$$

$$\left[\begin{array}{c} Ph_3P\overset{\frown}{-}CH_2R \\ \underset{O-H}{|}\;\overset{-}{O}H \end{array}\right] \longrightarrow Ph_3PO + RCH_3 + H_2O$$

occurs in the air to give, initially, a phosphine oxide and an aldehyde or ketone. The latter can then react with any remaining ylide to form alkene.

$$Ph_3P=CHR + O_2 \longrightarrow Ph_3PO + RCHO \xrightarrow{Ph_p=CHR} RCH=CHR + Ph_3PO$$

Where the substituent R is capable of conjugation with the P=C bond (e.g. R = CN, CO_2R', COR', aryl), the ylide is stabilized both against electrophilic and also against nucleophilic attack. Such 'stabilized' ylides are much less prone to oxidation, alcoholysis or hydrolysis. They can be generated, for example, by treating appropriate phosphonium salts with alkali metal alkoxides in alcohols. This is convenient, as lithium alkyls would react with groups such as CO_2R' or COR' leading to side products.

The Wittig reaction thus provides a mild synthetic approach to a wide range of olefinic products including those which contain functional groups. It proceeds in two stages, first to an intermediate betaine which subsequently breaks down to phosphine oxide and alkene.

$$R_3\overset{+}{P}-\overset{-}{C}\overset{R^1}{\diagdown}R^2 \xrightarrow{Step\ 1} R_3\overset{+}{P}-C\overset{R^1}{\underset{R^2}{\diagdown}} \xrightarrow{Step\ 2} R_3\overset{+}{P}\overset{R^1\ \ R^2}{\diagup\ \diagdown}C$$
$$\underset{O=C}{\overset{\delta-\ \ \delta+}{|}}\overset{R^3}{\diagdown}R^4 \qquad \underset{-O-C}{|}\overset{R^3}{\underset{R^4}{\diagdown}} \qquad \underset{O^-}{|}\ \underset{R^3}{C}\overset{}{\diagdown}R^4$$

betaine

On replacement of R = Ph by R = alkyl in a phosphorane, the positive charge on phosphorus is decreased on account of the inductive effect of the alkyl groups, but the negative charge at carbon is increased. Step 1 is therefore facilitated, but Step 2 becomes more difficult and rate determining. Betaines can often be isolated where R = alkyl. Introduction of electron-withdrawing groups R¹, R² causes the negative charge to be delocalized away from the nucleophilic carbon centre. Here betaine formation becomes rate determining and the polarity of the carbonyl group becomes important, so that while aldehydes usually react fairly readily, the less electrophilic ketones do not, or give only poor yields.

One way of overcoming this difficulty is to make R more electron withdrawing than phenyl. When trialkylphosphites react with alkyl halides the initially formed quaternary salt rearranges to a phosphonate ester (the Michaelis–Arbusov rearrangement). The driving force for the rearrangement is the formation of a strong P=O bond. Some of these reactions proceed sedately, but others can go almost explosively. The quaternary salts from triarylphosphites, however, generally rearrange only on strong heating, as the second stage involves nucleophilic attack on an aromatic ring.

Phosphonate esters derived from trialkyl phosphites and bromoesters, for example, are deprotonated by strong bases to give resonance stabilized ions which attack aldehydes or ketones as follows. This provides a useful synthesis of $\alpha\beta$-unsaturated esters.

The Wittig reaction is employed industrially as in the BASF syntheses for Vitamin A and β-carotene. Laboratory methods have to be modified for industrial practice. Triphenylphosphine is obtained by the Wurtz reaction (p. 126). The triphenylphosphine oxide recovered after reaction is reconverted into Ph_3P by treatment first with phosgene and then with phosphorus.

$$3Ph_3PO + 3COCl_2 \longrightarrow 3CO_2 + 3Ph_3PCl_2$$

$$3Ph_3PCl_2 + 2P \longrightarrow 3Ph_3P + 2PCl_3$$

Butyllithium is an unsuitable base on account of its sensitivity to air and moisture. In the process the phosphonium salt precursor (formed from the alcohol, Ph_3P and hydrogen halide), together with aldehyde in methanol, are treated with methanolic sodium methoxide at $-30°C$. After reaction the solution is neutralized with dilute sulphuric acid, the Vitamin A ester extracted into a hydrocarbon solvent and isolated after removal of solvent by distillation.

Generally the Wittig reaction yields a mixture of *cis* and *trans* geometrical isomers of an alkene. The stereochemical considerations are discussed elsewhere.

(b) ARSENIC YLIDES. Arsenic ylides can be prepared by deprotonation of quaternary arsonium salts.

$$R_3As + R'CH_2X \longrightarrow [R_3\overset{+}{As}CH_2R']X^- \xrightarrow{\text{base}} R_3As{=}CHR' ({\leftrightarrow} R_3\overset{+}{As}{-}\overset{-}{C}HR')$$

(R = alkyl, aryl; R' = H, alkyl, aryl, acyl, CO_2R, CN)

Potentiometric titrations indicate that they are much stronger bases than the corresponding phosphorus ylides. On the basis of $J(^{13}C{-}^1H)$ coupling constants derived from ^{13}C n.m.r. spectra it is concluded that the carbon atom in $Ph_3P{=}CH_2$ is in a trigonal planar environment (sp^2 C), whereas in $Ph_3As{=}CH_2$ it is probably pyramidal (sp^3 C; $J = 136.7$ Hz compared with 142 Hz in $Ph_3AsCH_3.I^-$). There therefore seems to be weaker $(2p{-}3d)\pi$ bonding in the arsenic compound, which means that the As—C bond is more polar in the sense $Ph_3\overset{\delta+}{As}{-}\overset{\delta-}{C}H_2$. This shows up in the enhanced reactivity, most noticeably of 'stabilized' arsenic ylides, compared with their phosphorus analogues. While the ylides PhE$=$CHR' (E = P, R' = CO_2Me, CN or COPh) do not react with ketones readily if at all, the arsenic ylides give yields of between 50–80% of the expected alkenes. With non-stabilized

139

The main Groups IV and V

arsenic ylides, however, the reactions can take another course leading to oxiranes (epoxides) or their rearrangement products. In this they resemble sulphonium or oxosulphonium ylides.

$$Ph_3As=CH_2 + Ph_2C=O \longrightarrow Ph_2C=CH_2\ (20\%) + Ph_2CHCHO\ (69\%)$$

$$Ph_3As=CHAr + Ar'CHO \xrightarrow[20°C]{C_2H_5OH} \underset{O}{\overset{Ar'}{\triangle}} + Ph_3As$$

These differences can be discussed in terms of the two step mechanism.

A major factor is the strength of the E=O bond, which provides the main driving force for path (a) in the case of phosphorus.

The carbanion $[Me_3Si-CH_2]^-$ is isoelectronic with the phosphorus ylide $Me_3\overset{+}{P}-\overset{-}{C}H_2$. The lithium derivative of the anion $Me_3SiCH_2^-$ can be prepared by treating tetramethylsilane with butyllithium/TMEDA at $-78°C$ in tetrahydrofuran.

$$Me_3Si-CH_3 \xrightarrow[TMEDA]{Bu^nLi} Me_3Si-CH_2Li$$

These α-silyl carbanions react with carbonyl compounds to give intermediates which readily eliminate the Me_3SiO^- group either spontaneously or on treatment with acids or bases.

Normally a mixture of *cis* and *trans* olefins results.

$$2Me_3SiO^- \xrightarrow{H_3\overset{+}{O}} (Me_3Si)_2O \quad \text{(The Peterson reaction)}.$$

The by-product, hexamethyldisiloxane, is volatile and is sometimes easier to remove from the olefinic product than solid Ph_3PO which is formed in the Wittig reaction.

140

4.9.8 *Catenated alkyls and aryls*

In 1760 L. Cadet de Gassicourt described the preparation of an evil smelling fuming liquid by distilling a mixture of As_2O_3 and potassium acetate. During the next century this product was further investigated in turn by two famous German chemists, Bunsen and von Baeyer. More recent work shows that Cadet's liquid largely consists of a mixture of cacodyl, $Me_2AsAsMe_2$ (56%), cacodyl oxide, $Me_2AsOAsMe_2$ (40%) and trimethylarsine (2.5%).

The compounds discussed in this section are of four main types.

(a) Diphosphines, diarsines and distibines, R_2MMR_2.
(b) Triphosphines and triarsines $(R_2M)_2MR$,
(c) Cyclopolyphosphines and polyarsines, $(RP)_n$ and $(RAs)_n$,
(d) More complicated polycyclic phosphines,

General methods of preparation are illustrated taking the diphosphines and diarsines as examples.

(i) $\qquad 2R_2MX + M' \xrightarrow[\text{liq. NH}_3]{\text{ether or}} R_2M\!-\!MR_2 + 2M'X \quad (M' = Na, Li)$

Mercury is also sometimes used to bring about this coupling, especially of trifluoromethyl derivatives $(CF_3)_2MI$ and CF_3MI_2.

(ii) Elimination of hydrogen halide between a hydride and a halide, with or without the addition of base.

$$R_2PX + R_2PH \xrightarrow[\text{reflux}]{\text{hexane}} R_2P\!-\!PR_2 + HX$$

(iii) Reaction between an alkali metal derivative and a halide. The alkali metal derivative is formed from the hydride or halide in solvents such as dioxan, tetrahydrofuran or liquid ammonia. The cleavage of arylphosphines or arsines by alkali metals in these solvents is also a particularly convenient route to them.

$$R_2PH \xrightarrow{M'} R_2PM' + \tfrac{1}{2}H_2 \qquad \text{(Also As)}$$

$$R_2PX \xrightarrow{M'} R_2PM' + M'X \qquad \text{''}$$

$$Ar_3P \xrightarrow[\text{or Li/THF}]{M'/NH_3} Ar_2PM' + M'Ar \qquad \text{''}$$

Then

$$R_2PM' + R_2PCl \longrightarrow R_2P\!-\!PR_2 + M'Cl \quad \text{(Also As)}$$

(iv) Elimination of amine between an amino derivative and a hydride.

$$R_2MNMe_2 + R_2MH \longrightarrow R_2M\!-\!MR_2 + Me_2NH \quad (M = P, As)$$

Compounds which contain homonuclear bonds decrease in stability as the radius of the central atom increases from P to Bi. Distibines R_2SbSbR_2 are all very reactive and easily decomposed. Apart from $Me_2BiBiMe_2$ and maybe

141

Ph$_2$BiBiPh$_2$ no compounds containing Bi—Bi bonds are known. We shall therefore concentrate our attention on the derivatives of phosphorus and arsenic.

Diphosphines and diarsines are readily oxidized in air, the tetramethyl compounds being spontaneously inflammable. The P—P or As—As bond is readily cleaved by a variety of reagents. Some of these reactions are exemplified by tetraphenyldiarsine.

When difunctional compounds RMX$_2$ (X = H, Hal, M', NR$_2'$) are used in the preparations listed above, cyclic oligomers (RP)$_n$ and (RAs)$_n$ are generally obtained. Polymeric forms are also known. Rings with three, four, five and six annular atoms have been obtained with a variety of alkyl, aryl and perfluoroalkyl substituents. There is evidence from ^{1}H n.m.r. measurements of larger rings. Polymethylarsine (MeAs)$_n$ prepared by the reduction of methylarsonic acid with H$_3$PO$_2$ exists in several forms. One is a yellow oil, m.p. 12°C, b.p. 118–120/ 1 mm, which is pentameric (MeAs)$_5$ in the solid state. The pentamer is also the major component in solution and in the vapour phase. The ring is puckered. A second purple-black form, m.p. 204°C, has an unusual ladder structure. Double chains of trans-As$_2$Me$_2$ units are joined by long As$\cdots$As links. The compound is both a semiconductor and a photoconductor. Unlike (MeAs)$_5$ which is rapidly oxidized, the ladder form is stable to air. The phenyl derivatives (PhP)$_n$ and (PhAs)$_n$ were originally formulated as PhM=MPh, analogous to azobenzene. They are now known to be either cyclic oligomers or chain polymers. X-ray diffraction has shown that the main form of arsenobenzene, m.p. 212°C, consists of six membered rings in the chair form with phenyl groups in equatorial positions. One form of phosphobenzene m.p. 195–9°C has a similar structure, and another, m.p. 148–150°C, is a cyclic pentamer.

An example of an arsenobenzene which is apparently a long chain polymer is Salvarsan or Arsphenamine. This results from reduction of 3-nitro-4-hydroxyphenylarsonic acid.

Salvarsan

Fig. 4.12 Some bicyclo and cage structures with phosphorus skeletons.

P_7R_5 P_7R_3 P_8R_6 P_9R_5 P_9R_3

This substance was discovered by Ehrlich in 1912 who soon found it to be moderately successful in the treatment of syphilis. It occupies an important place in the history of chemotherapy, as it was one of the first effective synthetic chemotherapeutic agents. It is still used in the treatment of sleeping sickness in its advanced stages.

There are probably only small energy differences between rings of different sizes. Even three membered species have been isolated in some cases, for example by the reaction

$$C_2F_5PI_2 \xrightarrow{\text{Hg}} (C_2F_5P)_n \quad (n = 3, 4, 5)$$

but they are usually readily converted into larger, less strained rings. Some kinetic stabilization of the small rings is achieved by employing bulky substituents such as *tert*-butyl.

Recently some bicyclo and cage structures based on P—P bonds have been obtained by reacting a mixture of $RPCl_2$ and PCl_3 with magnesium. Only the phosphorus skeletons are shown (Fig. 4.12). These discoveries show that the chemistry of catenated phosphorus derivatives, like that of catenated silicon compounds, is much more diverse than was imagined even ten years ago.

4.9.9 Pyridine and pyrrole analogues

Unlike silabenzene (p. 125), pyridine analogues of the Group V elements are quite robust, although their stability decreases from phosphorus to bismuth. The phosphorus compounds were discovered by Märkl who first made them from pyrylium salts.

The six-membered ring in 2,4,6-triphenylphosphorin is planar and the two P—C distances are equal (1.74 Å). This suggests electron delocalization associated with an aromatic system. This is consistent with the low basicity of the compound; it is not alkylated by methyl iodide or even by $Et_3O^+BF_4^-$. Alkyl derivatives can be prepared, however, *via* the resonance stabilized anion.

The parent heterocycles have been obtained from an organotin precursor (p. 108) followed by elimination of hydrogen chloride in the presence of a rather unusual base.

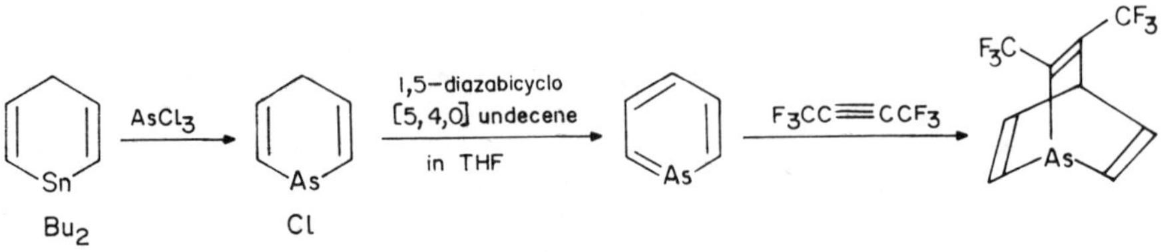

Arsabenzene (arsenin) is a colourless liquid which is rapidly attacked by air, turning bright red. Its photoelectron, u.v. and n.m.r. spectra all support an aromatic structure. In the 2,4,6-triphenyl derivative the ring is planar. The two As—C bonds are equal in length and so are the four C—C bonds. This indicates considerable electron delocalization. Nevertheless arsenins undergo 1,4-addition reactions with alkynes. 1-Arsanaphthalenes and 9-arsaanthracenes are also known.

The arsenic analogues of the pyrroles, the arsoles, however, are not aromatic. The arsenic atom in 9-phenyl-9-arsafluorene, for example, is pyramidally coordinated, even though the rest of the fused ring system is planar. While the barrier to inversion about arsenic is some $50 \, kJ \, mol^{-1}$ lower than in most tertiary arsines, resolution into optical isomers of unsymmetrically substituted arsafluorenes has still been achieved.

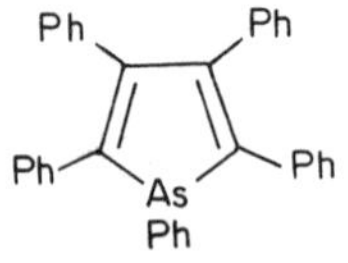

pentaphenylarsole

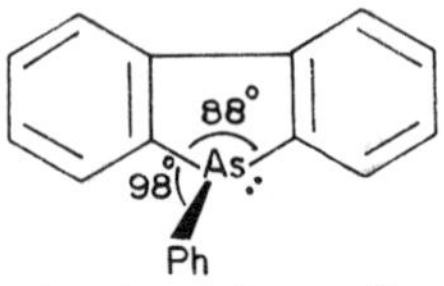

9-phenyl-9-arsafluorene

4.10 References

Ashe, A.J. III. (1978) The heterobenzenes. *Acc. Chem. Res.*, **11**, 153.

Aylett, B.J. (1979) *Organometallic Compounds*, (*Main Groups IV and V* Vol. 1, Part 2), 4th edn, Chapman and Hall, London.

144

Baudler, M. (1982) Chain and ring phosphorus compounds. *Angew. Chem. (Int. Edn)*, **21**, 492.

Brook, A.G. (1986) One thing leads to another – from silylcarbinols to silaethenes. *J. Organomet. Chem.*, **300**, 21.

Colvin, E. (1981) *Silicon in Organic Synthesis*, Butterworths, London.

Corriu, J.P. and Guérin, C. (1982) Mechanisms of nucleophilic substitution at silicon. *Adv. Organomet. Chem.*, **20**, 265.

Davies, A.G. and Smith, P.J. (1980) Organotin chemistry. *Adv. Inorg. Chem. Radiochem.*, **23**, 1.

Deschler, U., Kleinschmidt, P. and Panster, P. (1986) 3-Chloropropyltrialkoxysilanes. Key intermediates for commercial production of organofunctionalised silanes and polysiloxanes. *Angew Chem. (Int. Edn)*, **25**, 236.

Hellwinkel, D. (1983) Penta- and hexa-organyl derivatives of the main Group V elements. *Topics Curr. Chem.*, **109**, 1.

Kosolapoff, G. and Maier, L. (1976) *Organic Phosphorus Compounds*, Vols 1–7, Wiley-Interscience, New York.

Ng, S-W. and Zuckerman, J.J. (1985) Subvalent Group IV compounds. *Adv. Inorg. Chem. Radiochem.*, **29**, 297.

Pommer, H. (1977) Industrial applications of the Wittig reaction. *Angew. Chem. (Int. Edn)*, **16**, 423.

Raabe, G. and Michl, J. (1985) Multiple bonding to silicon. *Chem. Rev.*, **85**, 419.

West, R. (1986) The polysilane high polymers. *J. Organomet. Chem.*, **300**, 327.

West, R. and Barton, T.J. (1980) *J. Chem. Educ.*, **57**, 165, 334.

Problems

1. How do acids, halogens and mercury(II) salts react with tetraalkyltins? Indicate how each of the following affects the reactivity of the metal—carbon bond:
(a) Replacing Sn by Si, Ge or Pb.
(b) Introducing unsaturation into the hydrocarbon group.
(*University College, London*).

2. Indicate the nature of the organometallic product formed in each of the following reactions.

(a) $Me_3Bi + Br_2 \rightarrow$; $Me_3Sb + Br_2 \rightarrow$
(b) $Me_3As + EtI \rightarrow$; $(MeO)_3P + EtI \rightarrow$

(*University College, London*).

3. Give an account of the structure and bonding in the following compounds.
(a) $Me_3GeC_5H_5$ (b) Me_3SnF (c) Me_3SbBr_2 (d) Me_2AlPh (e) MeMgBr (f) MeK
(*University College, London*).

4. Describe synthetic routes to main-group 4 organometallic compounds, and compare the association properties of such species with those of related compounds from groups 1–3.

Dimethyltin difluoride is tetragonal, $a = 425$ and $c = 710\,pm$. Positions for non-hydrogen atoms have been determined as follows:

Sn	$(0,0,0)$
F	$(0,\frac{1}{2},0)$ and $(\frac{1}{2},0,0)$
C	$(0,0,z)$ and $(0,0,-z)$ $z = 0.29$

Describe the crystal structure of $(CH_3)_2SnF_2$, calculating the Sn—F and Sn—C bond distances.

$(CH_3)_2SnF_2$ decomposes without melting at *ca* 400°C, whilst $(CH_3)_2SnCl_2$ and $(CH_3)_2SnBr_2$ melt at *ca* 90°C and 75°C respectively. Suggest a possible reason for this difference in behaviour.
(*University of Edinburgh*).

5. Triethylphosphite $(C_2H_5O)_3P$ and benzyl chloride were heated together and gave a compound A $(C_{11}H_{17}O_3P)$. When A (in toluene) was treated sequentially with sodium hydride, benzaldehyde, and then dilute aqueous hydrogen chloride, the product B $(C_{14}H_{12})$ was obtained.

Formulate A and B and suggest a mechanism for the formation of each.
(*University of Strathclyde*).

6. A chemical method of detecting dative π-bonding to Group IV elements is to measure the relative donor properties of compounds in which such bonding is suspected. A guide to the Lewis basicity of such derivatives is provided by the shift (Δv) which they cause in the infrared spectrum of $CDCl_3$ through C—D....O or C—D...N hydrogen bonding. Discuss the results in the Table on the basis of inductive effects (electronegativities) of the Group IV elements and of their capacity to form $(p—d)\pi$ bonds with N, O or S. How is the extent of π-bonding reflected in the structures of the compounds?

Base	C	Si	Ge	Sn	Pb
$(Me_3M)_3N$	100	0	72	106	—
$(Me_3M)_2O$	33	13	55	84	—
$(Me_3M)_2S$	40	29	38	43	51

Relative donor properties of some organometallic bases. $\Delta v/cm^{-1}$ for $v(C—D)$ of $CDCl_3$

7. When dichlorodimethylsilane is treated with alkali metals in tetrahydrofuran, the main product X is a crystalline solid of composition, C, 41.4; H, 10.3; Si, 48.3%, and molecular weight 290. The proton n.m.r. spectrum of X measured in benzene consists of a single band. Compound X reacts with bromine to give a compound Y which contains 35.5% Br; and, when Y is allowed to react with sodium and chlorotrimethylsilane in toluene one of the products formed is a liquid

Z. The proton n.m.r. spectrum of Z consists of four bands with areas in the ratio $3:2:2:1$.

Suggest structures for X, Y, and Z.

(*University of St Andrews*).

8. Give structures for the unknown compounds in the following reaction sequence explaining your reasoning.

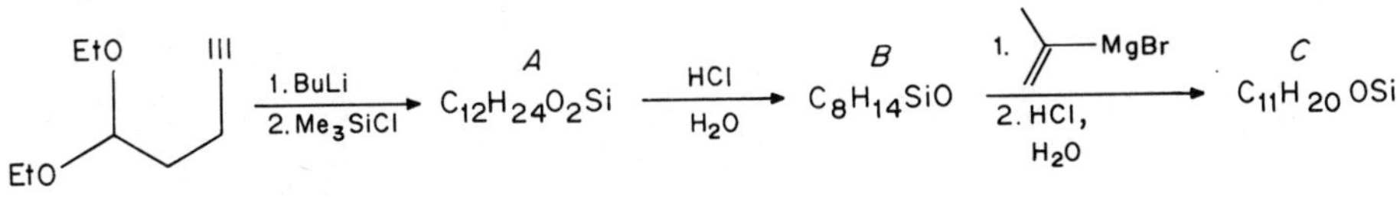

(*University of Leicester*).

9. In the following sequence one racemic diastereoisomer of F is formed. Give a structure for F and predict its relative stereochemistry from the subsequent transformations; give a structure for G.

Explain the reasoning behind your answers.

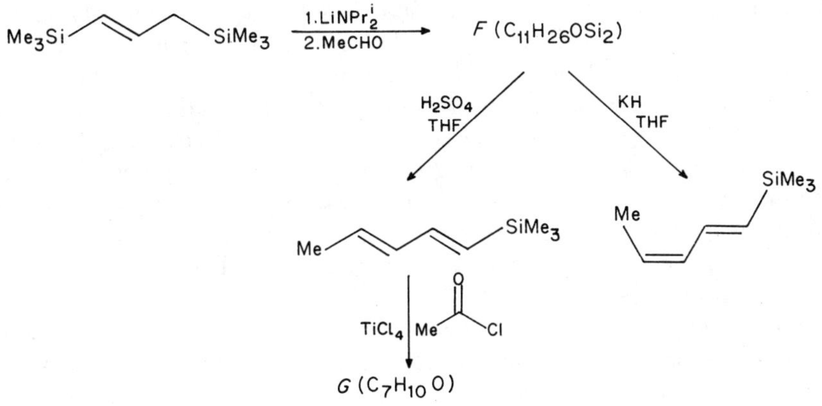

(*University of Leicester*).

Some transition metal chemistry relevant to organometallic chemistry

5.1 The 18-electron rule

The character of bonds between transition metal centres and attached groups or ligands can span the whole range from electrostatic to covalent. This is reflected in the approximations which have been used to discuss the coordinate bond. On the one hand crystal field theory starts from the assumption that the interactions between transition metal ions and ligands are entirely electrostatic. While this approach is useful in explaining magnetic and spectroscopic properties of aquo and ammine complexes, for example, especially when allowance is made for some covalency, it is far from successful in treating the essentially covalent compounds formed by ligands such as carbon monoxide, unsaturated hydrocarbons or alkyl and aryl phosphines.

In 1927 Sidgwick suggested that a coordinate bond arises from the donation of a pair of electrons from the ligand to the central metal atom. In a noble gas atom all the valence orbitals ns, np, $(n-1)d$ are filled, and a stable chemically inert entity results. Similarly, by filling all the valence orbitals of a transition metal centre by donation of an appropriate number of electrons from the ligands, unreactive complexes arise. Eighteen electrons are required from the metal and its associated ligands to attain this noble gas configuration. This is the 'effective atomic number (EAN)' or 18-electron rule. A consequence of the rule is that a metal should have a coordination number appropriate to its oxidation state. As we shall see below, this does apply in the case of metal carbonyls, but clearly does not hold for complexes such as the hexaaquo ions $M(H_2O)_6^{n+}$ ($n = 2, 3$) or ammines $M(NH_3)_6^{3+}$. The 18-electron rule is of great value, however, in rationalizing and predicting the stoichiometries and structures of essentially covalent complexes, such as those formed by organic ligands.

The 18-electron rule is essentially a statement about the '*kinetic stability*' of complexes. In this it resembles the octet rule in the chemistry of elements of the First Short Period, including carbon. In 18-electron compounds full use is made of

148

all the valence orbitals of the metal. There are no low-lying empty orbitals into which electrons can be promoted to initiate thermal decomposition or donated in the case of nucleophilic attack. Where such empty orbitals are present, the kinetic stability is usually greatly reduced.

Molecular orbital theory provides the most complete and general description of bonding because it can be applied to all situations ranging from electrostatic to fully covalent. Metal and ligand orbitals will combine to form molecular orbitals provided that (a) they are of comparable energy and (b) that they have the correct symmetry properties. Condition (a) means that only the valence orbitals need be considered. Where ligand orbitals are significantly lower in energy than the metal orbitals with which they interact, a polarized bond with appreciable ionic character results. Where both sets of orbitals are of similar energies, the interaction is essentially covalent (Fig. 5.1).

A qualitative m.o. diagram for an octahedral complex ML_6 is illustrated in Fig. 5.2. Interactions involving only the ligand σ-orbitals, which are directed along the metal–ligand axes, are considered here. The ns, np_z, $(n-1)d_{z^2}$ and $(n-1)d_{x^2-y^2}$ orbitals of the metal are directed along these axes and are of the correct symmetry to interact with the ligand σ-orbitals, but for the $(n-1)d_{xy}$, d_{yz} and d_{zx} (t_{2g}) set there is zero net overlap and these orbitals are non-bonding. Eighteen electrons are required to fill all the bonding and non-bonding m.o.s. This provides a theoretical justification of the 18-electron rule. The separation between the non-bonding t_{2g} set, which is purely metal-centred and the antibonding e_g^* pair which incorporates some ligand character, corresponds formally with the splitting t_{2g}/e_g (Δ_0) described by crystal field theory.

The metal t_{2g} orbitals are of the correct symmetry to enter into π-interactions with suitable ligand orbitals. If the latter are empty and high in energy this interaction leads to the stabilization of the t_{2g} set and an increase in Δ_0. Such a situation arises, for example, with the isoelectronic series of ligands CO, NO^+, N_2, CN^- which possess empty antibonding π^* orbitals. Phosphorus donors such as PF_3, $P(OR)_3$ and PR_3 can also act as acceptors by virtue of empty $3d$ orbitals on phosphorus (but see p. 160). Such π-bonding is an important source of stabilization in their complexes. If ligand π-orbitals are filled and lower in energy than metal t_{2g}, the latter are destabilized, leading to a decrease in Δ_0. This can arise, for example, with oxide, alkoxide or halide donors (Fig. 5.3).

Where there is a large energy gap (Δ_0), low spin 18-electron complexes are favoured, which are kinetically inert to ligand substitution. With a small separation, however, a wide range of electron configurations is to be expected. In the First Transition Series these complexes are often high spin and the 18-electron configuration has no special significance. It is therefore complexes of those ligands which occasion large d-orbital splittings which will obey the 18-electron rule (Table 5.1). Metal–ligand bonding in such complexes is largely covalent; the donor atoms are of medium electronegativity such as carbon, hydrogen, phosphorus or sulphur. Covalency is enhanced if the energies of the metal orbitals match well with those of the ligands. This occurs best with complexes of zero or

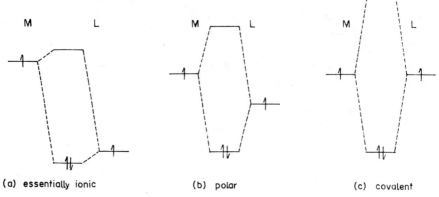

(a) **essentially** ionic   (b) polar   (c) covalent

Fig. 5.1 Interaction diagrams for metal and ligand orbitals.

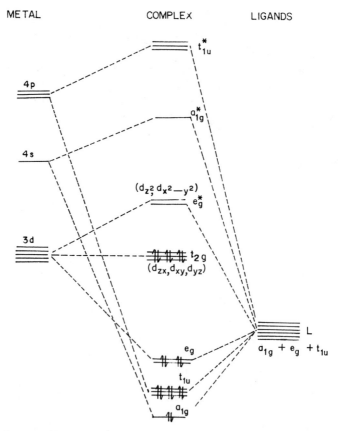

Fig. 5.2 Qualitative molecular orbital diagram for an octahedral complex, ML_6 σ-bonding only considered.

150

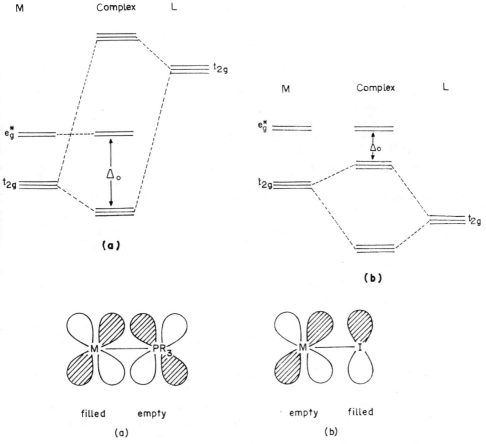

Fig. 5.3 Effect of π-bonding on separation. (a) π-acceptor ligands, (b) π-donor ligands.

Table 5.1. Scope of the 18-electron rule

	18-electron rule useful	Intermediate	18-electron rule not useful
Ligands	CO, NO, N_2, CS, RNC	H, CH_3, C_6H_5, CN	NH_3, H_2O, F, Cl, Br, I
			OR
	C_2H_4, unsaturated hydrocarbons R_3P especially PF_3, $P(OR)_3$	2, 2′-bipyridyl R_2S, RS, —S—	—O—(=O)
Compounds	Carbonyls, nitrosyls, organometallics phosphine complexes		Halides, oxides, aquo complexes alkoxides, ammines,

151

even negative charge where the metal is in a low formal oxidation state (the metal centre then having a higher than usual electronegativity). Moreover the effect is greatest if the ligand has empty orbitals of π-symmetry which can act as acceptors. The carbon monoxide molecule possesses these very characteristics. They are shared by many unsaturated hydrocarbons such as ethene and ethyne.

5.2 Transition metal carbonyls

Carbon monoxide is an extremely weak Lewis base towards conventional Lewis acids. It does not complex with the boron trihalides, although it does yield a weak adduct $H_3B \cdot CO$ with diborane. On the other hand it forms numerous complexes with transition elements. The source of this difference is that in the latter complexes not only is the weakly donating σ-orbital of CO involved, but also the π^* orbitals which can function as acceptors. The conventional explanation is that a 'synergic effect' exists in which the π interaction removes electron density from the metal, allowing σ donation from the ligand to be enhanced.

5.2.1 Range of compounds and structures

The most important binary carbonyls of the transition elements are listed in Table 5.2. Some physical properties are included for complexes of the First Transition Series. The corresponding compounds of the elements of the Second and Third series are broadly similar. Apart from $V(CO)_6$, which is a paramagnetic 17-electron complex with one unpaired electron, all binary metal carbonyls obey the 18-electron rule. Thus in $Cr(CO)_6$, the Cr atom formally provides six electrons (the number of electrons in its valence shell) and the six carbonyl ligands two each, so that $6 + (6 \times 2) = 18$. The stoichiometries and structures of the mononuclear compounds $Cr(CO)_6$ (octahedral, O_h), $Fe(CO)_5$ (trigonal bipyramidal, D_{3h}) and $Ni(CO)_4$ (tetrahedral, T_d) are determined by this requirement and by the symmetrical arrangement of the MCO σ-bonding pairs. $Mn_2(CO)_{10}$ and $Co_2(CO)_8$ have binuclear structures (Fig. 5.4). The sharing of the odd electrons between two 17-electron $Mn(CO)_5$ or $Co(CO)_4$ fragments respectively makes up the electron count of the metal to 18 in each case:

Mn	7 e	Co	9 e
5 CO	10 e	4 CO	8 e
Mn—Mn	1 e	Co—Co	1 e
	18 e		18 e

Vanadium hexacarbonyl is monomeric and is the only simple carbonyl which is paramagnetic at room temperature ($\mu_{\text{eff}} = 1.81$ B.M.). Steric effects have been held responsible, but this explanation is hard to believe, especially as a robust tetrahedral cluster $(Ph_3PAu)_3V(CO)_5$, in which vanadium is eight-coordinate,

Table 5.2 Binary carbonyls of the transition elements

First transition series	$V(CO)_6$ Blue-green pyrophoric crystals (μ_{eff} 1.81 B.M). Octahedral (O_h)	$Cr(CO)_6$ Colourless; sublimes; m.p. 150°C Air stable Octahedral (O_h)	$Mn_2(CO)_{10}$ Golden-yellow m.p. 154°C Air stable (D_{4h})	$Fe(CO)_5$ Yellow liquid b.p. 103°C Air stable Trigonal bipyramid (D_{3h}) $Fe_2(CO)_9$ Golden plates Air stable $Fe_3(CO)_{12}$ Black solid	$Co_2(CO)_8$ Orange-red crystals m.p. 51°C dec. Air sensitive $Co_4(CO)_{12}$ Black solid $Co_6(CO)_{16}$ Black solid	$Ni(CO)_4$ Colourless liquid b.p. 42°C Air sensitive; toxic, carcinogenic Tetrahedral (T_d)
Second transition series	–	$Mo(CO)_6$	$Tc_2(CO)_{10}$	$Ru(CO)_5$ $Ru_2(CO)_9^*$ $Ru_3(CO)_{12}$	$Rh_2(CO)_8^*$ $Rh_4(CO)_{12}$ $Rh_6(CO)_{16}$	–
Third transition series	–	$W(CO)_6$	$Re_2(CO)_{10}$	$Os(CO)_5$ $Os_2(CO)_9^*$ $Os_3(CO)_{12}$ higher clusters	$Ir_4(CO)_{12}$ $Ir_6(CO)_{16}$	–

*Unstable. For structures of bi- and polynuclear species, see Figure 5.4. Isolable compounds only listed.

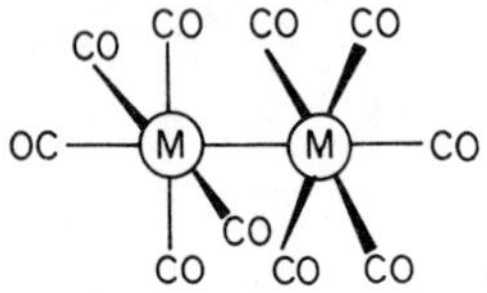

$M_2(CO)_{10}$, M=Mn,Tc,Re(D_{4d}) 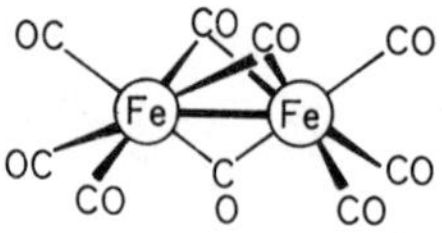 $Fe_2(CO)_9(D_{3h})$ 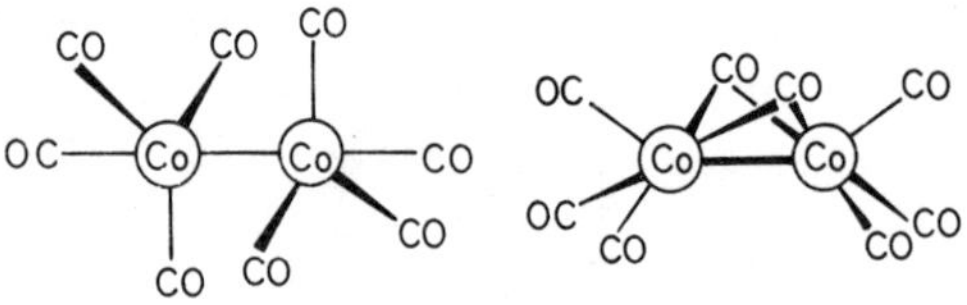 $Co_2(CO)_8$ two isomers (D_{3d},C_{2v})

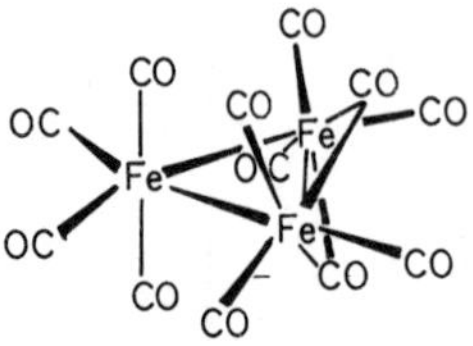

$Fe_3(CO)_{12}$ (C_{2v}) 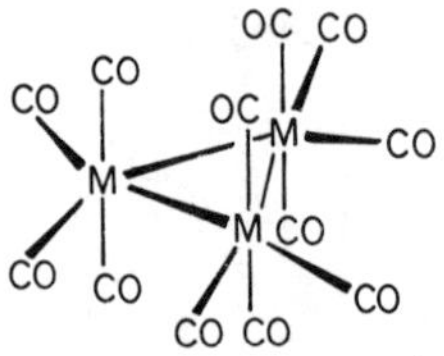 $Me_3(CO)_{12}$, M = Ru,Os(D_{3h})

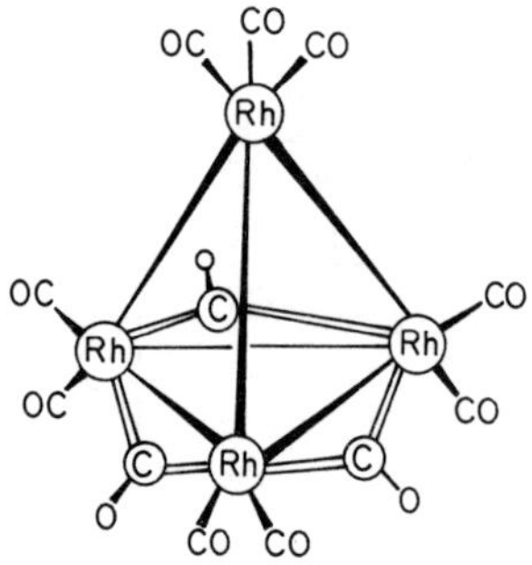

$Rh_4(CO)_{12}$ (Td) $Ir_4(CO)_{12}$(Td)

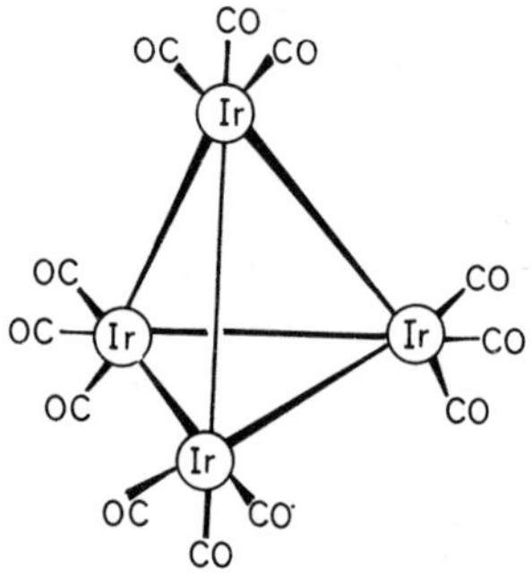

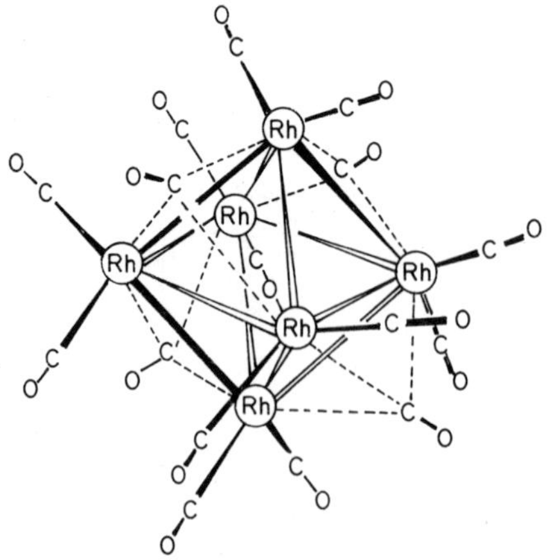

$Rh_6(CO)_{16}$(T$_d$)

Fig. 5.4 Structures of some bi- and polynuclear carbonyl complexes.

has been prepared from the reaction between $[V(CO)_5]^{3-}$ and Ph_3AuCl in THF. Cocondensation of vanadium vapour with carbon monoxide in a frozen matrix of noble gas affords $V(CO)_6$ and probably also the dimer. The metal–metal bond energies decrease from $88\,kJ\,mol^{-1}$ in $Co_2(CO)_8$ to $75\,kJ\,mol^{-1}$ in $Mn_2(CO)_{10}$. Extrapolation to vanadium would indicate an extremely weak bond in the dimer $V_2(CO)_{12}$. The molecular structures of some bi- and polynuclear carbonyls are shown in Fig. 5.4. Further examples are given in Chapter 11. These reveal some of the varied modes of bonding which the carbonyl ligand can adopt.

In general carbon monoxide can be considered to be a two electron donor as far as electron counting is concerned (see p. 191). It can act as a terminal ligand or as a bridge between two (μ_2-CO) or three (μ_3-CO) metal atoms. In nearly all known examples of bridge formation, the metal atoms are within bonding distance of

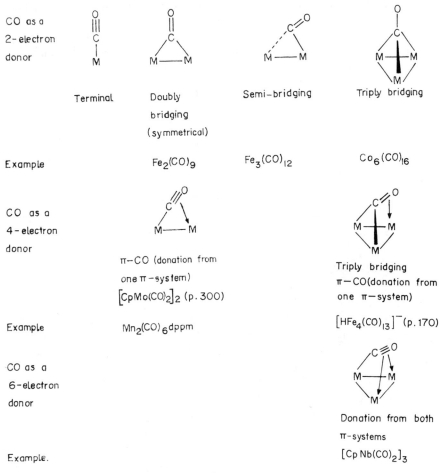

Fig. 5.5 Modes of bonding of the carbonyl ligand.

each other, that is, they are linked by some sort of metal–metal interaction. There are also cases in which one or both π systems of CO contribute to bridge formation, CO thus becoming an overall four or six electron donor respectively. These are still quite rare, and are not discussed further (but see p. 300).

5.2.2 Bonding

A qualitative molecular orbital diagram for the carbon monoxide molecule is given in Fig. 5.6. In its simplest form $C(2s)$ and $O(2s)$ are combined to form bonding $(2s\sigma)$ and antibonding $(2s\sigma^*)$ m.o.s. which have σ-symmetry about the internuclear CO axis (z axis). $C(2p_z)$ and $O(2p_z)$ are also directed along this axis, and give rise to m.o.s $(2p\sigma$ and $2p\sigma^*)$ of σ-symmetry. The $2p_y$ orbitals, however, interact to form orbitals of π-symmetry about the z axis; the orthogonal $2p_x$ orbitals behave in exactly the same way, so that doubly degenerate m.o.s $2p\pi$ and $2p\pi^*$ are formed.

In the point group $C_{\infty h}$ to which CO belongs, all orbitals of σ-symmetry can mix, regardless of whether they are derived from s or p orbitals, because they all belong

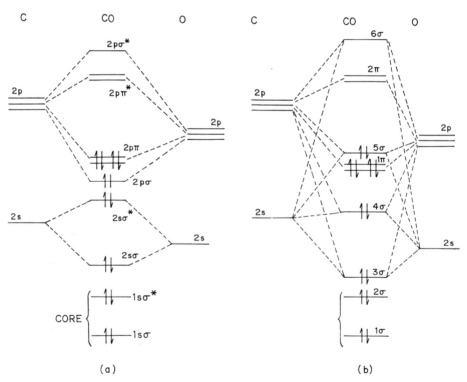

Fig. 5.6 Qualitative molecular orbital energy diagrams for CO.(a) Neglecting s, p mixing. (b) With s, p mixing.

to the same symmetry class (σ^+). A consequence of this mixing is that the highest occupied molecular orbital (HOMO) is a σ-orbital (5σ) (approximating to $2p\sigma$) rather than a pair of π-orbitals. (Molecular orbitals are numbered consecutively within a symmetry class, starting from the orbital of lowest energy. Thus for CO the HOMO is 5σ (Fig. 5.6) and the lowest unoccupied molecular orbital (LUMO) is 2π.)

The linear M—C—O bond has conventionally been discussed in terms of two interactions (Fig. 5.7). The first involves σ-donation of electrons from carbon (5σ) into suitable vacant metal orbitals which are directed along the bond axis. Secondly electrons are transferred from filled metal orbitals of π-symmetry into the empty antibonding m.o.s (2π) of the ligand. In carbonyls the π-component is

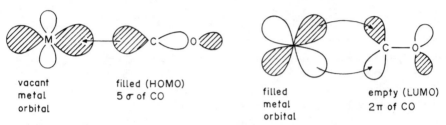

| vacant metal orbital | filled (HOMO) 5σ of CO | filled metal orbital | empty (LUMO) 2π of CO |

σ-donation from 5σ (HOMO) of CO into empty metal orbital directed along M–C–O axis.

π-back donation from filled metal orbital directed between the axes into 2π (LUMO) of CO.

Fig. 5.7 Conventional description of bonding interactions in a transition metal carbonyl linkage. The shading denotes the phase of the wave function.

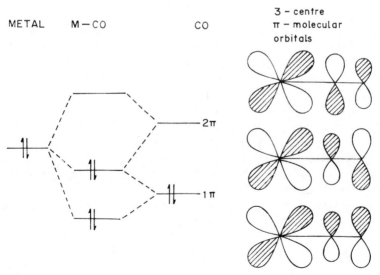

METAL   M—CO   CO   3 – centre
                                                     π – molecular
                                                     orbitals

2π

1π

Fig. 5.8 Three-centre π-molecular orbitals for M—C—O system. The shading denotes the phase of the wave function.

considered to be the more important; it is enhanced by electron donating substituents on the metal and by a zero or negative charge on the complex. It is also argued that the removal of electron density from the metal in this way permits more σ-electron transfer from carbon to metal to occur.

It is more correct to treat the M—C—O group as a three centre π-system (Fig. 5.8). Any interaction which involves the filled 1π orbitals of CO is thus taken into account. If, as is usual, 1π lies well below the metal orbitals this interaction will be small. The most important stabilizing effect is then still the formation of the central m.o. in the Figure, corresponding to the conventional picture of M $\rightleftharpoons$ CO back donation.

5.2.3 *Vibrational spectra of transition metal carbonyls*

Vibrational spectroscopy, especially in the infrared, provides a useful method of investigating the structures of carbonyl complexes. Moreover by providing information about bond force constants and hence bond orders within a molecule, insight can be gained into the relative donor/acceptor properties of carbon monoxide and other ligands.

Perhaps the simplest use of the technique is in distinguishing between terminal, doubly bridging and triply bridging carbonyls. As the CO bond order falls across this series, so the stretching frequency would also be expected to fall. In uncharged complexes terminal carbonyl groups give rise to bands in the region $2140-1800\,cm^{-1}$, doubly bridging groups at about $1850-1700\,cm^{-1}$, while triply bridging groups absorb even lower, at about $1550-1700\,cm^{-1}$. The infrared spectrum of $Fe_2(CO)_9$ (Fig. 5.9) clearly shows bands arising from the stretching vibrations of terminal and bridging carbonyl ligands. Two similar groups of bands are also seen for $Fe_3(CO)_{12}$, (2 semi-bridging COs) but only one appears in the spectrum of $Ru_3(CO)_{12}$, which has only terminal carbonyls. In the crystal $Co_2(CO)_8$ has the bridged C_{2v} structure. Infrared spectra of its solutions in hexane, however, reveal an equilibrium between this bridged species and an unbridged isomer, which exists preferentially in the staggered D_{3d} conformation. Rapid intramolecular exchange of carbonyl groups between different positions in a molecule is a feature of metal carbonyls. The ^{13}C n.m.r. spectrum of $Fe(CO)_5$ shows only one resonance even at the lowest temperatures attainable. Interchange of equatorial and axial carbonyl groups by a Berry pseudorotation mechanism is envisaged. Solutions of $Fe_3(CO)_{12}$ also give only one ^{13}C signal, even though bridging as well as terminal carbonyls are clearly indicated by infrared measurements. Again a rapid fluxional process must be occurring, in which the positions of all the carbonyl ligands become averaged.

The C—O stretching vibrations in metal carbonyl complexes absorb at lower frequencies than the corresponding vibration in CO itself ($2143\,cm^{-1}$). π-Back donation from metal into the π^* orbitals of carbon monoxide reduces the C—O bond order. The higher the electron density on the metal, the more the back donation and the greater reduction in C—O bond order and frequency. In

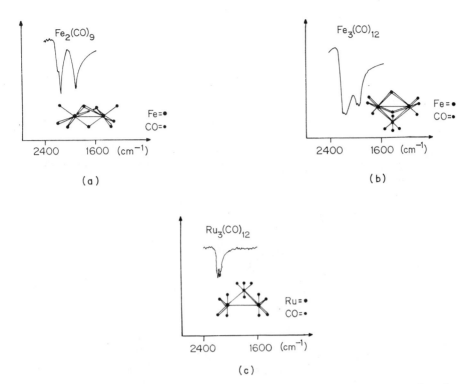

Fig. 5.9 Infrared spectra of (a) $Fe_2(CO)_9$ (b) $Fe_3(CO)_{12}$ and (c) $Ru_3(CO)_{12}$ in Nujol mulls.

molecular orbital parlance, high electron density at the metal means that the metal π-orbitals are of relatively high energy. There is therefore an especially strong interaction with the π^* orbitals of CO leading to considerable electron delocalization into these orbitals.

The effect of M—CO σ-donation on the CO bond order and frequency is less easy to predict. Nevertheless it is agreed that it is the back donation which exerts the dominant effect and trends in v(CO) frequencies can be understood in terms of it alone. A good demonstration of the relationship between electron density, the extent of back donation and v(C—O) frequency is provided by the positions of corresponding bands in infrared spectra of two isoelectronic series (Table 5.3).

As negative charge is added, M—CO(π^*) back donation increases and v(CO) decreases. The greater the back donation, however, the stronger does the M—C bond become, leading to an increase in frequency of the M—C stretching vibrations.

A phosphorus ligand, PR_3, can, like carbon monoxide, function both as a σ-donor through the lone pair on phosphorus and also as a π-acceptor. Where the substituents R are strongly electron withdrawing, the phosphine is only a poor σ-donor. It appears, however, that such ligands are quite good π-acceptors. Until

159

Table 5.3

	$Mn(CO)_6^+$	$Cr(CO)_6$	$V(CO)_6^-$	
$v(CO)(cm^{-1})$	2096	2000	1859	(T_{1u})
$v(MC)(cm^{-1})$	416	441	460	(T_{1u})
	$Ni(CO)_4$	$Co(CO)_4^-$	$Fe(CO)_4^{2-}$	$Mn(CO)_4^{3-}$	
$v(CO)(cm^{-1})$	2058	1883	1729	1670	(T_1)

Increasing back donation $M \overset{\frown}{\longrightarrow} CO(\pi^*)$ ⟶
CO bond order and $v(CO)$ decrease ⟶
MC bond order and $v(MC)$ increase ⟶

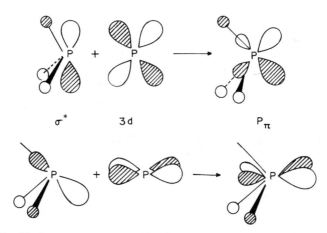

σ^* 3 d P_π

Fig. 5.10 Doubly degenerate PX_3 LUMO (C_{3v} local symmetry assumed). (From Orpen, A.G. and Connelly, N.G., 1985, *Chem. Commun.*, 1310.)

recently it was believed that the 3*d* orbitals of phosphorus were responsible for these acceptor properties. It was argued that a positive charge on phosphorus contracts the 3*d* orbitals and lowers their energy. This would hold especially for PF_3.

Recent calculations suggest that even in PF_3 the 3*d* orbitals may be too high in energy to fill the major role as π-acceptors. It seems that empty antibonding P—X σ^* orbitals of the phosphine are also involved. Indeed a combination of phosphorus 3*d* and P—X σ^* orbitals leads to well directed π-acceptor orbitals of the correct symmetry (Fig. 5.10). Removal of electron density from the metal centre will decrease donation into the acceptor orbitals. If the latter have P—X σ^* character, this decreased donation will cause a strengthening and hence a shortening of the P—X bond. This effect has been observed for several pairs of complexes which differ only in the total number of electrons (redox pairs). On oxidation of $CpCo(PEt_3)_2$ (18 electrons) to $[CpCo(PEt_3)_2]^+BF_4^-$ (17 electrons), for

Table 5.4 Band positions in the vibrational spectra of complexes $Ni(CO)_nL_{4-n}$, $L = P(OMe)_3$

	$Ni(CO)_4$	$Ni(CO)_3L$	$Ni(CO)_2L_2$	$Ni(CO)L_3$
Point Group	T_d	C_{3v}	C_{2v}	C_{3v}
$v(CO)$	A_1 2131 T_2 2058	A_1 2082 E 2006 2015	A_1 2025 B_1 1965	A_1 1964
$v(MC)$	A_1 368 T_2 423	A_1 420 E 455	A_1 458 B_1 495	A_1 499

The bands labelled A_1 arise from the symmetric breathing modes. For $Ni(CO)_4$, this mode is Raman, but not infrared active (change in polarizability but not in dipole moment). The bands labelled T_2 are due to the antisymmetric modes which are infrared, but not Raman active. For C_{3v}, C_{2v} all modes shown are both Raman and infrared active.

example, the P—Et bond decreases from 1.846(3) Å to 1.829(3) Å, while the Co—P distances increase from 2.218(1) Å to 2.230(1) Å.

This change in interpretation does not affect the general picture of a phosphine as a σ-donor and a π-acceptor ligand. The σ-donor power has been related to the pK_a values (p. 127). Trialkylphosphines therefore are quite strong σ-donors but trialkylphosphites are rather weak. The σ-donor ability increases across the series, $PF_3 < P(OMe)_3 < PPh_3 < PMe_3$. It is much more difficult to quantify the π-acidities. It is generally agreed, however, that the weakest σ-donors are the strongest π-acceptors and *vice versa*. Some people suggest that trialkylphosphines such as PBu_3 may lack all π-acceptor ability and may even be π donors.

Some band positions in the vibrational spectra of the complexes $Ni(CO)_nL_{4-n}$ ($L = P(OMe)_3$) are given in Table 5.4. On successive substitution of CO by $P(OMe)_3$ the position of $v(CO)$ falls, indicating increased back donation into the π^* orbitals of CO. This is consistent with the proposal that $P(OMe)_3$ is a stronger σ-donor than CO but a weaker π-acceptor. The trend in $v(MC)$ is consistent with this suggestion.

Tolman correlated the position of the A_1 band in the infrared spectra of complexes $Ni(CO)_3PR_3$ in dichloromethane with the electronic effect of the substituents R (Table 5.5). A reasonable correlation with Hammett σ-parameters was also found. The ligand behaviour of phosphorus donors are thus defined both by these electronic effects as well as by the steric effects measured by the cone angle (p. 169).

Where a molecule is fairly symmetrical different structures can be distinguished using vibrational spectroscopy. This method has been used, for example, to characterize geometrical isomers of substituted octahedral complexes. The point groups and symmetry classes of the normal $v(CO)$ vibrations for octahedral complexes $M(CO)_nL_{6-n}$ are listed in Table 5.6. Sometimes infrared spectroscopy

Table 5.5. Band positions of the A_1 stretching mode for the complexes $Ni(CO)_3PR_3$

R	$\nu(CO)$ (cm^{-1})		
$C(CH_3)_3$	2056.1	Best σ donor	Worst π acceptor
Me	2063.9		
Ph	2068.9		
OMe	2079.2		
OPh	2085.3		
F	2110.7	Worst σ donor;	best π-acceptor

Table 5.6. Infrared and Raman active carbonyl stretching modes for octahedral complexes $M(CO)_nL_{6-n}$

Molecule	Point group	IR active	Raman active	No. of IR bands	No. of Raman bands
$M(CO)_6$	O_h	T_{1u}	$A_{1g} + E_g$	1	2
$LM(CO)_5$	C_{4v}	$2A_1 + E$	$2A_1 + B_1 + E$	3	4
cis-$L_2M(CO)_4$	C_{2v}	$2A_1 + B_1 + B_2$	$2A_1 + B_1 + B_2$	4	4
$trans$-$L_2M(CO)_4$	D_{4h}	E_u	$A_{1g} + B_{1g}$	1	2
mer-$L_3M(CO)_3$	D_{2h}	$2A_1 + B_1$	$2A_1 + B_1$	3	3
fac-$L_3M(CO)_3$	C_{3v}	$A_1 + E$	$A_1 + E$	2	2
cis-$L_4M(CO)_2$	C_{2v}	$A_1 + B_1$	$A_1 + B_1$	2	2
$trans$-$L_4M(CO)_2$	D_{4h}	A_{2u}	A_{1g}	1	1
$L_5M(CO)$	C_{4v}	A_1	A_1	1	1

alone is sufficient to distinguish isomers. Thus cis-$L_2M(CO)_4$ affords four infrared active bands, whereas $trans$-$L_2M(CO)_4$ shows only one. In cases of doubt, however, it is always wise to check using Raman spectroscopy. If the ligands L are unsymmetrical, the symmetry of the molecule will depart to some extent from the ideal. This can lead to the observation of weak bands which would be excluded in the pure symmetry class. Moreover the degeneracy of vibrational modes can be lifted; doubly degenerate bands (E), for example, often split into two under these circumstances. The concept of 'local symmetry' within a molecule can also be useful. The pyramidal $M(CO)_3$ unit, which belongs to the point group C_{3v} gives rise to $3\nu(CO)$ fundamentals which are both infrared and Raman active ($A_1 + E$). This pattern, possibly with the E band split into two, appears in the spectra of all the complexes $C_7H_7V(CO)_3$, $C_6H_6Cr(CO)_3$, $C_5H_5Mn(CO)_3$, $C_4H_4Fe(CO)_3$ as well as fac-$L_3M(CO)_3$.

5.2.4 Mass spectra

Electron impact mass spectra of carbonyl complexes commonly show successive stepwise loss of all the carbonyl groups from the molecular ion (Fig. 5.11). The

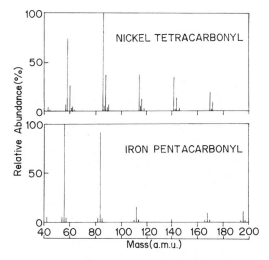

Fig. 5.11 Mass spectral cracking patterns of nickel tetracarbonyl and iron pentacarbonyl.

technique thus provides a means not only of determining the relative molecular mass of a volatile carbonyl, but also of counting the number of carbonyl ligands which it contains.

5.2.5 Bond dissociation energies

With respect to formation from the elements, metal carbonyls are exothermic compounds (Table 5.7). A measure of the strengths of the M—CO bonds is provided by the mean bond dissociation enthalpy $\bar{D}$(M—CO), which is derived from the enthalpy of dissociation:

$$M(CO)_n(g) \longrightarrow M(g) + nCO(g), \quad \text{given } \Delta H_f^{\ominus} \, (CO, g) = -110.5 \, \text{kJ mol}^{-1}$$

$\bar{D}$(M—CO) increases across the First Transition Series from Cr to Ni (except for Mn), and down Group VI (Cr < Mo < W). The increase in M—C bond strength down a transition metal group contrasts with the decrease down a group which is generally observed with main group elements (p. 5). M—CO bond distances also decrease across a transition series, following the contraction in covalent radius of the metal.

5.2.6 Bond distances in metal carbonyls

The carbon—oxygen bond length in the free CO molecule is 1.128 Å and in CO_2 1.171 Å. In terminal metal carbonyls the C—O distances lie between 1.14 and 1.16 Å, the slight lengthening compared with free CO being consistent with the reduction in bond order which occurs on complexing. This increase is so small,

163

Table 5.7. Thermochemical data and bond lengths in metal carbonyls

	$Cr(CO)_6$	$Mo(CO)_6$	$W(CO)_6$	$\frac{1}{2}[Mn_2(CO)_{10}]$	$Fe(CO)_5$	$\frac{1}{2}[Co_2(CO)_8]$	$Ni(CO)_4$
$\Delta H_f^{\ominus}[M(CO)n.g]$ (kJ mol^{-1})	-908	-915	-884	-798^*	-725	-593^*	-599
$\bar{D}$(M—CO) (kJ mol^{-1})	107	152	180	98	118	136	147
$\Delta H_f^{\ominus}[M.g]$ (kJ mol^{-1})	397	658	860	279	416	425	430
M—CO$_{terminal}$ (av.)(Å)	1.91	2.06	2.06	1.83	1.81	1.80	1.84

*Refers to $\frac{1}{2}$ mole.

however, that bond lengths do not provide a sensitive measure of electron distribution within the bond, infrared spectroscopy being much more informative. The M—C—O system in terminal carbonyls is approximately linear; the angles at carbon vary from about 165°–180°. In the gas phase intramolecular interactions between the M—C—O unit and other ligands may be responsible for small deviations from linearity. In the solid crystal packing forces may also contribute.

The C—O bonds in bridging carbonyl are somewhat longer than in terminal ligands. In $Fe_2(CO)_9$ (doubly bridging CO) the average distance is 1.176 Å and in $Rh_6(CO)_{16}$ (triply bridging) 1.201 Å. The M—C—M angle in a symmetrical bridge is rather acute, falling in the range 77–90°, compared with about 120° in a ketone. Bridging carbonyls are less common in complexes of the heavier transition elements than in those of the First Transition Series. It has been argued that the bridging mode is less favoured when the ligand has to span the larger metal–metal distance which obtains in the former compounds. In any case, the energy differences between bridging and terminal bonding modes are rather small (e.g. $Co_2(CO)_8$, p. 154), so it is difficult to predict which will occur in any particular case.

M—C bond distances are always significantly shorter to terminal than to bridging carbonyl ligands. There is a general parallel between M—CO(terminal) and the atomic radius of the metal. The radii of the corresponding elements at the beginning of the Second and Third Transition Series are almost equal, on account of the contraction which occurs across the lanthanides. Consequently the Mo—CO and W—CO bonds in the hexacarbonyls $Mo(CO)_6$ and $W(CO)_6$ are both 2.06 Å in length.

5.2.7 Metal carbonyls–Preparation

(a) DIRECT REACTION. The first binary carbonyl, $Ni(CO)_4$, was reported by Mond, Langer and Quincke in 1890. This volatile material is formed when carbon monoxide is passed over metallic nickel at or just above room temperature. It decomposes readily on warming. The Mond process for the extraction and refining of nickel depends on these reactions. Crude nickel oxide mixed with oxides of iron and copper is obtained by roasting the sulphide ore. Finely divided metal is produced by heating in hydrogen, which is obtained from water gas $(H_2 + CO)$, and the crude product reacted with CO at 50°C/1 atm to give volatile $Ni(CO)_4$. This is then passed over pellets of pure nickel at 230°C when it decomposes to metal of 99.97% purity, leaving CO to be recycled. A recent modification of the process which is operated in Canada employs slightly higher temperatures (130°C) and pressures (70 atm) for the carbonylation. Under these conditions a mixture of $Ni(CO)_4$ and $Fe(CO)_5$ is produced, which is separated by fractional distillation or decomposed directly to give an iron/nickel alloy. Nickel carbonyl is very toxic and strongly carcinogenic. It is present in low concentrations in tobacco smoke.

165

Transition metal chemistry

Nickel is the only metal to react directly with carbon monoxide at room temperature at an appreciable rate, although iron does so on heating under pressure. Cobalt affords $HCo(CO)_4$ with a mixture of hydrogen and carbon monoxide (p. 387). In general, therefore, direct reaction does not provide a route to metal carbonyls. The metal atom technique (p. 313) has been used to prepare carbonyls of other metals in the laboratory e.g. $Cr(CO)_6$, but it offers no advantages over the reduction method discussed below. When metal vapours are cocondensed with carbon monoxide in frozen noble gas matrices at very low temperatures (4–20K) the formation of carbonyl complexes is observed. These include compounds of metals which do not form any stable isolable derivatives e.g. $Ti(CO)_6$, $Nb(CO)_6$ and $Ta(CO)_6$ as well as $Pd(CO)_4$ and $Pt(CO)_4$. Vibrational spectra of the matrix show that coordinatively unsaturated species such as $Ni(CO)_n$ ($n = 1$–3) or $Cr(CO)_n$ ($n = 3$–5) are also formed under these conditions.

(b) REDUCTIVE CARBONYLATION. Binary metal carbonyls are usually prepared by reacting a metal halide (or other suitable complex such as an acetylacetonate) with carbon monoxide in the presence of a reducing agent, which both supplies electrons to the metal centre and acts as a halide ion acceptor. This is an example of a general method for making complexes in which the metal is in a low formal oxidation state:

$$\text{Metal halide} + \text{Ligand} + \text{Reducing agent} \longrightarrow \text{complex}$$

The metal halide (or other starting material) must be anhydrous when strong water sensitive reducing agents such as alkali metals are employed. The choice of reducing agent is based on experience; common examples include sodium, magnesium, triethylaluminium, sodium and benzophenone (for $Mn_2(CO)_{10}$) or carbon monoxide itself, sometimes in the presence of hydrogen (for $Co_2(CO)_8$). A polar organic solvent such as diethyleneglycol dimethylether ('diglyme') which dissolves the metal halide and the carbonyl product is often used and the reactions generally carried out in an autoclave under carbon monoxide pressure:

$$2MnCl_2 + 4Ph_2CO^-Na^+ + 10CO \xrightarrow[\substack{200°/ \\ 200\,atm.}]{THF} Mn_2(CO)_{10} + 4NaCl + 4Ph_2CO$$

These ether solvents also dissolve alkali metals to some extent, aiding the initial electron transfer. Often the first products of these reactions are carbonyl anions which are converted into the neutral species on acidification, probably via an unstable carbonyl hydride (p. 172).

$$VCl_3 + 4Na + 6CO \xrightarrow[160°C/200\,atm]{diglyme} [Na(diglyme)_2][V(CO)_6] + 3NaCl$$

$$CrCl_3 + 5Na + 5CO \xrightarrow[0-25°C/150\,atm]{diglyme} [Na(diglyme)_2]_2[Cr(CO)_5] + 3NaCl$$

$$\left\downarrow \begin{array}{l} 0°C/50\,atm \\ CO \\ H_2SO_4 \end{array}\right.$$

$$Cr(CO)_6$$

Niobium and tantalum halides yield salts of the 18 electron anions $Nb(CO)_6^-$ and $Ta(CO)_6^-$, but no neutral binary carbonyls have been isolated.

Some metal halides such as ruthenium trichloride afford carbonyl halides if the carbonylation is carried out in the absence of a halide acceptor.

$$2RuCl_3 \cdot 3H_2O + 7CO \xrightarrow[\text{1 atm}]{\text{MeOH}} Ru(CO)_3Cl_2 + 2HCl + CO_2 + 5H_2O \xrightarrow{\text{Zn, CO/10 atm}} Ru_3(CO)_{12}$$

$$2RhCl_3 \cdot 3H_2O + 3CO \xrightarrow[\text{1 atm}]{\text{MeOH}} [Rh(CO)Cl_2]_2 + 2HCl + CO_2 + 5H_2O$$

$$\bigg\downarrow \begin{array}{l} \text{Cu/MeOH} \\ \text{CO/1 atm} \end{array}$$

$$Rh_4(CO)_{12}$$

Further reduction to binary cluster carbonyls can be effected as shown.

Many metal carbonyls including $M(CO)_6$ (M = Cr, Mo, W), $Mn_2(CO)_{10}$, $Fe(CO)_5$, $Co_2(CO)_8$ and $Ni(CO)_4$ are commercially available at reasonable cost, so it is now seldom necessary to prepare them in the laboratory.

(c) PREPARATION OF POLYNUCLEAR CARBONYLS. Usually polynuclear carbonyls are prepared starting from the simpler derivatives. $Fe_2(CO)_9$ is obtained as insoluble orange plates by photolysis of a cooled solution of $Fe(CO)_5$ in acetic anhydride:

$$Fe(CO)_5 \xrightarrow{h\nu} Fe(CO)_4 + CO$$

$$Fe(CO)_4 + Fe(CO)_5 \longrightarrow Fe_2(CO)_9$$

$Co_4(CO)_{12}$ results when $Co_2(CO)_8$ is heated in an inert atmosphere:

$$2Co_2(CO)_8 \xrightarrow{50°C} Co_4(CO)_{12} + 4CO$$

Similarly higher osmium carbonyl clusters are obtained from the pyrolysis of $Os_3(CO)_{12}$ (p. 351). Other syntheses involve carbonyl anions as intermediates (p. 171).

5.2.8 Reactions of metal carbonyls

The chemistry of metal carbonyls is so vast that only a few examples of their reactions can be mentioned here.

(a) SUBSTITUTION OF CARBON MONOXIDE. Lewis bases, in particular phosphorus, arsenic and antimony donors, can substitute carbon monoxide in metal carbonyls. Two mechanisms can be envisaged for such reactions. In the first (a)

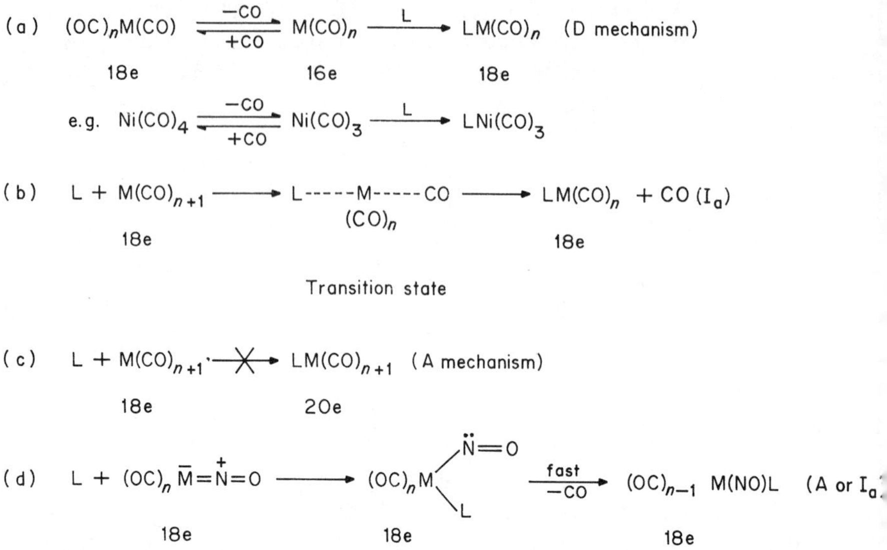

(a) $(OC)_nM(CO)$ $\underset{+CO}{\overset{-CO}{\rightleftharpoons}}$ $M(CO)_n$ $\xrightarrow{L}$ $LM(CO)_n$ (D mechanism)

$\qquad\qquad$ 18e $\qquad\qquad\qquad$ 16e $\qquad\qquad\qquad$ 18e

$\qquad$ e.g. $Ni(CO)_4$ $\underset{+CO}{\overset{-CO}{\rightleftharpoons}}$ $Ni(CO)_3$ $\xrightarrow{L}$ $LNi(CO)_3$

(b) $L + M(CO)_{n+1}$ $\longrightarrow$ $\underset{(CO)_n}{L\text{-----}M\text{-----}CO}$ $\longrightarrow$ $LM(CO)_n + CO\ (I_a)$

$\qquad\qquad$ 18e $\qquad\qquad\qquad\qquad\qquad\qquad\qquad\qquad$ 18e

$\qquad\qquad\qquad\qquad\qquad$ Transition state

(c) $L + M(CO)_{n+1}$ $\cdot\!\!\xrightarrow{\quad\times\quad}$ $LM(CO)_{n+1}$ (A mechanism)

$\qquad\qquad$ 18e $\qquad\qquad\qquad\qquad$ 20e

(d) $L + (OC)_n \overset{-}{M}\!\!=\!\!\overset{+}{N}\!\!=\!\!O$ $\longrightarrow$ $(OC)_nM\!\!\begin{smallmatrix}\nearrow\ddot{N}=O\\ \searrow L\end{smallmatrix}$ $\xrightarrow[-CO]{fast}$ $(OC)_{n-1}\ M(NO)L$ (A or I_a)

$\qquad\qquad$ 18e $\qquad\qquad\qquad\qquad\qquad$ 18e $\qquad\qquad\qquad\qquad\qquad$ 18e

Fig. 5.12 Mechanisms of substitution of CO in metal carbonyls.

initial rate determining dissociation of CO to give a coordinatively unsaturated intermediate is followed by fast uptake of the attacking ligand L (S_N1 or D mechanism). Alternatively an interchange associative mechanism (b) might operate (S_N2 or I_a) in which concerted replacement of CO by L occurs. A non-synchronous associative mechanism (A) involving addition of L to the metal centre followed by elimination of CO is precluded as it would entail a 20 electron intermediate (Fig. 5.12).

$\qquad$ The reactions of $Ni(CO)_4$ with phosphorus donors are first order in $Ni(CO)_4$ and essentially independent of the nature and concentration of the attacking nucleophile. Coordinating solvents such as THF, which can interact with and stabilize the 16 electron $Ni(CO)_3$ intermediate, accelerate the first step. The reactions are retarded by increase in pressure; the volume of activation for attack of triethylphosphite is positive, $+8\ cm^3\,mol^{-1}$ at $0°C$. All these observations point to a dissociative mechanism.

$\qquad$ $Ni(CO)_4$ reacts rapidly at room temperature to give a series of substituted derivatives $Ni(CO)_3L$, $Ni(CO)_2L_2$, $Ni(CO)L_3$ and NiL_4. The extent of reaction with excess L depends largely on the steric bulk of L. Tolman has classified this in terms of the cone angle Θ. This is the angle at the apex of a cone (Fig. 5.13) which just touches the substituents defined by van der Waals radii. It was found that the degree of substitution of carbonyl groups in $Ni(CO)_4$, on treatment with an eight-fold excess of L, is proportional to Θ for PR_3:R = H, 87°; F, 101°; OMe, 107°; Me, 118°; OPh, 130°; Ph, 145°; Pr^i, 160°; Bu^t, 182°.

$\qquad$ Substitution of CO in $Fe(CO)_5$ and in $Cr(CO)_6$ also follows the D mechanism, but

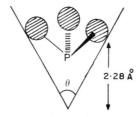

Fig. 5.13 The 'cone angle' Θ of a phosphine ligand. Metal atom (Ni) at apex of cone.

these reactions are much slower than for $Ni(CO)_4$. (At high concentrations of strongly basic phosphines, a second order term also appears in the kinetic equation for $Cr(CO)_6$.) Under reflux in a hydrocarbon solvent, $Fe(CO)_5$ yields first $Ph_3PFe(CO)_4$ and finally $(Ph_3P)_2Fe(CO)_3$. Photolysis of $Fe(CO)_5$ aids dissociation of CO, and the unsaturated $Fe(CO)_4$ can then be trapped by L. Other mild ways of generating $Fe(CO)_4$ include oxidation of one CO group by trimethylamine oxide and catalytic processes involving transition metal salts, which generate radical species which are labile to substitution:

$$R_3\overset{+}{N}\!\!-\!\!\overset{-}{O} \atop O\!=\!C\!=\!Fe(CO)_4 \longrightarrow R_3\overset{+}{N}\!\!-\!\!O\!-\!\underset{\underset{O}{\|}}{C}\!\!-\!\!\overset{-}{Fe}(CO)_4 \longrightarrow Fe(CO)_4 + CO_2 + R_3N$$

$$\downarrow L$$

$$LFe(CO)_4$$

Sequential replacement of carbon monoxide in octahedral carbonyls $M(CO)_6$ (M = Cr, Mo, W) occurs mainly by a series of dissociative substitutions. If the entering ligand L is a nitrogen, oxygen or halogen donor, a carbonyl group in the initial product $M(CO)_5L$ which is *cis* to L is labilized, so that further substitution leads preferentially to the *cis* isomer of $M(CO)_4L_2$. A carbonyl which is *cis* to both labilizing groups then dissociates and this leads in turn to the *fac*-isomer of $M(CO)_3L_3$. While good π-donors labilize *cis* carbonyl groups, they stabilize those in the *trans* position by increasing the electron density at the metal and hence π-back donation. Substitution often therefore stops at this stage.

Ligands with P, As, Sb or S donor atoms do not exert any stereospecific labilizing effect on carbonyl ligands, so that mixtures of isomers are often obtained in their direct reactions with the hexacarbonyls. They do however readily displace ligands such as CH_3CN or piperidine from their complexes without any rearrangement. Stereospecific synthesis of *cis* and *fac* isomers can therefore be carried out in two stages as follows:

$$Mo(CO)_6 + 2C_5H_{11}NH(xs) \longrightarrow cis\text{-}Mo(CO)_4(C_5H_{11}NH)_2 \overset{2L}{\longrightarrow} cis\text{-}Mo(CO)_4L_2$$

$$Mo(CO)_6 + 3CH_3CN(xs) \longrightarrow fac\text{-}Mo(CO)_3(CH_3CN)_3 \overset{3L}{\longrightarrow} fac\text{-}Mo(CO)_3L_3$$

$$(L = PR_3, P(OR)_3).$$

169

Transition metal chemistry

Carbonyl ligands in nitrosyl carbonyls $Mn(CO)_4(NO)$, $Fe(CO)_2(NO)_2$ and $Co(CO)_3(NO)$ are substituted by an S_N2 (A or I_a) mechanism (Fig. 5.12(d)). This occurs because a lone pair can be localized on nitrogen in the intermediate (or transition state). The nitrosyl group can act either as a formal three-electron

(linear $\bar{M}=\overset{+}{N}=O$) or as a one-electron donor (bent $M-N\overset{\nearrow O}{}$). (see p. 294 for substitution of cyclopentadienyl metal carbonyls).

(b) REACTION OF HARD BASES.   Pentacarbonyliron reacts with amines such as pyridine, not by substitution, but by disproportionation. A cluster anion is produced:

$$5Fe(CO)_5 + 6py \xrightarrow{\text{reflux}} [Fe(py)_6][Fe_4(CO)_{13}] + 12CO$$
$$(PPN^+ = Ph_2P\overset{+}{=}N=PPh_2) \qquad \longrightarrow [PPN]_2[Fe_4(CO)_{13}]$$

X-ray diffraction of the PPN salt reveals a tetrahedral cluster of iron atoms (Fig. 5.14(c)). One triangular face is capped by a triply bridging CO group; around each of the sides of this same face there is an edge-bridging carbonyl. On protonation the tetrahedral cluster opens out to a butterfly framework (cf. p. 350); the protonated CO ligand is now σ-bonded to the fourth iron atom. Further reduction under acid conditions affords methane; it has been suggested that this

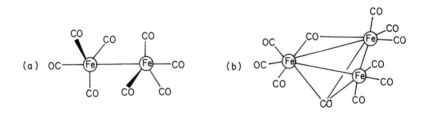

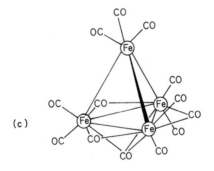

Fig. 5.14 Structures of (a) $Fe_2(CO)_8^{2-}$, (b) $Fe_3(CO)_{11}^{2-}$, (c) $Fe_4(CO)_{13}^{2-}$ showing three doubly bridging and one triply bridging CO groups (O atoms omitted).

170

sequence of reactions could mimic those occurring at metal surfaces in the Fischer-Tropsch process (p. 393).

Such anionic metal carbonyls are very common. Many are clusters (p. 353). One method of preparation is to treat the neutral precursor with hydroxide ion, or to use a reducing agent. In either case one CO ligand is formally replaced by an electron pair. Hydroxide is thought first to attack the carbonyl group, followed by release of carbon dioxide (as HCO_3^- or CO_3^{2-}) and transfer of electrons to the metal:

$$(OC)_4 Fe \overset{\delta+}{—}\overset{\delta-}{C} = O \xrightarrow{\quad} \left[(OC)_4 Fe—C \overset{O}{\underset{O—H}{\diagdown}} \right]^- \xrightarrow{\ 3OH^-\ }$$

$$(OC)_4 Fe^{2-} + CO_3^{2-} + 2H_2O$$

$Fe(CO)_4^{2-}$ is the conjugate base of the carbonyl hydride $H_2Fe(CO)_4$. This shows two neutralization steps in aqueous solution with $pK_1 = 4.0$ and $pK_2 = 12.7$. The sodium salt $Na_2Fe(CO)_4$ is a useful reagent in organic chemistry. Single crystal X-ray diffraction of $Na_2Fe(CO)_4 . \frac{3}{2}$ dioxan reveals six coordinate Na^+ ions which interact with the oxygen atoms of adjacent $Fe(CO)_4^{2-}$ units. Similar ion pairing through $Na \cdots O$ contacts also occurs in solvents such as tetrahydrofuran.

Other controlled syntheses afford $Fe_2(CO)_8^{2-}$ from $Fe_2(CO)_9$ and $Fe_3(CO)_{11}^-$ from $Fe_3(CO)_{12}$:

$$Fe_2(CO)_9 \xrightarrow[Et_4N^+I^-]{KOH/} [Et_4N]_2[Fe_2(CO)_8] \quad \text{(red-orange crystals)}$$

$$Fe_3(CO)_{12} + 4OH^- \longrightarrow [Fe_3(CO)_{11}]^{2-} + CO_3^{2-} + 2H_2O$$

In contrast to $Co_2(CO)_8$, $Fe_2(CO)_8^{2-}$ has no bridging carbonyl ligands. In crystalline $[PPN]_2[Fe_2(CO)_8]$ the ion adopts a staggered D_{3d} conformation νCO 1920, 1852 cm^{-1}; $\nu(Fe—Fe)$, 170 cm^{-1} (Raman). $Fe_3(CO)_{11}^{2-}$ has one face bridging and one edge bridging CO group. The structure is related to that of $Fe_4(CO)_{13}^{2-}$ by removal of the face bridging $Fe(CO)_3$ group (Fig. 5.14).

A laboratory preparation of $Fe_3(CO)_{12}$ entails treatment of $Fe(CO)_5$ with alkali to give $HFe(CO)_4^-$, oxidation with freshly prepared MnO_2, followed by acidification. It is likely that the reaction proceeds via $HFe_3(CO)_{11}^-$.

(c) REDUCTION TO CARBONYL ANIONS.   In the last Section we saw that a carbonyl ligand in a neutral metal carbonyl can often be replaced by an electron pair by reduction. $Cr(CO)_6$, however, can be reduced electrochemically or by using the graphite intercalation compound C_8K in THF to the dimeric anion $Cr_2(CO)_{10}^{2-}$; that is, by a one electron reduction followed by coupling of the resulting $Cr(CO)_5^- \cdot$ radicals. Further reduction with sodium in liquid ammonia affords $Cr(CO)_5^{2-}$:

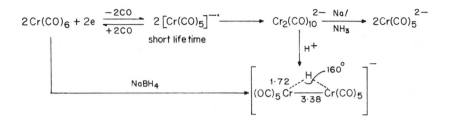

The metal—metal bonds in $M_2(CO)_{10}$ (M = Mn, Tc, Re) and in $Co_2(CO)_8$ are readily cleaved by sodium amalgam in THF:

$$\tfrac{1}{2}Mn_2(CO)_{10} \xrightarrow[\text{THF}]{\text{Na/Hg}} Mn(CO)_5^- \xrightarrow{2\varepsilon} Mn(CO)_4^{3-}$$
$$\xrightarrow{H_3PO_4} HMn(CO)_5 \quad (pK \sim 7)$$

$$\tfrac{1}{2}Co_2(CO)_8 \longrightarrow Co(CO)_4^- \longrightarrow HCo(CO)_4 \quad (pK \sim 2)$$

The resulting anions are the conjugate bases of carbonyl hydrides. $HCo(CO)_4$ is a strong acid, but $HMn(CO)_4$ is much weaker. Some carbonyl hydrides such as $HCo(CO)_4$ and $H_2Fe(CO)_4$ are very unstable volatile materials, decomposing below $0°C$. $HMn(CO)_5$ is somewhat more robust (dec. $80°C$). It has a distorted octahedral structure; the equatorial CO groups are bent towards the hydrogen. In favourable circumstances hydrogen atoms attached to transition metals can now be located quite precisely by X-ray diffraction in spite of their low scattering effect on X-rays compared with heavy metal atoms. Neutron diffraction, when available, provides a more accurate method, as scattering is effected by the nuclei, and hydrogen presents a fairly high cross section. Spectroscopic methods, in particular infrared and n.m.r. spectroscopy are invaluable in the initial characterization of a transition metal hydride. Terminal M—H stretching vibrations (1700–2250 cm^{-1}) fall in about the same region as terminal carbonyl vibrations (for $HMn(CO)_5$, $v(Mn—H) = 1783$ cm^{-1}). Confusion between the two can be overcome by comparing the spectrum of the suspected hydride with that of the corresponding deuteride when $v(M—H)/v(M—D) \sim \sqrt{2}$. In their proton n.m.r. spectra transition metal hydrides typically show resonances well to high field of tetramethylsilane ($- 7$ to $- 24\delta$; $HMn(CO)_5$, $- 7.5\delta$). These shifts are very characteristic of this class of compound. Most organic compounds resonate in the region 10 to 0δ. The observation of coupling with metal nuclei with $I = \tfrac{1}{2}$ such as ^{103}Rh, ^{195}Pt, ^{187}Os or ^{89}Y (p. 401) can provide evidence for direct M—H bonding in specific cases.

Many carbonyl anions are strong nucleophiles. Some reactions of $Mn(CO)_5^-$ illustrate this (Fig. 5.15). They lead to various alkyl and aryl derivatives as well as complexes in which Mn is bonded to Main Group and transition elements.

172

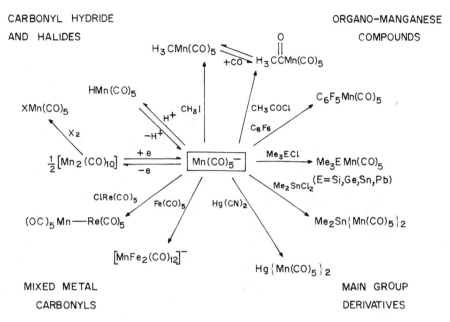

Fig. 5.15 Some reactions of $Mn(CO)_5^-$.

5.2.9 *Carbonyl halides*

Metal—metal bonds in transition metal carbonyls can be cleaved by halogens. $Mn_2(CO)_{10}$ affords the complexes $XMn(CO)_5$, $(X = Cl, Br, I)$ (cf. $Os_3(CO)_{12}$, p. 352). Halogens also attack metal carbonyls which do not possess M—M bonds with loss of carbonyl ligands e.g.

$$Fe(CO)_5 + X_2 \longrightarrow \textit{cis-}Fe(CO)_4X_2 + CO$$

$$Mo(CO)_6 + Cl_2 \longrightarrow \tfrac{1}{2}[Mo(CO)_4Cl_2]_2 + 2CO$$

$$2Fe_3(CO)_{12} + 3I_2 \longrightarrow 3Fe_2I_2(CO)_8 \quad (D_{4d})$$

Carbonyl halides of the platinum group metals, however, are often prepared by reacting the metal halide with carbon monoxide. $[Ru(CO)_3Cl_2]_2$ and $[Rh(CO)_2Cl]_2$ are mentioned above (p. 167). Platinum forms $Pt(CO)_2X_2$ and $[Pt(CO)X_2]_2$. The latter were the first carbonyl complexes of any element to be described. They were reported by the French chemist, P. Schutzenberger, in 1870.

5.3 Some general reactions of transition metal complexes

Figure 5.16 shows the structures of the 18-electron complexes ML_n ($n = 4$ to 7) the 16-electron species ML_{n-1} which would result from the dissociation of one ligand: L and some 14-electron species which would arise from the dissociation of

Transition metal chemistry

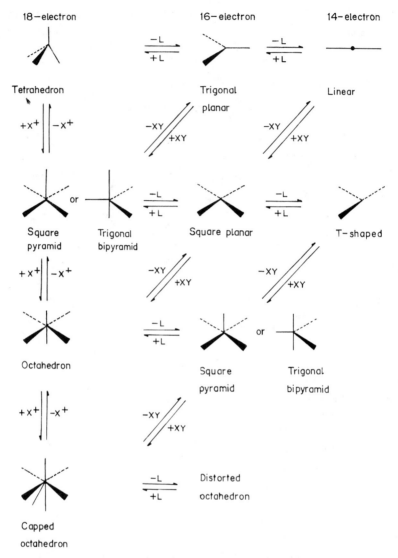

Fig. 5.16 Substitution, addition and elimination reactions of transition metal complexes. L = Lewis base, two electron ligand. X = one electron ligand, e.g. Cl, H, R, COR, SiR_3. XY includes homonuclear species, e.g. H_2, Cl_2.

a further ligand. The 16 and 14-electron species are 'coordinatively unsaturated', that is, they possess vacant coordination sites which are potentially available for occupation by ligands. These additional ligands must provide the electrons needed to reach a stable electron configuration. In general this requires a total of 18-electrons. Some square planar 16-electron complexes, especially those of Ni,

Pd, Pt and Au, however, often show little or no tendency to be converted into 18-electron products. In these cases the 16-electron configuration may often be viewed as being essentially 'saturated'. Sixteen electron complexes are common for (Co), Rh and Ir and for Ni, Pd and Pt. Even lower electron configurations (16, 14 and 12) are normal for Ag and Au, for which 18-electron complexes are rare.

Some general types of reaction which are followed by the essentially covalent transition metal complexes which are the subject of this book are listed in Table 5.8. The simplest way in which an unsaturated species ML_{n-1} can be become saturated is by addition of a Lewis base L, in other words by reversal of the dissociation mentioned above. This is shown in Fig. 5.16 by horizontal arrows ($\rightleftharpoons$) (Table 5.8, 1 (a), 1 (b)). Examples are provided by the substitution of carbon monoxide in $Ni(CO)_4$ or $Fe(CO)_5$ (p. 168) which follows a dissociative (D) mechanism, 1(a) followed by 1(b). Ligand substitution in 16-electron square planar complexes of Pd, Pt and Au for example, normally follows an associative mechanism, 1(b) followed by 1(a). An intermediate reaction type, synchronous interchange of ligands, in which the leaving group moves away as the incoming group enters (I_d, I_a), involves no change in electron configuration. 18-Electron complexes never react by an associative mechanism, as the first step would require the formation of a 20-electron intermediate. Where overall second order kinetics are observed, an interchange process is likely. This probably accounts for the small ligand dependent contribution to the rate of substitution of the carbonyls $M(CO)_6$ (M = Cr, Mo, W) (Rate = $k_1[M(CO)_6] + k_2[M(CO)_6][L]$) which is observed especially with strong σ donors such as trialkylphosphines.

Coordinatively unsaturated transition metal complexes may therefore behave as *electrophiles* (Lewis acids) by virtue of empty low lying orbitals centred on the metal atom. Complexes, however, especially those with 18-electron configurations, may also possess electrons in essentially non-bonding d-orbitals. Where the ligands are rather electronegative (N, O, halogen donors), or where the complex carries an overall positive charge, such orbitals lie rather low in energy. With relatively electropositive donor atoms (P, As, C, H donors) or in anionic species (e.g. $Mn(CO)_5^-$) the energy of these d-orbitals is raised and the compound becomes 'electron rich'. Such complexes can behave as bases; this property is noted above for carbonyl anions (p. 173) which are the conjugate bases of hydridocarbonyls. Replacement of CO by phosphites or phosphines increases the base strength; while $HCo(CO)_4$ is a strong acid (pK ~ 2), $HCo(CO)_3PPh_3$ is rather weak (pK ~ 7). $HCo\{P(OMe)_3\}_4$ is converted into the anion only by very strong bases such as KH in tetrahydrofuran, while $HCo(PMe_3)_4$ cannot be deprotonated at all. Neutral complexes ML_n (L = $P(OR)_3$ or PR_3) are quite strong bases. In some cases crystalline salts such as $HFeL_5^+X^-$ have been isolated, especially with large weakly coordinating anions (X = BF_4, PF_6) which have little tendency to displace a ligand from the coordination sphere.

A few adducts between metal complexes and Lewis acids such as boron trifluoride (p. 179) have been isolated.

175

Table 5.8. General reactions of transition metal complexes

Type of reaction	General equation	Change in no. of electrons in valence shell	Change in coordination number
1(a) Dissociation of Lewis base	$L_{n-1}M:L \rightarrow L_{n-1}M + :L$	-2	-1
(b) Association of Lewis base [reverse of 1(a)]	$L_{n-1}M + :L \rightarrow L_{n-1}M:L$	$+2$	$+1$
1(a) & 1(b) Substitution by dissociative mechanism(D)	$\left.\begin{array}{l} L_{n-1}M:L \rightarrow L_{n-1}M + :L \\ L_{n-1}M + :L' \rightarrow L_{n-1}M:L' \end{array}\right\}$	0	0
1(c) Synchronous substitution (I_a, I_d)	$L': + ML_n \rightarrow L':ML_{n-1} + :L$	0	0
1(b) & 1(a) Substitution by associative mechanism(A)	$\left.\begin{array}{l} L_{n-1}M + :L' \rightarrow L_{n-1}M:L' \\ L_{n-1}ML' \rightarrow L_{n-2}ML' + :L \end{array}\right\}$	0	0
2(a) Dissociation of Lewis acid	$A:ML_n \rightarrow A + :ML_n$	0	-1
(b) Association of Lewis acid (includes protonation) [reverse of 2(a)]	$A + ML_n \rightarrow A:ML_n$	0	$+1$
3(a) Nucleophilic substitution by complex	$L_nM: + X:Y \rightarrow [L_nM:X]^+ [Y:]^-$ or $[L_nM:]^- + X:Y \rightarrow L_nM:X + [Y:]^-$	0 0	$+1$ $+1$
4(a) Addition to coordinatively unsaturated complex	$L_{n-1}M: + X:Y \rightarrow L_{n-1}M\overset{X}{\underset{Y}{\cdots}}$	$+2$	$+2$
(b) Elimination from coordinatively saturated complex [reverse of 4(a)]	$L_{n-1}M\overset{X}{\underset{Y}{\cdots}} \rightarrow L_{n-1}M: + X \div Y$	-2	-2
5 Free radical pathways: (a) Radical addition to coordinatively unsaturated complex	$(Q\cdot + X - Y \rightarrow Q - Y + X\cdot)$ $X\cdot + :ML_{n-1} \rightarrow [X:\dot{M}L_{n-1}] \overset{\cdot Y}{\underset{\cdots}{\longrightarrow}} X:M L_{n-1}\overset{X}{\underset{Y}{\cdots}} + X\cdot$	$+1$ $+1$ ___ $+2$	$+1$ $+1$ ___ $+2$
(b) Initiated by complex as radical	$L_nM\cdot + X \div Y \rightarrow L_nM:X + Y\cdot$ $L_nM\cdot + Y\cdot \rightarrow L_nM:Y$	$+1$ $+1$	$+1$ $+1$

Electron rich complexes are strong nucleophiles. This is illustrated by some reactions with methyl iodide, which may be compared with attack by a tertiary amine:

$$Fe[P(OMe)_3]_5 \xrightarrow[ii)NH_4PF_6]{i)MeI} [MeFe\{P(OMe)_3\}_5]^+ [PF_6]^-$$

$$Ni(PMe_3)_4 \xrightarrow[CH_3CN]{MeI} [MeNi(PMe_3)_4]^+ [I]^-$$

$$Me_3N \xrightarrow{MeI} Me_4N^+ I^-$$

In these reactions (2(a), 2(b), 3(a) in Table 5.8), the coordination number of the metal increases by one, while the electron configuration remains the same. The structural relationships are shown in Fig. 5.16 by vertical arrows ($\Updownarrow$).

Halogens behave as electrophiles towards nucleophilic complexes. There are two common modes of reaction, e.g.

$$M(CO)_4L_2 + X_2 \longrightarrow [M(CO)_4L_2X]^+ X^-; \quad (M = Mo, W)$$
18e,   18 e
six-coordinate   seven-coordinate

$$M(CO)_4L_2 \underset{+CO}{\overset{-CO}{\rightleftharpoons}} M(CO)_3L_2 \xrightarrow{X_2} M(CO)_3L_2X_2$$

16 e,   18 e
five-coordinate   seven-coordinate
(unsaturated)

The first is analogous to the reaction just described ($\Updownarrow$), but the second involves expulsion of one carbonyl ligand. This probably proceeds by dissociation of CO to give an unsaturated 16-electron intermediate, followed by addition of X_2. (A synchronous mechanism could also operate.) The general reaction pathway is along the route horizontal ($\rightleftharpoons$) then diagonal ($/\!\!/$) in Fig. 5.16.

When counting electrons it is recommended that all bonds are treated as covalent. Thus a single bond M—X (X = H, Cl etc.) is considered as being derived from M· and X·, rather than from M^+ and $:X^-$ or from $M:^-$ and X^+. In this way possible confusion over assignment of a formal oxidation state to the metal is avoided. Ligands (X = H, F, Cl, Br, I, CN, alkyl, aryl, acyl etc.) which form single two-electron bonds are therefore treated as one-electron donors.

Examples:

$HCo(CO)_4$: $1(H) + 9(Co) + (4 \times 2)(4CO) = 18$
$Mn(CO)_4^-$: $7(Mn) + 10(5CO) + 1$ (negative charge) $= 18$
$Fe(CO)_4I_2$: $8(Fe) + 8(4CO) + (2 \times 1)(2I) = 18$
$HNi(PMe_3)_4^+$: $1(H) + 10(Ni) + 8(4PMe_3) - 1$ (positive charge) $= 18$

177

5.3.1 *Addition and elimination*

A coordinatively unsaturated species can thus achieve saturation by addition of two one electron ligands, derived from a molecule X—X or X—Y. Although many reactions fall into this broad classification and may appear to be similar, their mechanisms may be quite different. This problem is addressed below by considering three complexes which show particularly rich chemistry and which have played an important part in the development of this area.

It has been the convention when discussing such reactions to assign a formal oxidation state to the metal. The oxidation state is defined as the charge left on the metal when ligands are removed in their closed shell configurations i.e. L as :L and X as $:X^-$. The five coordinate 18-electron complexes FeL_5, CoL_4X and NiL_3X_2 thus contain Fe(0), Co(I) and Ni(II) respectively, as do their 16-electron counterparts FeL_4, CoL_3X and NiL_2X_2. The metal species Fe^0, Co^+ and Ni^{2+} all have eight valence electrons (d^8). Using this convention, addition of XY e.g. $FeL_4 \rightarrow FeL_4XY$ leads to an increase in oxidation state of the metal by two units, that is, the metal loses two electrons ($d^8 \rightarrow d^6$). Consequently this type of process has been termed 'oxidative addition' and the reverse reaction 'reductive elimination'.

There are several objections to this terminology. Confusion can arise in defining oxidation state. Although transition metal hydrides are essentially covalent, one has to decide whether H is removed as $M^+ :H^-$ or as $M:^- H^+$. The convention has been to adopt the former. Under this convention a protonation involves an increase in the oxidation state of the metal by two units. This is rather artificial; proton transfer is not usually treated as a redox reaction, so why should it become one in the field of transition metal chemistry? Again in a nucleophilic substitution at carbon e.g. $I^- + CH_3Br \rightarrow ICH_3 + Br^-$, the 'oxidation states' of carbon, iodine and bromine are invariant, whereas in the analogous reaction $Co(CO)_4^- + CH_3Br \rightarrow H_3CCo(CO)_4 + Br^-$ the cobalt has to be 'oxidized' from the -1 to the $+1$ state.

The reactions which have been grouped together under the heading 'oxidative addition' vary greatly in mechanism. Some are substitutions of an electrophile by a nucleophilic complex. Others e.g. $Ir(CO)Cl(PPh_3)_2 + H_2$ (p. 180) are considered to be a concerted process in which the complex acts both as a nucleophile and as an electrophile. Some involve free radical pathways. The lumping together of all these processes under the one heading has tended to obscure these differences.

The concept of 'oxidation state' is a legacy of classical coordination chemistry where ligands are essentially σ donors (Cl^-, H_2O, NH_3), the complexes have a fair degree of ionic character and crystal field theory provides a valid springboard for discussion. It is less helpful in describing the essentially covalent complexes formed by π-acceptor ligands, hydrogen and organic groups.

(a) ADDITIONS TO SQUARE PLANAR 16-ELECTRON COMPLEXES: VASKA'S COMPOUND. Vaska's compound, $(Ph_3P)_2Ir(CO)Cl$ is a 16-electron square planar complex. It is conveniently prepared by reducing Na_3IrCl_6 in boiling

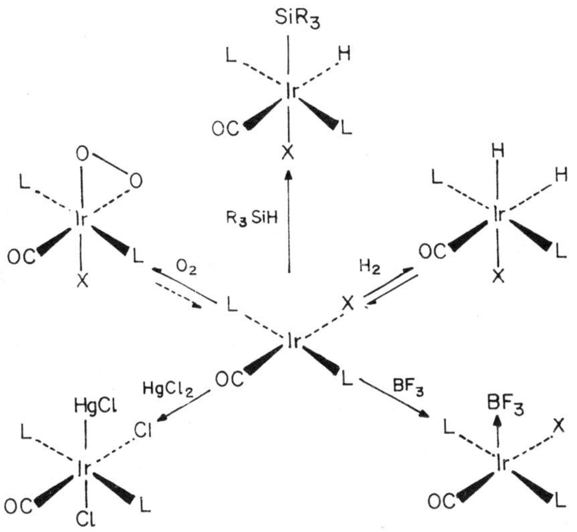

Fig. 5.17 Additions to Vaska's compound and related complexes. L = PPh$_3$ or similar. X = Cl, Br, I.

ethanol with formaldehyde in the presence of excess triphenylphosphine. The formaldehyde also is the source of the carbonyl ligand. The metal centre is basic. The complex is protonated, even by carboxylic acids, which form octahedral derivatives in which H and OCOR are *trans* to each other. This suggests a two stage reaction, initial protonation being followed by coordination of the anion:

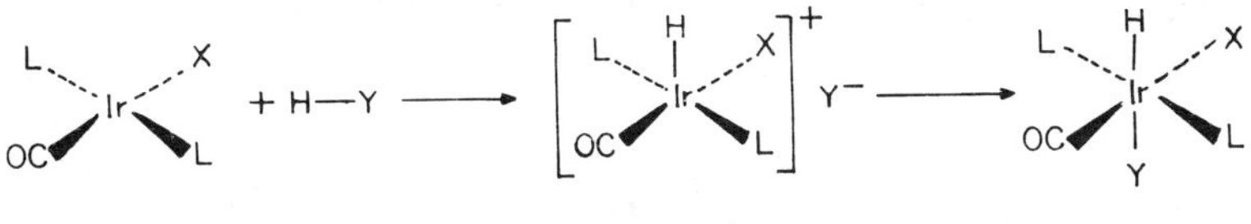

The equilibrium constant (*K*) for the addition of benzoic acid to Vaska's type compounds depends on the electron density at iridium. Thus electron attracting ligands lower the basicity and electron donating ones raise it (Order of *K* where X = I > Br > Cl; L = Me$_3$P > Me$_2$PPh > MePPh$_2$ > PPh$_3$).

Vaska's compound forms adducts with the Lewis acids BF$_3$, BCl$_3$ and BBr$_3$. With HgCl$_2$ and SnCl$_4$ adducts may be formed initially, but transfer of chlorine to iridium occurs and the final products are octahedral (Fig. 5.17).

Evidence on the reactions with methyl halides is consistent with S$_N$2 attack by iridium at carbon. The reactions are first order each in CH$_3$X and Ir(CO)XL$_2$. The most basic iridium complexes react fastest. In non-polar solvents *trans* addition products result; this is consistent with the mechanism:

179

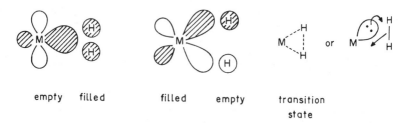

not detected

Most other alkyl halides, however, follow a free radical pathway (see general scheme in Table 5.8) rather than the S_N2 mechanism.

Among other essentially electrophilic substrates which add to Vaska's compound are halogens and acyl halides. Dioxygen forms a complex in which the O—O unit forms part of a three membered ring; it may be considered to be a peroxo compound. The addition can be reversed by flushing the solution with nitrogen. This is an example of an oxygen carrying system which may be compared with naturally occurring oxygen carriers such as haemoglobin.

An exciting discovery, reported by Vaska and Di'Luzio in 1962, was that molecular hydrogen reacts with a solution of $(Ph_3P)_2Ir(CO)Cl$ at 1 atm pressure and 25°C to give a *cis*-dihydride. At that time transition metal hydride complexes were rare and many were unstable. The formation of a hydride from hydrogen under such mild conditions was quite new. It has since been shown that such reactions are typical of electron rich coordinatively unsaturated complexes. The *cis* stereochemistry of the adduct has been established by infrared and ¹H n.m.r. spectroscopy. There are two bands of similar intensity in the infrared spectrum which can be assigned to Ir–H stretching of a *cis* IrH_2 grouping. The Ir–H protons give rise to signals at $-7.3\,\delta$ and $-18.4\,\delta$ in the region expected for transition metal hydrides. Both are doublets of triplets which arise through coupling to two equivalent ³¹P $(I = \frac{1}{2})$ nuclei and to the other (inequivalent) proton.

The addition of hydrogen to Vaska's compound is probably a concerted reaction. The hydrogen molecule is normally susceptible neither to nucleophilic nor to electrophilic attack. The metal atom however has both nucleophilic and also electrophilic character. Interaction is thought to occur between a filled metal orbital and the antibonding H_2 m.o. and also between an empty metal orbital and the bonding H_2 m.o. Similar mechanisms may be followed in additions of other relatively non-polar species such as R_3SiH.

empty   filled   filled   empty   transition state

Recently complexes of molecular hydrogen have been discovered. One of these was prepared by the reaction:

$$W(CO)_3L_2 \xrightleftharpoons{H_2/1atm}$$

16e

$$L = PCy_3$$

It forms yellow crystals which lose hydrogen readily; it must therefore be kept under hydrogen. The structure, determined at $-100°C$ by X-ray diffraction and at room temperature by neutron diffraction, shows that direct H—H bonding is retained. Moreover, in the n.m.r. spectra of the HD complex, coupling between ^{1}H and ^{2}D nuclei is observed ($J_{HD} = 33.5$ Hz; cf. HD(g), 43.2 Hz). Protonation of the polyhydride IrH$_5$L$_2$ (L = PCy$_3$) affords a cationic species in which two H$_2$ molecules are bonded to the metal centre.

$$IrH_5L_2 \xrightleftharpoons[NEt_3]{H^+} [IrH_2(H_2)_2L_2]^+ \xrightarrow[-2H_2]{MeCN} [IrH_2(MeCN)_2L_2]^+$$

(b) TETRAKIS(TRIPHENYLPHOSPHINE)PALLADIUM AND -PLATINUM. Nickel, palladium and platinum form neutral tetrahedral phosphine complexes ML$_4$ (L = PR$_3$). They are prepared by reduction of metal salts in the presence of excess ligand. The platinum and palladium complexes, M(PPh$_3$)$_4$, have been most thoroughly studied. They tend to dissociate in solution to three-coordinate and even to two-coordinate species. The extent of dissociation depends on the steric bulk of the phosphine (p. 168). While no dissociation of Pd(PMe$_3$)$_4$ can be detected in solution by ^{31}P n.m.r., Pd(PPh$_3$)$_4$ forms Pd(PPh$_3$)$_3$ and a trace of Pd(PPh$_3$)$_2$. The very bulky ligand PPhBu$_2^t$ affords a monomeric 14-electron complex

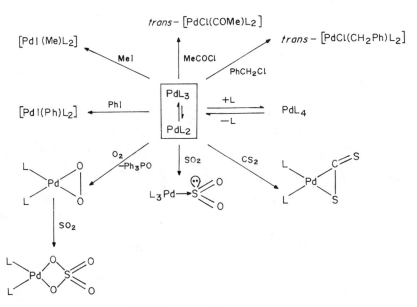

Fig. 5.18 Some reactions of Pd(PPh$_3$)$_3$, L = PPh$_3$.

Pd(PPhBu$_2^t$)$_2$ which shows little tendency to add further ligands or undergo addition reactions.

The complexes ML$_4$ add molecules X—Y in which two ligands are lost and a square planar product results:

$$ML_4 \rightleftharpoons ML_3 \rightleftharpoons ML_2 + XY \rightarrow MX(Y)L_2$$

Acids HX react with Pt(PPh$_3$)$_3$ to give salts [PtH(PPh$_3$)$_3$]$^+$ X$^-$ if X$^-$ is weakly coordinating, but otherwise Ph$_3$P is displaced, yielding neutral *trans* derivatives:

$$Pt(PPh_3)_3 \underset{KOH}{\overset{HX}{\rightleftharpoons}} [PtH(PPh_3)_3]^+X^- \underset{+PPh_3}{\overset{-PPh_3}{\rightleftharpoons}} trans\ PtHX(PPh_3)_2 \quad X = CN, Cl$$

Some reactions of Pd(PPh$_3$)$_4$ are given in Fig. 5.18. The platinum derivative behaves similarly. Sulphur dioxide acts as a Lewis acid. The reactions with alkyl halides usually follow the S$_N$2 mechanism, although free radical pathways can also participate. The reactions of aryl halides, which are normally rather unreactive towards nucleophiles are of interest. The palladium complexes catalyse the arylation of alkenes (the Heck reaction) by the following cycle:

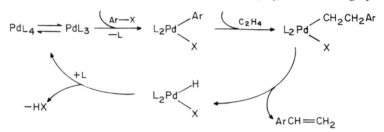

(c) WILKINSON'S COMPLEX – CATALYTIC HYDROGENATION OF ALKENES AND ALKYNES. Wilkinson's compound, RhCl(PPh$_3$)$_3$, is readily prepared by heating RhCl$_3$.3H$_2$O under reflux in ethanol with an excess of triphenylphosphine. It is usually obtained as red-violet crystals, but there is also a metastable orange modification. The molecule is distorted from ideal square planar geometry, presumably on account of steric interactions. Many of its reactions involve addition to the coordinatively unsaturated rhodium centre. The products are sometimes 18-electron species, but often loss of one phosphine ligand occurs, so that a 16-electron derivative results (as with CO or C$_2$H$_4$) (Fig. 5.19).

Dilute solutions ($\sim 10^{-3}$ mol dm^{-3}) of Wilkinson's complex in organic solvents such as ethanol/benzene actively catalyse the hydrogenation of alkynes and alkenes at ambient temperature and pressure (1 atm, H$_2$). Traces of oxygen must be excluded. Hydrogenation rates are dependent on steric effects. The following general order has been established: 1-alkenes > *cis*-2-alkenes > *trans*-2-alkenes > *trans*-3-alkenes. An exception is ethene itself, which forms a rather stable complex (Fig. 5.19) so that higher temperatures are required. Arenes, esters, ketones, carboxylic acids, amides and nitro compounds are unaffected, but aldehydes are slowly decarbonylated.

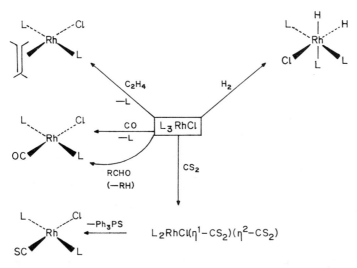

Fig. 5.19 Some reactions of Wilkinson's compound, L_3RhCl, $(L = PPh_3)$.

Tolman proposed two general rules relating to organometallic reactions and homogeneous catalysis.

Rule 1: The only metal-containing species which can exist in significant concentration possess either 16 or 18-electron configurations.

Rule 2: Organometallic reactions, including catalytic ones, proceed by elementary steps which involve 16 or 18-electron intermediates.

Some exceptions have already been noted; with Pd or Pt, for example, the 16-electron square planar complexes behave to some extent as if they are 'saturated' and 14-electron species are sometimes observed and are likely reaction intermediates. Moreover odd-electron free radical intermediates also sometimes intervene. Nevertheless the rules provide a very useful starting point for formulating possible mechanisms for catalytic cycles.

Hydrogenation using Wilkinson's catalyst has been studied in great detail and some features are still not understood. A simplified scheme, based on Tolman's rules is given in Fig. 5.20.

The dihydride L_3RhH_2Cl arises from *cis* addition of H_2 to L_3RhCl and has the structure shown in Fig. 5.19. The catalytically active species, however, has only two phosphine ligands and may be five-coordinate. Although isomer A is of lower energy than B, the latter is thought to be involved in the actual hydrogenation. Addition of alkene is followed by hydrogen transfer, the slowest stop of the whole cycle, to give a rhodium alkyl hydride (C). Elimination of alkane from C is very fast. A problem with the scheme appears at this point. Does C give a 14-electron species L_2RhCl (in contravention of the rules) or does a ligand enter? It has been suggested that the L_2RhCl is stabilized by coordination of a solvent molecule; this is supported by the observation that hydrogenation is about twice as fast in

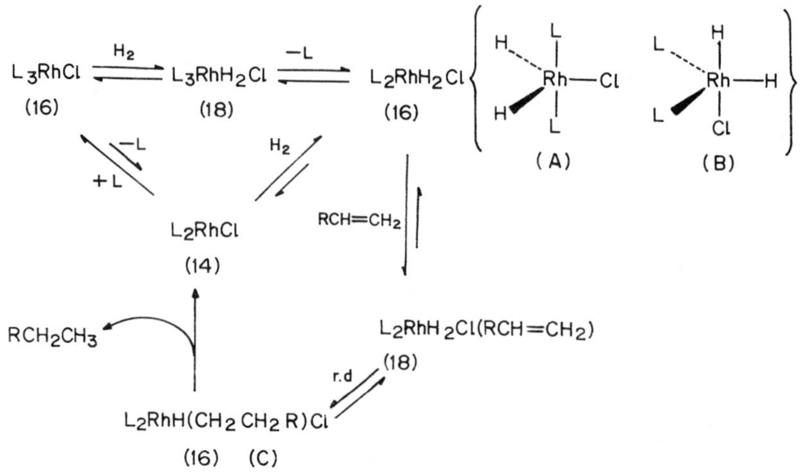

$L_3RhCl \xrightleftharpoons{H_2} L_3RhH_2Cl \xrightleftharpoons{-L} L_2RhH_2Cl$

(16) (18) (16)

L_2RhCl

(14)

$RCH=CH_2$

RCH_2CH_3

$L_2RhH_2Cl(RCH=CH_2)$

(18)

r.d

$L_2RhH(CH_2CH_2R)Cl$

(16) (C)

(A) (B)

Fig. 5.20 Simplified catalytic cycle for hydrogenation of an alkene using Wilkinson's catalyst. (16), (18) mean 16-, 18-electron etc. Unsaturated species, e.g. L_2RhCl may have additional solvent ligands. r.d. = rate determining step. $L = PPh_3$.

benzene/ethanol as in pure benzene. Whether or not solvent coordination is important the inner cycle is favoured from kinetic evidence. Hydrogen addition to L_3RhCl is strongly inhibited by excess triphenylphosphine. While the equilibrium $L_3RhCl \rightleftharpoons L_2RhCl + L$ lies well the left ($K \sim 10^{-5} \text{ mol dm}^{-3}$) the concentration of L_2RhCl could still be kinetically significant.

(d) ASYMMETRIC HYDROGENATION. The conversion of an achiral precursor stereospecifically into one enantiomer, rather than into a racemic mixture, is often a desirable goal in the pharmaceutical industry. Enantiomers can behave very differently in biological systems. Thus S-asparagine has a bitter taste, whereas the R-isomer is sweet. Thalidomide presents a particularly disastrous example. While the R-isomer aids sleep,the S-isomer is teratogenic. The very building blocks used by nature are themselves chiral. Only L-amino acids occur in proteins and D-sugars in DNA.

One way of accomplishing an asymmetric synthesis is to use a chiral reagent or catalyst. Rhodium complexes which contain optically active tertiary phosphines as ligands (p. 127) are, like Wilkinson's compound, active catalysts for the hydrogenation of alkenes. Moreover, in favourable cases, one enantiomer can be produced selectively in high optical purity. The drug L-DOPA which is used to control Parkinson's disease is made by hydrogenation of an acyl-aminopropenoic acid (Monsanto process). The catalyst contains a chiral phosphine such as DIPAMP. In spite of the cost of the catalyst the process is much more viable than one in which resolution of a racemic mixture of D- and L-isomers is required. Rhodium can be recovered, so that the main outlay is in synthesizing the chiral phosphines.

184

Synthesis of DIPAMP

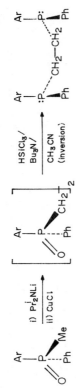

$$\text{Ar}-\underset{\underset{O}{\|}}{P}\overset{Me}{\underset{Ph}{\diagdown}} \xrightarrow[\text{ii) CuCl}]{\text{i) Pr}^i_2\text{NLi}} \left[\text{Ar}-\underset{\underset{O}{\|}}{P}\underset{Ph}{\diagdown}\text{CH}_2\right]_2 \xrightarrow[\substack{\text{CH}_3\text{CN} \\ \text{(inversion)}}]{\text{HSiCl}_3/ \\ \text{Bu}_3\text{N}/} \underset{Ph}{\overset{Ar}{\diagdown}}:P\text{---CH}_2\text{---CH}_2\underset{Ph}{\overset{Ar}{\diagup}}P:$$

DIPAMP. (Ar=2−MeOC$_6$H$_4$)

Hydrogenation

$$\text{Ar'CH}=\underset{\text{NHCOCH}_3}{\overset{\text{CO}_2\text{H}}{\text{C}}} \xrightarrow[\substack{\text{DIPAMP, 50°C} \\ \text{MeOH,H}_2(3\text{atm})}]{[\text{RhCl(1,5-C}_8\text{H}_{12})]_2/} \text{Ar'CH}_2\overset{*}{\underset{\text{NHCOCH}_3}{\text{CH}}}\text{CO}_2\text{H} \longrightarrow \text{Ar'CH}_2\overset{*}{\underset{\text{NH}_2}{\text{CH}}}\text{CO}_2\text{H}$$

L−DOPA
(98% optical purity)

Ar'= HO

MeO

Phosphines in which the seat of optical activity is in the attached organic groups rather than at phosphorus have also been used in asymmetric hydrogenation. They are easier to synthesize but the optical purity of the hydrogenated products is usually not so high.

(e) DEHYDROGENATION OF ALKANES. Can an alkane be dehydrogenated catalytically under mild homogeneous conditions? There are several reasons why this does not normally occur. Alkanes are poor ligands which cannot compete with better ligands such as Ph_3P which are present. Anyway, the chelate effect favours orthometallation of such ligands (p. 223) in preference to the addition of alkane. Moreover the reaction

$$C_2H_6(g) \rightleftharpoons C_2H_4(g) + H_2(g) \qquad \Delta H^\ominus = 137\,kJ\,mol^{-1}; \ \Delta S^\ominus = 121\,JK^{-1}\,mol^{-1}$$

is thermodynamically unfavourable below about $850°C$. In principle, however, it can be brought about by adding a hydrogen acceptor such as another alkene. Unfortunately this usually coordinates to the metal in preference to an alkane. Sometimes the alkene itself is dehydrogenated.

Crabtree has used 3, 3-dimethylbutene as an acceptor in conjunction with the cationic iridium complexes $[IrH_2S_2L_2]^+BF_4^-$ (S = e.g. acetone, CH_2Cl_2; $L = Ph_3P$). The bulky *tert*-butyl group in the alkene makes it a poor ligand. Orthometallation of triphenylphosphine does seem to occur, but it is reversible. Stoichiometric reaction of cycloalkanes is observed on heating in dichloromethane. Unlike some other compounds which activate alkanes (p. 224), these complexes are not electron rich. They are cationic and do not contain strongly donating ligands.

Felkin has shown that $(Ph_3P)_2ReH_7$ behaves similarly. Some reactions are catalytic, but with low turnover numbers.

5.4 References

Albers, M.O. and Coville, N.J. (1984) Reagent and catalyst induced substitution reactions of metal carbonyls. *Coord. Chem. Revs.*, **53**, 227.

Atwood, J.D. (1983) Ligand effects on organometallic substitution reactions. *Acc. Chem. Res.*, **16**, 350.

Basolo, F. (1985) Associative substitution reactions of metal carbonyls. *Inorg. Chim. Acta*, **100**, 33.

Brunner, H. (1986) Enantioselective catalysis by transition metal complexes. *J. Organomet. Chem.*, **300**, 39.

Connor, J.A. (1977) Thermochemical studies on organotransition metal carbonyls. *Topics Curr. Chem.*, **71**, 71.

Cross, R.J. (1985) Ligand substitution of square-planar molecules. *Chem. Soc. Revs.*, **14**, 197.

Geoffroy, G.L. and Wrighton, M.S. (1979) *Organometallic Photochemistry*, Academic Press, New York.

Hieber, W. (1970) A personal account of work by the father of carbonyl chemistry. *Adv. Organomet. Chem.*, **8**, 1.

Horwitz, C.P. and Shriver, D.F. (1984) C and O-Bonded carbonyls. *Adv. Organomet. Chem.*, **23**, 219.

Nixon, J.F. (1985) Trifluorophosphine complexes. *Adv. Inorg. Chem. Radiochem.*, **29**, 41.

Pearson, R.G. (1985) The transition metal-hydrogen bond. *Chem. Revs.*, **85**, 41.

Tolman, C.A. (1972) The 16- and 18-electron rules. *Chem. Soc. Revs.*, **1**, 337.

Problems

1. What are the structures of each of the following complexes? Show how either the 18 or the 16-electron rule applies in each case.

(a) $IrCl(CO)(PPh_3)_2$ (b) $IrCl_2H(CO)(PPh_3)_2$
(c) $Fe(CO)_2(NO)_2$ (d) $PtClH(PEt_3)_2$

(*University of York* (modified)).

2. Give the structures of the organometallic products of the following reactions, indicating a plausible mechanism for each.

(a) $PhMn(CO)_5 + Me_3NO + Ph_3P \xrightarrow{CH_2Cl_2}$

(b) $Fe(CO)_5 + aq.\ alkali \longrightarrow$

(c) $Re(CO)_6^+ + H_2{}^{18}O \longrightarrow$

(d) $(\eta^5\text{-}C_5H_5)Mn(CO)_3 + Ph_3P \xrightarrow[\text{THF}]{h\nu}$

(e) $MeCOMn(CO)_5 \xrightarrow[\text{vacuum}]{\text{heat under}}$

(*University of Sheffield*).

3. A white product in the form of long needles was obtained from the reaction between CO and $PtCl_2$ at 120° and at high pressure. The molecular weight was found to be 322 and the percentage of chlorine was determined as 21.8.

(a) Find the formula of the compound and suggest the types of isomers which may be found with such a formula.

(b) Comment on the probable relative MC and CO bond lengths in these isomers.

(c) Give the point groups of the isomers and, in each case, the irreducible representations of the vibrations of the compound. How may vibrational spectra be used to distinguish between the isomers?

(d) Comment on the observed dipole moment of 4.9 D.
(*University of East Anglia*).

4. The mass spectrum of a yellow, diamagnetic and volatile solid A reveals a molecular ion at m/z = 390 and a fragmentation pattern consisting of sequential losses of 28 mass units until m/z = 110. Treatment of A with chlorine affords a white crystalline solid B whose i.r. spectrum in the region 1700–2100 cm^{-1} is quite similar, but not identical, to that of A. Reaction of B with AlCl$_3$ under 10 atm CO pressure affords, after appropriate work-up, a water soluble salt C which, on treatment with NH$_4$PF$_6$, gives the crystalline salt D. Reaction of D with LiCl in acetone caused regeneration of B. Treatment of A with sodium amalgam in THF affords an air-sensitive salt E which, on acidification with H$_3$PO$_4$, gives a colourless volatile material F. The ^{1}H n.m.r. spectrum of F exhibits a single resonance at $\delta = -4$ ppm, and the i.r. spectra of B and F are very similar. On heating F, a non-condensible gas and A are formed. Reaction of E with methyl iodide affords a white solid G whose ^{1}H n.m.r. spectrum consists of a single resonance near $\delta = 0.2$ ppm. If the formation of G is carried out under 2 atm CO pressure a similar species H is formed, but the ^{1}H n.m.r. spectrum of H reveals that there is a single resonance near $\delta = 1.2$ ppm. The i.r. spectrum of H is similar to that of G but has an additional absorption at 1740 cm^{-1}.

Identify A, B, C, D, E, F, G and H and account for the sequence of reactions.
(*University of Sheffield*).

5. A metal M forms a carbonyl $M_x(CO)_y$ which contains 18.40% of carbon and in the mass spectrometer shows a molecular ion, the isotopic pattern of which shows its most intense peak at m/z = 652. The carbonyl reacts with triphenylphosphine to produce two compounds; A, which has a metal—phosphorus ratio of 1:1, and B, for which the corresponding ratio is 2:1. A reacts with bromine to form a single product C, whereas B with bromine produces C and D. D, which is identical with the sole product from the reaction of $M_x(CO)_y$ with bromine, reacts with triphenylphosphine to produce E, which is isomeric with C. C and D, but not E, have a C$_4$ axis of symmetry. All the compounds are diamagnetic.

Draw conclusions from each piece of evidence, deduce the identity of M, the values of x and y and the structures of A, B, C, D and E.
(*University of Exeter*).

6. Explain the following observations:
(a) The infrared spectrum of F$_3$SiCo(CO)$_4$ shows three bands at 2128, 2073 and 2049 cm^{-1}, which are assigned to carbonyl stretching modes. The corresponding bands in the spectrum of Me$_3$SiCo(CO)$_4$ are at 2100, 2041 and 2009 cm^{-1}. (Assume that the complexes have trigonal bipyramidal geometry. Treat the SiX$_3$ group as a sphere.)
(b) Hydrogen adds to the complex *trans* IrCl(CO)(PMe$_2$Ph)$_2$ [ν(Ir—Cl), 311 cm^{-1}] to give a product Q[ν(Ir—Cl), 249 cm^{-1}]. The ^{1}H n.m.r. spectrum of Q

shows peaks at δ/relative intensity, multiplicity: $\delta 7$–8, 10, multiplet; $\delta 1.89$, 6, triplet; $\delta 1.86$, 6, triplet; $\delta - 7.58$, 1, triplet of doublets; $\delta - 18.36$, 1, triplet of doublets. (I $^{31}P = \frac{1}{2}$).
(*University of Southampton*).

7. P visited a neighbouring University to lecture on his recent discovery of a new octahedral carbonyl complex of formula $M(CO)_2L_4$. He said that it was the *cis* isomer because an infrared spectrum of a solution of it showed two bands in the carbonyl stretching region. His rival Q said that, in his opinion, the *trans* isomer was more likely and that the method of preparation described by P would give a product contaminated with unchanged starting material, the octahedral complex $M(CO)_6$.

Explain why Q's suggestion fits the infrared evidence, and show how the argument could be settled using Raman spectroscopy.
(*University of Leicester*).

8. M belongs to the First Transition Series. It forms a carbonyl of empirical formula $M(CO)_5$ (*A*). Treatment of *A* with iodine in CS_2 produces *B*, empirical formula $MI(CO)_5$. On heating in petroleum spirit to $120°C$, *B* yields *C*, empirical formula $MI(CO)_4$. In hot pyridine (py), compound *C* gives *D*, molecular formula *fac* $Mpy_2I(CO)_3$. The compounds *A*, *B*, *C* and *D* all exhibit strong absorptions in their infrared spectra at about $2000 \, cm^{-1}$, but none near $1750 \, cm^{-1}$.

Identify element M. Propose and draw structures for the compounds *A*, *B* and *C*, showing your reasoning in detail. Why does *D* have a *fac* geometry?
(*University of Leicester*).

9. Suggest structures for the compounds *A* to *E* in the following reactions. Discuss fully the data given and show how it is consistent with your structural assignments.

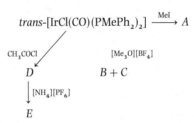

Analysis data for *B* and *E* gave:
B: 44.4% C, 10.0% F
E: 13.5% F
Other studies for *A*, *B*, *C*, *D* and *E* gave:
A: ^1H n.m.r. resonances at $\delta 7.43$ (multiplet, 20H), $\delta 1.79$ (triplet, 6H), and $\delta 0.88$ (triplet, 3H); $v_{IrCl} = 300 \, cm^{-1}$; $v_{CO} = 2050 \, cm^{-1}$.
B and E are 1:1 electrolytes in CH_3CN.
C: showed a molecular ion in its mass spectrum at 46 amu.

189

D: ^{1}H n.m.r. resonances at $\delta 7.45$ (multiplet, 20H), $\delta 1.80$ (triplet, 6H) and $\delta 1.7$ (singlet, 3H); $v_{IrCl} = 243$ and $315\,cm^{-1}$, $v_{CO} = 2049$ and $1618\,cm^{-1}$.

E: ^{1}H n.m.r. resonances at $\delta 7.40$ (multiplet, 20H), $\delta 1.89$ (triplet, 6H) and $\delta 0.89$ (triplet, 3H); $v_{IrCl} = 310\,cm^{-1}$, $v_{CO} = 2045$ and $2000\,cm^{-1}$.

Discuss the mechanism of formation of A and indicate how MeCl would have reacted with *trans*-[IrI(CO)(PMePh$_2$)$_2$].

(*University of Southampton*).

10. Addition of D$_2$ to Z-butenedioic acid (maleic acid) catalysed by (Ph$_3$P)$_3$RhCl gives RS-1,2-dideuteriosuccinic acid. On the other hand E-butenedioic acid (fumaric acid) gives RR- + SS-1,2-dideuteriosuccinic acid. (Succinic acid is 1,4-butanedioic acid.)

When 1-hexene is reduced by a mixture of H$_2$ and D$_2$ in the presence of (Ph$_3$P)$_3$RhCl, only CH$_2$DCHDC$_4$H$_9$ and CH$_3$CH$_2$C$_4$H$_9$ but no CH$_2$DCH$_2$H$_4$H$_9$ or CH$_3$CHDC$_4$H$_9$ are produced.

What can be deduced about the mechanism of the hydrogenation from these results?

6 Organometallic compounds of the transition elements. Classification of ligands and theories of bonding

6.1 Classification of ligands

While organometallic compounds of the Main Group elements are most conveniently classified from the position of the metal in the Periodic Table, it is useful to classify d-block transition metal complexes from the organic ligands which they contain. In Table 6.1 ligands are listed according to the number of electrons which they formally contribute to the metal—ligand bonding. While there is usually no doubt as to the total number of valence electrons in a complex, confusion in counting electrons can arise if attempts are made to take into account any polarization of the metal—ligand bonds, by assigning an oxidation state to the metal, for example. A metal—hydrogen bond can be thought of as arising from M^- and H^+, $M\cdot$ and $H\cdot$ or M^+ and H^-. In all cases the same essentially covalent bond results, which contains the same number of electrons (two), independent of the final polarization $M^{\pm}\!-\!H^-$ or $M\!-\!H^+$. It is therefore recommended that in counting electrons, all metal—ligand bonds are treated as covalent. Thus an M—H bond is considered as arising from a metal atom $M\cdot$ and the radical $H\cdot$, so that $H\cdot$ is formally a one electron donor. Metal alkyls and less conventionally metal halides are treated similarly. A metal cyclopentadienyl bond $M\!-\!C_5H_5$ is also considered as arising from $M\cdot$ and a cyclopentadienyl radical $C_5H_5^{\cdot}$ rather than from M^+ and $C_5H_5^-$, making C_5H_5 formally a five electron ligand.

Ligands are also classified by their hapto number (η^n) (from Greek ἅπτω, I fasten). The hapto number is the number of ligand atoms within bonding distance of the metal. For unsaturated hydrocarbon ligands the maximum hapto number is equal to the number of carbon atoms in the unsaturated system. Each 'hapto' carbon atom formally contributes one electron to the metal—ligand bonding, so that here the hapto number and the electron number are the same. (This may not apply when heteroatoms are present e.g. in $(\eta^5\text{-}C_4H_4S)Cr(CO)_3$ thiophene is a six-electron ligand).

Table 6.1 A classification of ligands

Number of electrons	Name of ligand class	Examples of group
1	-yl	η^1-alkyl, aryl cf. Cl·, ·CN, H· η^1-alkenyl M—CH=CH$_2$; Terminal M—COR η^1-alkynyl M—C≡CR; M—NR$_2$; M—OR η^1-allyl M—CH$_2$CH=CH$_2$
2	-ene	η^2-ethene cf. CO, N$_2$ η^2-ethyne; Lewis bases in general, R$_3$N, R$_3$P, H$_2$O etc. Carbene (R$_2$CM) or alkylidene (R$_2$C=M)
3	-enyl	η^3-allyl; NO, various bidentate groups e.g. acac η^3-benzyl
4	-diene	η^4-butadiene η^4-cyclobutadiene η^4-alkyne (p. 245)
5	-dienyl	η^5-cyclopentadienyl η^5-pentadienyl η^5-cyclohexadienyl
6	-triene	η^6-benzene, η^6-cycloheptatriene
7	-trienyl	η^7-cycloheptatrienyl
8	-tetraene	η^8-cyclooctatetraene

Many isolable transition metal complexes which contain organic groups, CO or hydrogen as ligands obey the 18-electron rule. The application of this rule to metal carbonyls is discussed in Chapter 5. To count electrons in an organometallic compound the number of electrons in the valence shell of the metal is added to the total number of electrons contributed by the ligands (Table 6.1). If the complex bears a charge, this is subtracted from the metal electron count. Thus iron in a cation of charge $+1$ contributes $8 - (+1) = 7$, in a neutral compound 8 and in an anion of charge -1, $8 - (-1) = 9$ electrons.

Two series of iso-electronic 18-electron complexes are the sandwich compounds

$$(\eta^6\text{-C}_6\text{H}_6)_2\text{Cr} \quad (\eta^5\text{-C}_5\text{H}_5)(\eta^6\text{-C}_6\text{H}_6)\text{Mn} \quad (\eta^5\text{-C}_5\text{H}_5)_2\text{Fe}$$
$$6+6+6 \qquad\qquad 5+6+7 \qquad\qquad 5+5+8$$
$$(\eta^4\text{-C}_4\text{Ph}_4)(\eta^5\text{-C}_5\text{H}_5)\text{Co} \quad (\eta^4\text{-C}_4\text{Ph}_4)_2\text{Ni}$$
$$4+5+9 \qquad\qquad 4+4+10$$

and the 'piano-stool' tricarbonyls

$$(\eta^7\text{-C}_7\text{H}_7)\text{V(CO)}_3 \quad (\eta^6\text{-C}_6\text{H}_6)\text{Cr(CO)}_3 \quad (\eta^5\text{-C}_5\text{H}_5)\text{Mn(CO)}_3 \quad (\eta^4\text{-C}_4\text{H}_4)\text{Fe(CO)}_3$$
$$7+5+6 \qquad\qquad 6+6+6 \qquad\qquad 5+7+6 \qquad\qquad 4+8+6$$
$$(\eta^7\text{-C}_7\text{H}_7)\text{Cr(CO)}_3^+ \quad (\eta^6\text{-C}_6\text{H}_6)\text{Mn(CO)}_3^+ \quad (\eta^5\text{-C}_5\text{H}_5)\text{Fe(CO)}_3^+ \quad (\eta^5\text{-C}_5\text{H}_5)\text{Cr(CO)}_3^-$$

Some ligands can bond in more than one way to transition elements, particularly those with more than one double bond or with extended π-systems.

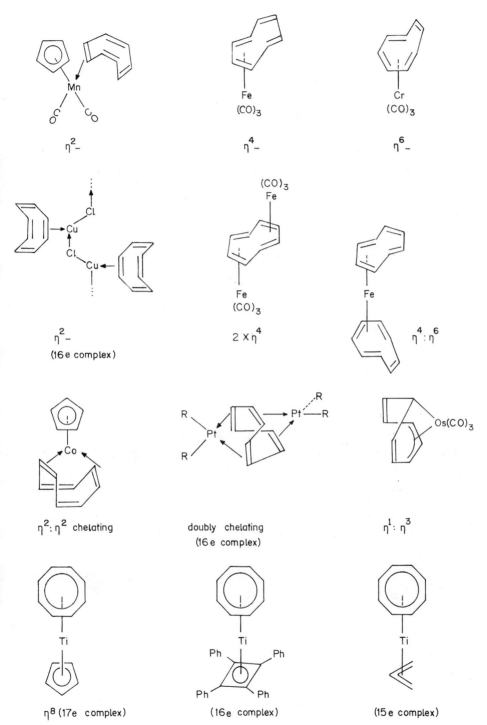

Fig. 6.1 Some complexes of cyclooctatetraene.

Organometallic compounds of the transition elements

The 18-electron rule is often of value in predicting likely structures for their complexes. 1, 3-Butadiene, for example forms the iron carbonyls $(C_4H_6)Fe(CO)_4$ and $(C_4H_6)Fe(CO)_3$. In the former the butadiene is η^2, linked to iron through only one of the double bonds, whereas in the latter it is η^4 bonded.

	C_4H_6	2 e
	Fe	8 e
Fe	4 CO	8 e
$(CO)_4$		18 e

	C_4H_6	4 e
	Fe	8 e
Fe	3 CO	6 e
$(CO)_3$		18 e

Cyclooctatetraene is an extremely versatile ligand in this respect. Some examples of the different modes of bonding it exhibits are illustrated in Fig. 6.1. The room temperature ^{1}H n.m.r. spectrum of $(C_8H_8)Fe(CO)_3$ shows only one signal, but the symmetrical η^8 ligand this suggests would give an electron count of 22. X-ray diffraction, however, reveals a η^4 structure consistent with the 18-electron rule. The explanation is that the $Fe(CO)_3$ group is rapidly switching its position between the four double bonds, so that an averaged n.m.r. signal is observed.

Most of the examples in Fig. 6.1 obey the 18-electron rule although some exceptions are shown. The platinum compound has a 16-electron configuration, common for many planar complexes late in the transition series (p. 174). Complexes of the early transition elements also often depart from the rule; the metal valence orbitals are relatively high in energy (especially $4p$) leading to greater ionic character in the bonding.

For some compounds it is possible to write down two or more structures which obey the 18-electron rule. A decision between these can be made only in the light of further evidence. The structure of $C_8H_8Fe(CO)_3$ could be written with an η^4-cyclooctatetraene ligand (as determined by X-ray diffraction) or with an η^2, η^2-chelating ligand (as found in $C_5H_5CoC_8H_8$). In fact both isomers of the cyclooctadiene analogue, $(\eta^4$-1, 3-$C_8H_{12})Fe(CO)_3$ and $(\eta^2:\eta^2$-1, 5-$C_8H_{12})Fe(CO)_3$ are known, although the latter isomerizes quite easily to the former. Dipolar structures can usually be excluded. Ready isomerization between the η^6 and η^5 forms of some fluorenyl complexes, however, is observed.

Fe
$(CO)_3$

Extreme dipolar
representation of the
bonding in $C_8H_8Fe(CO)_3$
(not correct)

Cr
$(CO)_3$

η^6

Cr$^-$
$(CO)_3$

η^5

Isomers of [fluorenyl $Cr(CO)_3$]$^-$

6.2 Molecular orbital theory

The bonding of unsaturated organic ligands to transition metals is conveniently discussed in terms of molecular orbital theory. For significant interaction to occur (a) metal and ligand orbitals must be of similar energies (this energy requirement is met by metal ns, np and $(n-1)d$ and ligand $C(2p)\pi$ orbitals) and (b) the symmetry properties of the metal and ligand orbitals must be the same. By considering the symmetry of a molecule and by applying group theory, it is possible to deduce which orbitals of the metal and which of the ligands fulfil the second condition. We concentrate first on one symmetry element, an axis of rotation which passes through the metal atom and lies perpendicular to the plane of the ligand. This is defined as the z-axis. Metal and ligand orbitals can be classified as having σ, π or δ symmetry (the last applies only if the rotational symmetry of the molecule is > 2) with respect to this axis. The phase of the wave function of a σ-orbital does not change sign on rotation through $180°$ about this symmetry axis. With a π-orbital the sign changes once, while with a δ-orbital it changes twice. This division into σ, π and δ-orbitals is further illustrated in Fig. 6.7.

As only ligand and metal orbitals of the same symmetry can combine to form bonding and antibonding molecular orbitals it is often possible through symmetry considerations to develop a qualitative picture of the bonding in a molecule. We confine ourselves here to the valence orbitals of the metal and the p molecular orbitals of the ligand. Orbitals of similar energy interact more strongly than those of disparate energy. Estimates of the energies of the interacting metal and ligand orbitals are obtained from spectroscopic measurements (UV/visible spectra, UV-photoelectron spectra). The ordering of molecular orbitals in a complex is deduced from the results of molecular orbital calculations in conjunction with spectroscopic (especially UV-photoelectron spectra) and magnetic data. Even though more and more powerful computers are becoming available, molecular

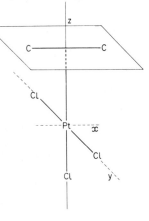

Fig. 6.2 Structure of the anion $[C_2H_4PtCl_3]^-$.

Table 6.2. Symmetry of metal orbitals with respect to rotation about z-axis (metal–ligand axis).

Symmetry	Metal Orbitals
σ	s, p_z, d_{z^2}
π	p_x, p_y, d_{zx}, d_{yz}
δ	$d_{x^2-y^2}$, d_{xy}

orbital calculations on large molecules still involve considerable approximations. The general principles of bonding in η-complexes, however, have been clarified using fairly simple computational methods and symmetry arguments. The discussion which follows is based on these results.

6.2.1 The ethene—metal bond

The anion of Zeise's salt, $K^+[(C_2H_4)PtCl_3]^-$, has the structure shown in Fig. 6.2.

The platinum atom lies in a square planar coordination sphere; the C=C double bond axis lies perpendicular to this plane. The anion has C_{2v} symmetry; the z-axis is placed coincident with the C_2 axis. In the following discussion it is assumed that the electrons which form the σ-bond skeleton of the ethene play no part in the bonding to the metal. The ethene—metal bond is formed by interaction of $2p_z$ ($2p\pi$) orbitals of the carbon atoms and metal orbitals of suitable symmetry. It must, however, be stressed that this is often a gross simplification.

The different symmetry allowed combinations of orbitals are shown in Fig. 6.3. Because of the low rotational symmetry (C_2) the weak interaction involving the

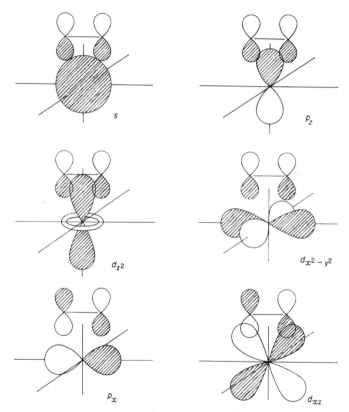

Fig. 6.3 Diagrammatic representation of types of orbital overlap between ethylene $2p_z$ atomic orbitals and ns, np and $(n-1)d$ orbitals of a transition metal.

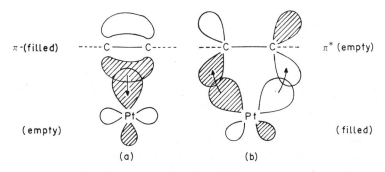

π-(filled) π^* (empty)

(empty)   (filled)

(a)   (b)

Fig. 6.4 Conventional representation of the metal—alkene bond. (a) Alkene → M σ-donation; (b) M alkene π-back donation.

$d_{x^2-y^2}$ orbitals is of σ-rather than of δ-symmetry.

An older but still useful view of the bonding is given in Fig. 6.4. This shows σ-donation from the filled bonding ($p_z\pi$) molecular orbital of ethene into an empty hybrid metal orbital and back-donation from a filled hybrid metal orbital into the empty antibonding ($p_z\pi^*$) m.o. of ethene. This description resembles the situation in metal carbonyls (p. 157). Like CO ethene is too weak a σ-donor to form adducts with typical Lewis acids such as BMe$_3$, which complex with ammonia. Unlike ammonia, however, CO and C$_2$H$_4$ possess empty orbitals of low energy which can accept electrons from filled d-orbitals of transition elements. It is difficult to evaluate the separate contributions from σ- and π-components in a metal—alkene bond. Both removal of electrons from the bonding m.o. of ethene and also electron donation into the antibonding m.o. lead to a weakening of the C=C bond. In agreement with this the frequency of the C=C stretching vibration falls by 60 to 150 cm^{-1} on complexing. In K$^+$[C$_2$H$_4$PtCl$_3$]$^-$ this band is at 1511 cm^{-1} in the infrared spectrum; in the Raman spectrum of free ethene the corresponding band is at 1623 cm^{-1}. Moreover the C=C bond length in Zeise's salt is 1.354 Å (cf. 1.337 Å in ethene).

The relative proportions of alkene → M σ-donation and M → alkene π-back-donation are influenced by the electron density at the metal centre. In Zeise's salt the σ-component is thought to be dominant. Where electron withdrawing ligands

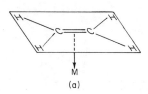

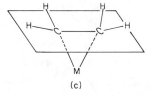

(a)   (c)

Planar alkene
(sp^2 carbon)
σ-donation dominant

Alkene substituents bent
up out of plane (sp^3 carbon)
π-back donation dominant.

Fig. 6.5 Two extreme representations of the metal—alkene bond.

197

such as Cl or CO are present the σ-component will be enhanced and the π-component diminished relative to cases where the ligands are electron donating. Alkenes with electron attracting substituents such as F, CN or COOH are poorer σ-donors and better π-acceptors than the parent alkenes. The lengthening of the C=C bond in an alkene on complex formation is particularly marked where strong π-back donation is expected. This is well illustrated by measurements on the complex $(\eta^5\text{-}C_5H_5)Rh(C_2H_4)(C_2F_4)$. A high degree of π-back donation is accompanied by some bending back of substituents out of the plane of the alkene away from the metal. In the extreme (e.g. $(Ph_3P)_2Pt\{C_2(CN)_4\}$) the structure approaches that of a metallacyclopropane (Fig. 6.5).

(a) ROTATION OF AN ALKENE ABOUT THE METAL—LIGAND BOND. The ethene—metal bond thus consists of a σ- and a π-component. In this it formally resembles the C=C double bond in ethene itself. Rotation about a C=C double bond is restricted; Z and E geometrical isomers can be isolated at ambient temperatures as the activation energy barrier which separates them is about $150\text{--}250\,kJ\,mol^{-1}$, depending on substituents. The energy barriers to rotation of an alkene about the metal–alkene axis, however, are in general much lower than this, falling in the range $30\text{--}110\,kJ\,mol^{-1}$. The σ-component of the bonding, being cylindrically symmetrical about the rotation axis, does not contribute to this barrier. The π-component, however, can be formed by interaction of either $p_x\pi^*$ with d_{zx} or $p_y\pi^*$ with d_{yz}. If d_{zx} and d_{yz} are degenerate in the complex, any linear combination of the two orbitals is possible and this leads to equal metal–alkene

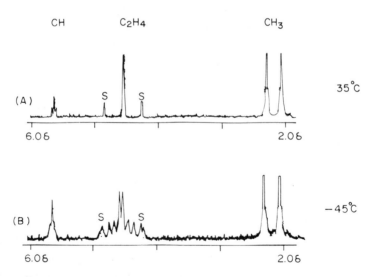

Fig. 6.6 100 MHz ^{1}H n.m.r. spectra of PtCl(C_2H_2) (acac). S. denotes ^{195}Pt–^{1}H satellites. (From C.E. Holloway, G. Hulley, B.F.G. Johnson and J. Lewis, *J. Chem. Soc.* (A), (1969), 53.)

π overlap whatever the orientation of the alkene. This is predicted for (alkene)$M(CO)_4$, for which rotational barriers are small ($< 40\,\text{kJ}\,\text{mol}^{-1}$) arising essentially from steric effects. Usually, however, d_{zx} and d_{yz} are not of equal energy and their respective overlap with $p_x\pi^*$ and $p_y\pi^*$ differs. This gives rise to a definite preference for one of the two mutually perpendicular conformations in which overlap is maximized.

(a) Square planar ML_3 (alkene): (d^8) (b) Trigonal planar ML_2 (alkene): (d^{10})
Alkene $\perp^r$ to plane of PtL_3 Alkene in plane of PtL_2

Evidence for restricted rotation about the metal–alkene axis in complexes is derived from measurements of n.m.r. spectra over a range of temperature. The ^{1}H n.m.r. spectra of $[PtCl(C_2H_4)(acac)]$ at $-45°C$ and at $35°C$ are shown in Fig. 6.6. At $-45°C$ the ethene gives rise to an AA'BB' pattern, as expected from the complex in its 'frozen' conformation similar to (a). At $-28°C$ the AA'BB' multiplet coalesces and reemerges at higher temperatures as a sharp singlet. The environment of the ethene protons is averaged on rotation through the metastable conformation similar to (b). The activation energy barrier to rotation ($\Delta G^{\ddagger}$) is about $50\,\text{kJ}\,\text{mol}^{-1}$. Platinum has an isotope ^{195}Pt (33% natural abundance) which has nuclear spin $I = \frac{1}{2}$. The ^{195}Pt–^{1}H coupling appears as satellites S. These satellites are present over the complete temperature range of the experiment. This indicates an intramolecular process in which the ethene remains bound to platinum throughout. If the alkene were to dissociate, satellite structure would be lost above the coalescence temperature.

6.2.2 Bonding between other unsaturated hydrocarbons and transition elements

The bonding between other hydrocarbon ligands and transition metals can be described in similar m.o. terms. In the discussion which follows it is assumed that the hydrocarbon σ-skeleton does not contribute to bonding with the metal. After formation of this skeleton each carbon atom has an unused $2p_z$ orbital, which combines with the other $2p_z$ orbitals of the delocalized system to form an equal number of π-molecular orbitals. The shapes of these m.o.s and their energies, calculated using the simple Hückel approximation, are given in Figs. 6.7 and 6.8. These m.o.s can be classified as of σ, π or δ symmetry with respect to rotation about the z-axis. These ligand orbitals can combine only with metal orbitals of the same symmetry.

From Fig. 6.8 it can be seen that the energies of the π- and δ-m.o.s of the ligand

199

	Classified as σ-symmetry	Classified as π-symmetry	Classified as δ-symmetry
3-electron η^3-allyl			
4-electron η^4-cyclobutadiene			
5-electron η^5-cyclopentadienyl			

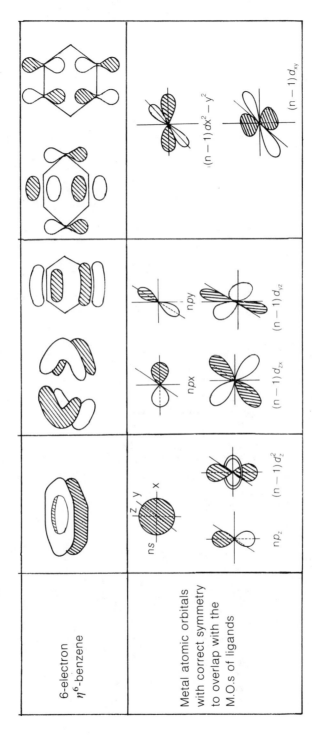

6-electron η^6-benzene			
Metal atomic orbitals with correct symmetry to overlap with the M.O.s of ligands	ns $(n-1)d_{z^2}$ np_z	np_x $(n-1)d_{zx}$ np_y $(n-1)d_{yz}$	$(n-1)dx^2-y^2$ $(n-1)d_{xy}$

Fig. 6.7 Representation of ligand π-molecular orbitals and their possible interactions with metal orbitals. The highest energy benzene m.o. is omitted as it makes little or no contribution to bonding.

201

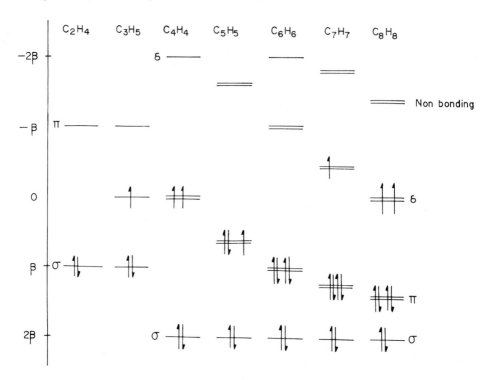

Fig. 6.8 Hückel energies of ligand orbitals.

decrease with increasing size of the delocalized system. Their occupancy also increases. The π m.o. in C_2H_4 is empty, so that it is a π-acceptor. In benzene, however, both π-m.o.s are filled, so that this ligand acts as a π-donor as well as a σ-donor. δ-Interactions are rather unimportant for small ligands where the relevant orbitals are of high energy compared with metal $d_{x^2-y^2}$ and d_{xy}, but they play a bigger part in the binding of larger ligands such as η^8-C_8H_8 to metals.

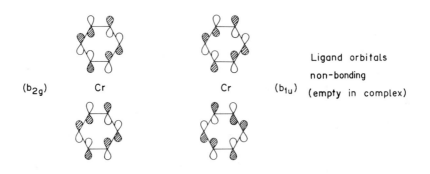

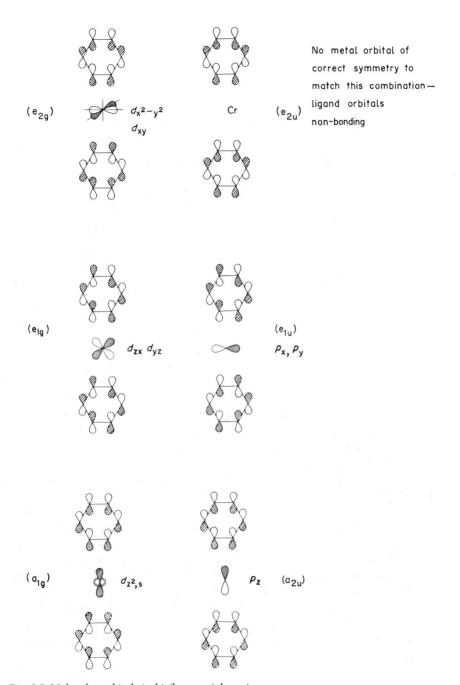

(e_{2g}) $d_{x^2-y^2}$ d_{xy} Cr (e_{2u})

No metal orbital of correct symmetry to match this combination — ligand orbitals non-bonding

(e_{1g}) d_{zx} d_{yz} p_x, p_y (e_{1u})

(a_{1g}) $d_{z^2, s}$ p_z (a_{2u})

Fig. 6.9 Molecular orbitals in bis(benzene)chromium.

Organometallic compounds of the transition elements

The allyl ligand possesses no rotation symmetry axis coincident with the z-axis. While the description here is therefore only approximate, it gives a useful pictorial representation.

(a) BONDING IN BIS(BENZENE)CHROMIUM. In this section, the symmetry arguments outlined above are developed further for the complex bis(benzene)chromium. This molecule has a sandwich structure in which the rings adopt an eclipsed conformation. The vertical six-fold rotation z-axis (C_6), six vertical planes of symmetry (σ_v) each of which includes a C_2 axis, and the horizontal plane (σ_h) characterize the point group as D_{6h}. In addition to the symmetry elements mentioned, attention is drawn to a centre of symmetry (i), situated at the chromium atom. The twelve $C(2p_z)$ orbitals, six on each ring, transform according to $a_{1g} + a_{2u} + e_{1g} + e_{1u} + e_{2g} + e_{2u} + b_{2g} + b_{1u}$. Reference to the character table of D_{6h} reveals that the metal orbitals transform as follows: σ-orbitals, d_{z^2}, s (a_{1g}), p_z (a_{2u}); π-orbitals, d_{zx}, d_{yz} (e_{1g}), p_x, p_y (e_{1u}); δ-orbitals, $d_{x^2-y^2}$, d_{xy} (e_{2g}). The symbols g and u refer to the parity on inversion through the metal atom. s- and d-orbitals have even (gerade) and p-orbitals odd (ungerade) parity. The ligand m.o.s on the two rings can be combined either to form sets of even a_{1g} (σ), e_{1g} (π), e_{2g} (δ) and b_{2g} (non-bonding) or odd a_{1u} (σ), e_{1u} (π), e_{2u} and b_{1u} (non-bonding) parities. The interactions between the metal and ligand orbitals which are permitted on symmetry grounds are illustrated in Fig. 6.9.

The next step is to construct a *qualitative* molecular orbital energy level diagram (Fig. 6.10). It must be stressed that this approach is qualitative; molecular orbital calculations might be expected to yield a different order for the molecular orbitals, this relative order varying depending on the assumptions made in the method of calculation chosen. Calculations of good quality would also not neglect the carbon s, p_x and p_y orbitals and the hydrogen $1s$ orbitals which are involved mainly in holding together the skeletons of the benzene rings.

The metal d-orbitals are split, d_{zx} and d_{yz} (e_{1g}) becoming antibonding through π-interaction with filled π_2 ligand orbitals, and $d_{x^2-y^2}$, d_{xy} (e_{2g}) somewhat bonding through δ-interaction with empty π_3 ligand orbitals. d_{z^2} should be weakly antibonding through overlap with the low-lying filled a_{1g} (σ) combination of π_1 ligand orbitals. There will be some mixing with the high lying metal $4s$ orbital, which is also of a_{1g} symmetry, however, so that the metal-centred orbital (approximately d_{z^2}) seems to be essentially non-bonding. The 18 valence electrons, six from chromium and six from each of the benzene rings, fill all the bonding and non-bonding (i.e. a_{1g}) orbitals. Bis(benzene)chromium is readily oxidized in solution to an air-stable 17-electron cation $(C_6H_6)_2Cr^+$. The ease of removal of one electron suggests that the highest occupied m.o. is fairly high lying.

UV-photoelectron spectroscopy provides a means of studying the energy levels of molecules experimentally. Simple interpretation of PE spectra relies on Koopmans' theorem which states that the same order of molecular orbitals

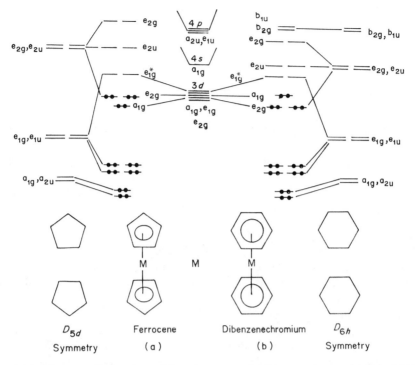

Fig. 6.10 Schematic illustrations of the important orbital interactions in (a) Fe(η-C$_5$H$_5$)$_2$ and (b) Cr(η-C$_6$H$_6$)$_2$. (After Mingos, D.M.P., 1982, *Comprehensive Organometallic Chemistry*, Vol. 3, Ch. 1.)

applies both to the neutral molecule and to the molecular ion. Molecular orbital calculations of good quality reveal that this is not always true. Nevertheless the PE spectrum of bis(benzene)chromium (Fig. 6.11) can be assigned on the basis of the m.o. diagram in Fig. 6.10. The first sharp band A is assigned to ionization from the essentially non-bonding d_{z^2} (a_{1g}) orbital. The second band B is of approximately double the intensity of the first and is broad. This is consistent with ionization from the doubly degenerate e$_{2g}$ orbital (reason for intensity) which is bonding in character (reason for breadth). The large group of ionizations at ca. 9–10 eV are from essentially ligand orbitals e$_{1u}$, e$_{1g}$, while those at still higher energies are associated with the a$_{2u}$ and a$_{1g}$ orbitals and the σ-skeleton (C—H and C—C bonds) of the ligands.

If Koopmans' theorem is not obeyed, the order of m.o.s indicated by the spectrum will not correspond to that present in the neutral molecule. Even so, much useful information can be derived from studying the photoelectron spectra of series of closely related complexes. In this way the effect of changing ligand substituents for example, on the relative energies of molecular orbitals can be investigated.

Organometallic compounds of the transition elements

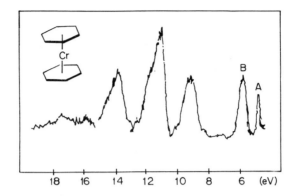

Fig. 6.11 He(I) UV-photoelectron spectrum of bis(benzene)chromium.

An ionization from a non-bonding m.o. gives rise to a sharp line, because the dimensions of the neutral molecule and the resulting ion are essentially the same. This means that a vertical transition from the zero vibrational level of the neutral molecule on ionization leads to the ion in its zeroth vibrational level also (Franck-Condon principle). Ionization from a bonding m.o. leads to an ion of increased dimensions, the potential energy surface of which does not lie vertically above that of the neutral molecule. Vibrational fine structure, which for a large molecule is not resolved, therefore accompanies the ionization peak. This effect is seen in the spectrum under discussion.

6.2.3 *Fragment analysis*

Molecules which contain two or more different types of ligand may be of low overall symmetry. While it is of course possible to perform molecular orbital calculations on such structures, the results may not give a very clear picture of the bonding, especially to the non-theoretician. A useful approach, largely developed by Hoffmann, is called 'fragment analysis'. This involves breaking the molecule down into fragments and calculating the shapes and energies of the orbitals associated with these fragments. Those orbitals which are not involved in holding together the atoms of the fragment, that is, the orbitals which are available for bonding to other ligands to form a real molecule, are called 'frontier orbitals'.

This approach is first illustrated with reference to the common grouping $M(CO)_3$. A metal hexacarbonyl such as $Cr(CO)_6$ possesses an octahedral structure. The σ-bonding skeleton can be considered as being derived from six equivalent d^2sp^3 hybrid metal orbitals and the σ-ligand orbitals. A pyramidal (C_{3v}) $M(CO)_3$ fragment can be derived from an octahedron by removal of three facial ligands, leaving three low lying acceptor orbitals directed towards the vacated ligand sites. These may be resolved (Fig. 6.12) into a hybrid orbital along the z-

206

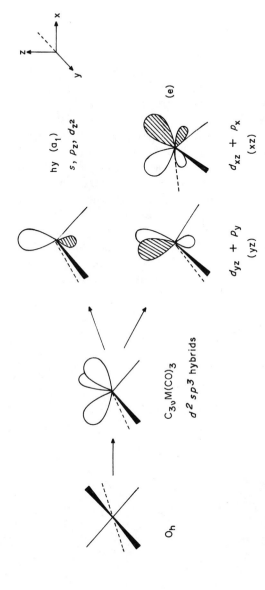

Fig. 6.12 Frontier orbitals of an M(CO)₃ fragment.

Organometallic compounds of the transition elements

axis with σ-symmetry with respect to this axis, and a degenerate pair of two π-orbitals, which are essentially the metal d_{zx} and d_{yz} orbitals with some p character Molecular orbital calculations using the extended Hückel (EHMO) method on $M(CO)_3$ fragments lead to the general energy level diagram shows in Fig. 6.13.

It will be noted that 18 electrons are required to fill the nine orbitals shown in Fig. 6.13 for the $M(CO)_3$ fragment. When $M = Cr$, six come from M and six from the three COs, requiring six more from the ligand L (e.g. C_6H_6). When $M = Fe$, eight come from M and six from the three COs, so that four more are needed from L (e.g. C_4H_4 or C_4H_6). This is merely a restatement of the 18-electron rule. Fragment analysis, however, enables us to see clearly the nature of the interactions between the frontier orbitals of the $M(CO)_3$ group (labelled xz, yz and hy in Fig. 6.12) and the available $2p_z\pi$ orbitals of the ligand. As an example consider the interaction between the frontier orbitals of the $Mn(CO)_3$ group and the $2p\pi$ m.o.s of the C_5H_5 ligand to form the molecule $(\eta^5\text{-}C_5H_5)Mn(CO)_3$. In the conformation shown, there is a plane of symmetry passing through the metal atom and one CO group, which bisects the C_5H_5 ligand and also the angle between the other two carbonyl ligands. This defines the xz plane; the point group is C_s. The metal and ligand orbitals can be classified as symmetric (a′) or antisymmetric (a″) with respect to reflection across this plane, depending on whether or not the wave function changes sign on carrying out this symmetry operation.

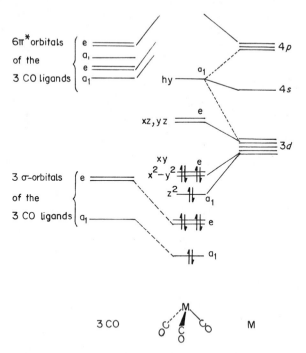

Fig. 6.13 Interaction diagram for an $M(CO)_3$ fragment. Orbital occupancy for $Cr(CO)_3$ shown.

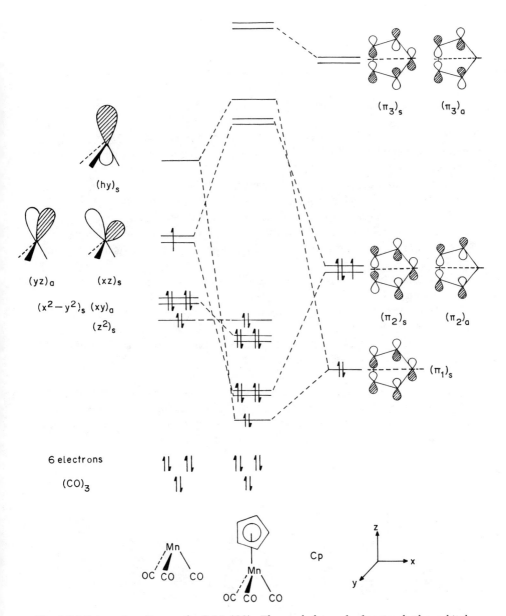

Fig. 6.14 Interaction diagram for CpMn(CO)$_3$. The symbols s and a denote whether orbital is symmetric or antisymmetric with respect to reflection across the xz plane of symmetry (Cp = C$_5$H$_5$).

Frontier orbitals	Symmetric (a')	Antisymmetric (a'')
Metal	hy, 'xz'	'yz'
Ligand	$(\pi_1)_s$, $(\pi_2)_s$	$(\pi_2)_a$

The main orbital interactions between C_5H_5 and $Mn(CO)_3$ are between $(\pi_1)_s$ and 'hy', $(\pi_2)_s$ and 'xz' and $(\pi_2)_a$ and 'yz'. A partial energy level diagram (Fig. 6.14) illustrates this. Involvement of the frontier orbitals only gives a rather oversimplified picture; it is usual to extend it somewhat by including overlap of 'xy' and 'x² − y²' with the π_3 m.o.s of C_5H_5 as shown. A substantial energy gap between the highest occupied and lowest unoccupied m.o.s is predicted. This is consistent with the lack of reactivity towards substitution of CO by other ligands which is a property of this molecule. A similar energy diagram for $CpCoL_2$, however, (Fig. 6.15) shows that the metal atom possesses a non-bonding pair of electrons. The consequence of this in the chemistry of $CpCoL_2$ is referred to below (p. 295).

Calculations have shown that the number, symmetry properties, spatial extent and energies of the frontier orbitals of the fragments $M(CO)_3$, $M(C_6H_6)$ and $M'(C_5H_5)$ are very similar. (If electron occupancy is considered M' has to provide one more electron than M for the groups $M(C_6H_6)$ and $M'(C_5H_5)$ to be isoelectronic, e.g. if $M = Cr$, $M' = Mn$.) Such fragments are termed 'isolobal'. In fact all hydrocarbon metal fragments $M(C_nH_n)(n = 4, 5, 6, 7)$ are essentially isolobal.

Fragment analysis is particularly useful in predicting the favoured conformations of molecules. In cyclohexadienyltricarbonylmanganese there are two extreme orientations of the ligand with respect to the $Mn(CO)_3$ fragment. Molecular orbital calculations as well as n.m.r. measurements on derivatives over

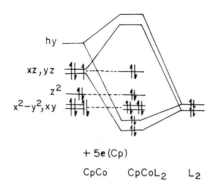

$+ 5e$ (Cp)

CpCo   CpCoL$_2$ L$_2$

Fig. 6.15 Schematic interaction diagram showing the bonding in $CpCoL_2$. Here L is a σ-donor only; no π-bonding to L is included. $C_6H_6FeL_2$ or $C_6H_6RuL_2$ resemble $CpCoL_2$.

210

a range of temperature indicate that conformation (A) is more stable than (B) by about $50\,kJ\,mol^{-1}$.

(A)   (B)

6.3 References

Albright, T.A. (1982) *Acc. Chem. Research*, **15**, 149.

Albright, T.A. and Whangbo, M-H. (1985) *Orbital Interactions in Chemistry*, Wiley, Chichester.

Green, J.C. (1981) Gas phase photoelectron spectra of *d* and *f* block organometallic compounds. *Struct. Bonding*, **43**, 37.

Hoffmann, R. (1982) Nobel lecture. *Angew. Chem. (Int. edn.)*, **21**, 71.

Mingos, D.M.P. (1982) *Comprehensive Organometallic Chemistry*, (Vol. 3, Chapter 1).

Muetterties, E.L., Bleeke, J.R., Wucherer, E.J. and Albright, T.A. (1982) Structural, stereochemical and electronic features of arene metal complexes. *Chem. Revs*, **82**, 499.

Stone, F.G.A. (1984) Isolobal relationships. *Angew. Chem. (Int. edn.)*, **23**, 89.

Problems

1. Apply the 18-electron rule to predict the compounds which are likely to be formed by the following combinations of metals and ligands. Mononuclear, neutral compounds only are required.

(a) η^5-Cp, CO, V (b) η^5-Cp, CO, Co (c) η^5-Cp, NO, Ni (d) H, PF_3, Fe (e) 1,3,5-cycloheptatriene, η^5-Cp, Mn (f) 1,3-butadiene, CO, Fe (g) allyl (C_3H_5), CO, Co (h) pentadienyl ($H_2C\cdots CH\cdots CH\cdots CH\cdots CH_2$), CO, Mn.

2. Suggest likely structures for each of the cations
$[(\eta^5\text{-Cp})Mo(C_8H_8)]^+$; $[(\eta^5\text{-Cp})Mo(C_8H_8)(PMe_3)]^+$;
$[(\eta^5\text{-Cp})Mo(C_8H_8)(Me_2PCH_2CH_2PMe_2)]^+$.

3. Using the 18-electron rule, or the 16-electron rule where appropriate, predict likely structures for the following compounds:

(a) $Mo(C_6H_6)(CO)_3$ (b) $[Rh(1,5\text{-COD})Cl]_2$
(c) $[Rh(1,5\text{-COD})(C_6H_6)]^+$ (d) $Rh(1,5\text{-COD})(acac)$
(e) $Rh(C_5H_5)\{C_6(CF_3)_6\}$ (f) $(Azulene)Mo_2(CO)_6$
(g) $(Azulene)Mn_2(CO)_6$ (h) $(Azulene)Fe_2(CO)_5$
$(1,5\text{-COD} = 1,5\text{-cyclooctadiene}; acac = acetylacetonate (2,4-pentanedionate)$.

4. Sketch the structures of each of the following molecules, clearly indicating the ways in which the ligands are attached to the metal.

$Cr(CO)_5py$ $Cr(CO)_3py$ $(C_8H_8)Mo(CO)_3$ $(C_8H_8)Fe(CO)_3$
$FeCp_2(CO)_2$ $Fe_2Cp_2(CO)_4$ (py = pyridine)
(*University College, London*).

5. The magnetic moments (μ_{eff}) of $Ti(C_6H_6)_2$, $V(C_6H_6)_2$ and $Cr(C_6H_6)_2$ at 298K are 0.0, 1.68 and 0.0 B.M. respectively. Discuss these data with reference to the molecular orbital diagram given on p. 205. Predict how the average metal—carbon bond distances in $Ti(C_6H_6)_2$ and $V(C_6H_6)_2$ would compare with that in $Cr(C_6H_6)_2$.

6. Consider the bonding in bis(cyclobutadiene)nickel, $Ni(C_4H_4)_2$. To what point group does the molecule belong? Classify ligand and metal orbitals in terms of their symmetry properties and construct a qualitative molecular orbital energy level diagram for the compound (consider metal 3d, 4s and 4p orbitals and ligand C $2p_z$ orbitals only).

7. For *each* of the following complexes:
$Ir_3(CO)_3(\eta^5\text{-}C_5H_5)_3$; $Fe(NO)(CO)_3(C_3H_5)$; $Fe_2(CO)_6(C_3H_5)_2$; $Mo_2(CO)_4(HC_2H)(\eta^5\text{-}C_5H_5)_2$; $Re(CO)(C_7H_7)(C_8H_8)$, where C_3H_5 = allyl, C_7H_7 = cycloheptatrienyl and C_8H_8 = cyclooctatetraene.

(a) Use the '18-electron rule' to predict two possible structures.
(b) Describe how i.r. and/or n.m.r. spectroscopy could be used to distinguish between the structures.
(c) Discuss any features of interest.

8. Describe the accepted model for the bonding of ethene and of ethyne to transition metals. Show how this model accounts for the structural, spectroscopic and reactivity variations of these coordinated ligands.

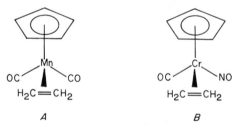

A   B

The proton n.m.r. spectra of the ethene complexes A and B are temperature dependent. Ignoring the cyclopentadienyl singlets, A exhibits two multiplets ($\delta1.93$ and $\delta2.82$) at $-123°C$ but a singlet ($\delta2.37$) at $-40°C$. The low temperature spectrum of B exhibits four multiplets due to the ethene protons, collapsing to two multiplets at room temperature. Account for these observations and predict whether the ^1H n.m.r. spectra of the ethyne analogues of A and B, i.e. $(C_5H_5)Mn(CO)_2(C_2H_2)$ and $(C_5H_5)Cr(NO)(CO)(C_2H_2)$ would be temperature dependent.
(*University of Southampton*).

Organotransition metal chemistry. Alkyl and alkylidene derivatives. Complexes of alkenes and alkynes.

7.1 Alkyls and aryls

7.1.1 *Introduction*

At the time when the alkyl and aryl derivatives of the Main Group elements were being discovered and their chemistry developed, little progress was made in the search for corresponding compounds of the transition elements. In general, reactions of Grignard reagents with anhydrous transition metal halides yielded mixtures of hydrocarbons and metal-containing residues from which no organometallic compounds could be isolated. The few exceptions were thought to be anomalies, rather than indications of a whole new area of knowledge yet to be explored. In 1907 Pope and Peachey in Cambridge, for example, isolated trimethylplatinum halides, $[Me_3PtX]_4$, from the reaction of methyl Grignard reagents with platinum(IV) halides. These air-stable, thermally robust materials have since been shown to have the cubane structure in which the platinum atoms attain an 18-electron configuration through triply bridging halogen atoms.

Alkylgold(III) derivatives were studied by C.S. Gibson in the 1930s and 1940s. Iododimethylgold, $(Me_2AuI)_2$, is formed by the action of methylmagnesium iodide on a cooled suspension of $[py_2AuCl_2]Cl$ in pyridine. It forms colourless crystals, melting to a dark red liquid which detonates violently. It is dimeric in benzene. Silver cyanide yields tetrameric $[Me_2AuCN]_4$.

Treatment of chromium(III) chloride with phenylmagnesium bromide under an atmosphere of carbon monoxide provided a route to chromium hexacarbonyl (Job-Cassels reaction). When this reaction was carried out under nitrogen it was possible to isolate organochromium complexes, but these are now known to be η^6-arene compounds, formed via unstable chromium phenyls (see p. 314).

On account of such rather discouraging results, there grew up a belief that transition metal alkyls and aryls were inherently unstable. It was suggested that transition metal—carbon σ-bonds must be very weak (thermodynamic instability). By implication an important mechanism for decomposition was thought to be homolysis to alkyl radicals, which led to mixtures of hydrocarbons (kinetic factors).

In the 1950s and 1960s, however, more transition metal alkyls and aryls were discovered. These were of two general types. First there were relatively stable compounds such as $MeMn(CO)_5$, $CpMo(CO)_3Me$, Cp_2TiPh_2 and $trans[(Et_3P)_2PtMe_2]$ which contain π-bonding ligands such as carbon monoxide, unsaturated hydrocarbons or alkyl phosphines. The enhanced stability was attributed to an increase in the separation of filled and empty molecular orbitals brought about by interaction between metal orbitals and acceptor orbitals on the ligand (p. 149). The stabilization was thus considered to be essentially thermodynamic. Secondly, some unstable alkyls which contain no supporting ligands were prepared (sometimes called 'binary' or 'homoleptic' alkyls). When titanium(IV) chloride is treated with methyllithium in ether at $-78°C$, for example, tetramethyltitanium is formed. This can be codistilled with ether under vacuum at $-30°C$ giving a yellow solution which decomposes at room temperature. Pure Me_4Ti decomposes slowly even at $-70°C$ and is very air sensitive. The lability of tetramethyltitanium is in striking contrast to the inertness of tetramethylsilane, which it might be expected to resemble. The latter is air stable and resists heating to 450°C.

7.1.2 Bond energies

It is now known that the lability of many transition metal alkyls and aryls is not due to the presence of weak metal—carbon bonds, but because pathways for decomposition are available which have low activation energies. That is, the property is kinetic rather than thermodynamic in origin. There have been rather few good measurements of bond enthalpies in transition metal alkyls and aryls (Table 7.1), but those which have been made indicate that the bonds are of similar strength to those found in many compounds of the Main Group elements. In contrast to Main Group chemistry, however, it seems that metal—carbon bond energies increase down a Group e.g. Cr < Mo < W. Once this was realized, the mechanisms of decomposition were investigated and ways devised of preparing robust compounds in which some of these pathways are blocked.

Table 7.1. Some M—C bond enthalpies (kJ mol^{-1})

$D(M-CH_2CMe_3)$		$D(M-Me)$		$D(M-Et)$	
TiR$_4$	188	TaMe$_5$	261	SiEt$_4$	287
ZrR$_4$	227	WMe$_6$	161	GeEt$_4$	243
HfR$_4$	224			SnEt$_4$	195
				PbEt$_4$	130

$R = CH_2CMe_3$

7.1.3 Preparation of alkyls and aryls of the transition elements

Some of the most widely used methods for preparing alkyl(aryl) complexes are listed below.

i) Reaction of a transition metal halide MX with an alkyl of another element M'R (often an organolithium or a Grignard reagent)

$$MX + MR' \longrightarrow MR + M'X$$

e.g. $WCl_6 + 6MeLi \xrightarrow{Et_2O} WMe_6 + 6LiCl$ Neutral, 'homoleptic' compounds

$CrCl_3 + 6MeLi \xrightarrow[\text{ii) dioxan}]{\text{i)Et}_2\text{O, }-18°C} Li_3CrMe_6 . 3$ (dioxan) Anionic complexes

η^5-Cp$_2$TiCl$_2$ + 2MeLi $\longrightarrow$ η^5-Cp$_2$TiMe$_2$ + 2LiCl   Complexes with

$trans[(Et_3P)_2PtCl_2] \xrightarrow{MeMgCl} trans(Et_3P)_2PtClMe$ π-bonding ligands.

$\xrightarrow{2MeLi} trans(Et_3P)_2PtMe_2$

While the Grignard reagent effects only mono-substitution at platinum, the more reactive lithium compound yields the dimethyl derivative. Use of mild alkylating agents can sometimes be used to advantage to give partial substitution. This is illustrated by the methyltitanium system. The very low thermal stability of some of these complexes should be noted.

$TiCl_4 \xrightarrow{Me_2AlCl} MeTiCl_3 \xrightarrow{Me_2Zn} Me_2TiCl_2$

deep violet solid   Black, decomp.
melts at 29°C to $-10°C$
yellow liquid

$\downarrow$ 4MeLi

TiMe$_4$

decomp. $-70°C \xrightarrow[\text{ii) dioxan}]{\text{i) MeLi/Et}_2\text{O}} LiTiMe_5 . 2$ dioxan $\longrightarrow$ MeTiCl$_3$(dmpe)
(p. 214)   (p. 223)

(Dioxan = (dioxane ring)) ; dmpe = bis(1, 2-dimethylphosphino)ethane

215

ii) Reaction of an anionic metal complex with an alkyl halide (or related electrophile e.g. *p*-toluenesulphonate).

$$M^- + RX \longrightarrow MR + X^-$$

As most anionic complexes also contain π-acid ligands such as CO, C_5H_5 or PR_3 this method is most useful in preparing alkyls supported by such groups e.g.

Carbonyl complexes	$Mn(CO)_5^- + MeI \longrightarrow MeMn(CO)_5 + I^-$	(p. 173)
	$Fe(CO)_4^{2-} + MeI \longrightarrow MeFe(CO)_4^- + I^-$	(p. 171)
	$Co(CO)_4^- + MeI \longrightarrow MeCo(CO)_4 + I^-$	(p. 178)
Cyclopentadienyl carbonyls	$CpFe(CO)_2^- + MeI \longrightarrow CpFe(CO)_2Me + I^-$	(p. 297)
Nitrogen bases	$[(dmgH)_2Co(py)]^- + MeI \longrightarrow (dmgH)CoMe(py) + I^-$	(p. 232)

(dmgH = dimethylglyoximato)

iii) Addition of an alkyl halide to a coordinatively unsaturated metal complex.

$$M + RX \longrightarrow M{\overset{R}{\underset{X}{\diagup\diagdown}}} \quad (e.g. \text{ Vaska's compound})$$

$$ML + RX \longrightarrow M{\overset{R}{\underset{X}{\diagup\diagdown}}} + L (e.g. (Ph_3P)_3 Pt)$$

$(L = PMe_3)$

The metal complex in these reactions generally behaves as a strong nucleophile. All the starting complexes possess a lone pair of electrons in a non-bonding orbital centred on the metal. (In some cases, e.g. Vaska's complex and alkyl halides other than methyl iodide a radical chain mechanism competes successfully with nucleophilic substitution, S_N2.)

iv) Addition of a transition metal hydride to an alkene.

$$M-H + {\diagdown \atop \diagup}C=C{\diagdown \atop \diagup} \rightleftharpoons M-\overset{|}{\underset{|}{C}}-\overset{|}{\underset{|}{C}}-H$$

This type of reaction is also observed in Main Group chemistry e.g. hydroboration (p. 66), hydroalumination (p. 80), hydrostannation (p. 108). It and its reversal are central to catalytic cycles such as those involved in the hydrogenation (p. 182) and hydroformylation (p. 387) of alkenes. Hydrozirconation (p. 290) provides another example of its use in stoichiometric organic synthesis. Its reverse

216

corresponds to β-hydrogen transfer, one of the most important pathways for the decomposition of transition metal alkyl derivatives, which is discussed below.

7.1.4 Decomposition pathways

In Table 7.2 pathways for decomposition of alkyl and aryl derivatives of the transition elements are summarized. Only the initial step of the decomposition is shown. The metal-containing species produced in this step can sometimes be isolated or detected spectroscopically, but it may also decompose further.

(a) β-HYDROGEN TRANSFER (β-ELIMINATION).   β-Hydrogen transfer is well recognized in Main Group chemistry as a pathway for decomposition of alkyl derivatives (e.g. Be, p. 54, B, p. 66 and Al, p. 80). It also provides a major mechanism for transition metal alkyls. Chatt found that the ethyl platinum complex below eliminates ethene on heating, leaving an isolable hydride. The latter takes up ethene again under pressure to regenerate the ethyl complex. This shows that the reaction is reversible.

$$trans[(Et_3P)_2PtCl(CH_2CH_3)] \; \underset{95°C/40\,atm}{\overset{180°C}{\rightleftharpoons}} \; trans[(Et_3P)_2PtClH] + C_2H_4$$

(16e)

For the elimination to proceed, a vacant coordination site at the metal centre is required. In the example quoted the square planar platinum complex has only a 16-electron configuration, so it is coordinatively unsaturated. In 18-electron compounds this site has to be provided by dissociation or by change of hapto number of a ligand.

The equilibrium involving β-hydrogen transfer has been observed directly by proton n.m.r. spectroscopy. The complex $[Ru(\eta^6\text{-}C_6H_6)(\eta^2\text{-}C_2H_4)(PMe_3)]$ is a strong base which is readily protonated to give a hydride $[RuH(\eta^6\text{-}C_6H_6)(C_2H_4)(PMe_3)]^+$. At $-78°C$ separate signals from the ethyl complex and the alkene hydride with which it is in equilibrium are observed, as the interconversion is slow. As the temperature is raised the signals from the individual species coalesce and emerge as an averaged spectrum. The ethyl compound can be trapped by addition of trimethylphosphine, which occupies the free coordination site which is required for the hydrogen transfer to proceed. It is

217

Table 7.2. Mechanisms of decomposition of η^1-alkyls

α-Hydrogen transfer (α-elimination)	$[M]-\underset{\underset{H}{\vert}}{\overset{\overset{H}{\vert}}{C}}-H \rightleftharpoons [M]{=}\underset{H}{\overset{CH_2}{<}}$	further decomposition
β-Hydrogen transfer (β-elimination)	$[M]-\underset{\underset{R}{\vert}}{\overset{\overset{R}{\vert}}{C}}-\underset{\underset{R}{\vert}}{\overset{\overset{R}{\vert}}{C}}-H \rightleftharpoons [M]-\overset{R\ R}{\underset{H\ R\ R}{||}}C$	may lose alkene; hydride may decompose further
Cyclometallations: γ-Hydrogen transfer	$[M]\overset{CH_2}{\underset{H-CH_2}{<}}\overset{R}{\underset{P}{C}} \rightleftharpoons [M]\overset{CH_2}{\underset{H-CH_2}{<}}\overset{R}{\underset{R}{C}}$	
δ-Hydrogen transfer	$[M]\overset{CH_2}{\underset{H-CH_2}{<}}\bigcirc \rightleftharpoons [M]\overset{CH_2}{\underset{H-CH_2}{<}}\bigcirc$	
cf. orthometallation	$[M]\overset{PPh_2}{\underset{H}{<}}\bigcirc \rightleftharpoons [M]\overset{PPh_2}{\underset{H}{<}}\bigcirc$	
Intramolecular elimination of hydrocarbon	$[M]\overset{R}{\underset{R}{<}} \longrightarrow [M]+\overset{R}{\underset{R}{|}}$ $[M]\overset{R}{\underset{H}{<}} \longrightarrow [M]+\overset{R}{\underset{H}{|}}$	
Binuclear (intermolecular) elimination[†]	$[M]-R+H-[M] \longrightarrow [M]-[M]+R-H$	
Free radical (homolytic) fission	$[M]-R \longrightarrow [M]^{\cdot}+R^{\cdot} \longrightarrow$	

[†]e.g. $(OC)_4Os{-}\overset{H}{C}H_3 + H{-}Os(CO)_4 \longrightarrow CH_4 + (OC)_4Os{-}\overset{H\ \ CH_3}{Os}(CO)_4$

now believed that the main reason that ligands such as $\eta^5\text{-}C_5H_5$, phosphines and CO often stabilize alkyls and aryls in 18-electron (and sometimes 16-electron) complexes is that they firmly occupy available coordination sites.

Many such sites, however, are present in binary compounds such as TiR_4. Having realized that β-hydrogen transfer is an important pathway for decomposition, people set out to prepare compounds in which the alkyl groups lack β-hydrogen atoms. Methyls, although still unstable, are generally more robust than ethyls, but benzyls are more inert than either. Thus $Ti(CH_2Ph)_4$ decomposes only at its melting point (91°C), but $Ti(CH_2CH_3)_4$ has never been prepared at all. The TiCC bond angles at the α-carbon of the benzyl groups in $Ti(CH_2Ph)_2$ are only about 90°, whereas in $Sn(CH_2Ph)_4$ normal tetrahedral angles are adopted. The tight angles in the former molecule bring the benzene rings within about 2.8 Å of the metal atom. Some interaction between the π-system and titanium is therefore indicated and this may help to stabilize the compound.

The neopentyl (Me_3CCH_2—) and trimethylsilylmethyl (Me_3SiCH_2—) groups have proved particularly useful in blocking β-hydrogen transfer. Not only do they lack β-hydrogen atoms, but their bulk probably also restricts sterically both intra- and intermolecular decomposition.

$$MCl_4 + 4LiCH_2EMe_3 \xrightarrow[-78°C]{Et_2O} M(CH_2EMe_3)_4 \quad (M = Ti, V; E = C, Si)$$

$$CrCl_3 + 4LiCH_2EMe_3 \xrightarrow{THF} Li[Cr(CH_2EMe_3)_4] \xrightarrow[Na/Hg]{O_2(air)} Cr(CH_2EMe_3)_4$$

$Ti(CH_2CMe_3)_4$ is yellow solid, m.p. 105°C with decomposition, which sublimes at $40°C/10^{-3}$ mm Hg. It is monomeric in solution in benzene. It is thermally quite stable decomposing in solution at 60°C to CMe_4 and a black solid with a half life of 14 h. The deep green paramagnetic liquid $V(CH_2CMe_3)_4$ is spontaneously inflammable in air and must be stored under nitrogen or argon at $-30°C$. The initial product from chromium(III) chloride is a dark blue solution which contains the anion $[Cr(CH_2CMe_3)_4]^-$. This is oxidized in air to purple $Cr(CH_2CMe_3)_4$ ($\mu_{eff} = 2.9$ B.M.), in which chromium has two unpaired electrons. The unusual tetravalent state is paralleled in the alkoxides $Cr(OR)_4$ and amino compounds $Cr(NR_2)_4$.

While the tetrakis compounds MR_4 are obtained from Me_3ECH_2Li, tris complexes MR_3' ($M = Ti, V, Cr; R' = CH(SiMe_3)_2$) result from reactions between MCl_3 and the very bulky $(Me_3Si)_2CHLi$. Here steric effects apparently determine the valency (cf. Sn, p. 123). Norbornyl compounds MR_4 ($M = Ti, V, Cr, Fe, Co$) are even more inert to thermal decomposition and to oxidation than the neopentyls. β-Hydrogen transfer from the norbornyl group to the metal is strongly disfavoured as this would result in the formation of a strained alkene with a double bond at a bridgehead carbon atom, contrary to Bredt's rule. (There is also no α-hydrogen). The bulk of the group is probably responsible for the great resistance to oxidation observed for these complexes.

219

Organotransition metal chemistry

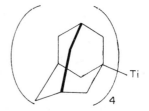

tetrakis(norbornyl) titanium   tetrakis (adamantyl)titanium

Tetrakis(adamantyl)titanium, m.p. 235°C, prepared by heating titanium(IV) chloride, sodium and 1-chloroadamantane in cyclohexane, could be decomposed for analysis only at 170°C by a mixture of concentrated nitric and hydrofluoric acids in the presence of hydrogen peroxide! The contrast with tetramethyl-titanium is amazing.

Transfer of a β-hydrogen atom to give a benzyne (or orthophenylene) complex as an intermediate can occur in the decomposition of some η^1-aryls. Certainly aryl complexes with ortho-substituents (e.g. 2,4,6-trimethylphenyl, 'mesityl' or 2,6-dimethylphenyl, 'o-xylyl') are more robust than their unsubstituted analogues.

$$trans[CoBr_2(PEt_2Ph)_2] + 2MesMgBr \longrightarrow trans[CoMes_2(PEt_2Ph)_2]$$
golden-yellow crystals,
planar cobalt, 15-electron,
μ_{eff} 2.5 B.M.

The related diamagnetic 16-electron nickel complexes $trans[NiMes_2(PR_3)_2]$ and $trans[NiXMes (PR_3)_2]$ (X = Cl, Br) can be heated unchanged in boiling benzene and can be stored in air in the crystalline form. (Mes = 2,4,6-trimethylphenyl). An orthophenylene complex has actually been isolated from the decomposition of a phenyltantalum derivative.

$$\text{(structure)} \xrightarrow{-CH_4} \text{(structure)}$$

(b) α-HYDROGEN TRANSFER (α-ELIMINATION). A possible decomposition pathway for alkyls which lack a β-hydrogen atom, but which possess an α-hydrogen is termed α-hydrogen transfer or α-elimination.*

---

*The second term is somewhat misleading as there is no loss of material from the coordination sphere in the elementary step $M—CH_3 \rightarrow M(=CH_2)H$. In the case of β-

220

The product of the initial step is an alkylidene (hydride), $M(=CH_2)H$. In all known cases this intermediate is unstable and decomposes further. If there are other alkyl groups R in the molecule, a hydrocarbon RH can be lost in an intramolecular process. Intermolecular loss of RH is also a possibility.

$$R_nMCH_2R \xrightarrow[\text{transfer}]{\alpha-H} R_nM\overset{H}{\underset{CHR}{<}} \xrightarrow{-RH} R_{n-1}M=CHR$$

$$\xrightarrow{?} \text{may decompose further}$$

The production of CMe_4 and $SiMe_4$ from the decomposition of $Ti(CH_2CMe_3)_4$ and $Ti(CH_2SiMe_3)_4$ respectively (p. 219) is consistent with this mechanism. Further evidence comes from the isolation of stable alkylidene complexes of tantalum. Partial alkylation of $TaCl_5$ with neopentyllithium affords $(Me_3CCH_2)_3TaCl_2$. Further reaction with 2 mol $LiCH_2CMe_3$ in pentane does not give $Ta(CH_2CMe_3)_5$ as expected, although $TaCl_5 + 5MeLi$ do give unstable $TaMe_5$. Instead the products are the orange complex $(Me_3CCH_2)_3Ta=CHCMe_3$ and CMe_4. The molecular structure of the related complex $Cp_2Ta(=CH_2)CH_3$ show a short $Ta=CH_2$ distance consistent with a metal—carbon double bond (p. 232). These results can be explained by the following mechanism, in which the incipient metal hydride produced by α-hydrogen transfer and another neopentyl group are expelled as tetramethylmethane in a concerted process (note that tantalum cannot exceed a valency of five).

$$(Me_3CCH_2)_3TaCl_2 \xrightarrow{Me_3CH_2Li} (Me_3CCH_2)_4TaCl \xrightarrow[\text{and loss of alkane}]{\alpha-H \text{ transfer}}$$

$$\text{not isolated}$$

$$(Me_3CCH_2)_2\overset{Cl}{\underset{Me_3CCH_2--H}{Ta}}=CHCMe_3 \xrightarrow[-LiCl]{Me_3CCH_2Li} (Me_3CCH_2)_3Ta=CHCMe_3$$

Evidence for the equilibrium $[M]-CH_3 \rightleftharpoons [M]H(=CH_2)$ has been obtained by trapping the intermediate $[M]H(=CH_2)$ with a phosphine. The complex $[Cp_2W(CH_3)(C_2H_4)]^+ PF_6^-$ was heated in acetone with PMe_2Ph and the changes monitored using proton n.m.r. spectroscopy. The first product (A) (Fig. 7.1) arises by nucleophilic addition of the phosphine to the ethene ligand in the cation. This may be compared with the similar reactions of alkene iron complexes described on p. 241. On further heating (A) loses ethene producing a vacant coordination site. This opens the way to α-hydrogen transfer from the η^1-methyl group. The resulting alkylidene hydride is trapped as a phosphonium salt (B). On prolonged heating this is converted into the thermodynamically most stable

---

hydrogen transfer the alkene hydride $M(CH_2=CH_2)H$ is often unstable to loss of alkene, so the term β-elimination is understandable. Both terms 'α- and β-elimination' are, however, well entrenched in the literature.

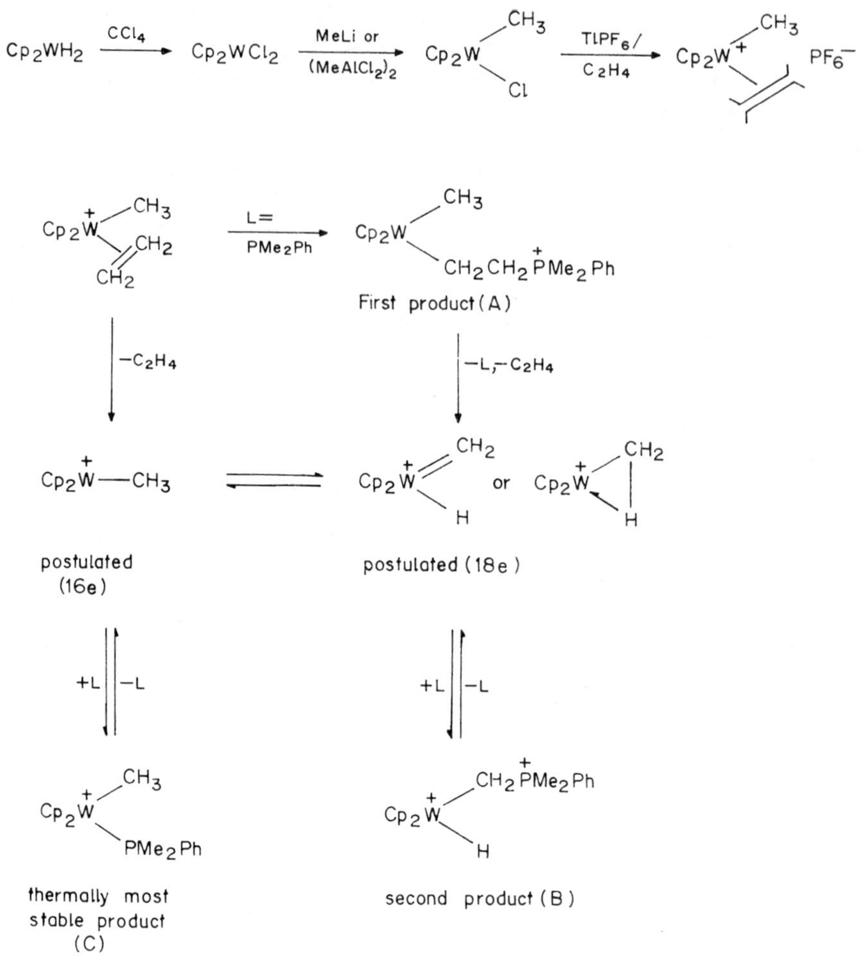

Fig. 7.1 Reversible α-hydrogen transfer.

product (C). This experiment indicates that in this case α-hydrogen transfer is the first step in the decomposition of the methyl complex and that like β-hydrogen transfer the process is reversible.

The molecular structures of the complexes (dmpe)TiCl$_3$R (R = Me, Et), determined by X-ray diffraction, shed light on pathways which lead to α- and β-hydrogen transfer (dmpe = Me$_2$PCH$_2$CH$_2$PMe$_2$) (Fig. 7.2). In the ethyl complex one of the β-hydrogen atoms is situated within bonding distance of the metal atom. The structure thus models the incipient formation of a metal(alkene)hydride envisaged in β-hydrogen transfer. The methyl compound similarly reflects the commencement of α-hydrogen transfer. The hydrogen atom

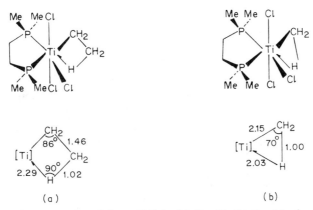

Fig. 7.2 Molecular structures of dmpe TiCl$_3$R; (a) R = Et, (b) R = Me showing agostic hydrogen interactions.

is bonded to the metal by a three-centre two-electron bond, which is common in the chemistry of Main Group elements (see p. 33), but has only recently been recognized in transition metal complexes. The interacting hydrogen atom has been called an 'agostic' hydrogen (Greek ἀγοστόσ, flat of the hand, arm). Alkylidene complexes are discussed further in Section 7.2.

(c) γ- AND δ-HYDROGEN TRANSFER; CYCLOMETALLATIONS. Transfer of γ- or δ-hydrogen atoms to the metal can trigger decomposition *via* metalla-cyclobutane or -cyclopentane intermediates. These reactions are examples of a much wider class termed cyclometallations. These can be intra- or intermolecular.

$$e.g. \ (Ph_3P)_3 \ IrCl \xrightarrow[\text{decalin}]{\text{heat in}}$$
(16 e)

(18e)

Such reactions, which can be thought of as addition of a carbon—hydrogen bond to a coordinatively unsaturated metal centre, are promoted by an increase in the bulk of ligands in the coordination sphere. In part this forces the C—H bond close to the metal centre; a weak interaction of this type has been found by X-ray crystallography in (Ph$_3$P)$_3$RuCl$_2$.

(d) INTRAMOLECULAR ELIMINATION OF ALKANE. Complexes which possess at least two alkyl substituents in a *cis* configuration (or an alkyl and a hydride substituent) can eliminate alkane in a concerted intramolecular process.

$$[M]\begin{subarray}{l}\diagup R\\\diagdown R\end{subarray} \longrightarrow [M] + \begin{subarray}{c}R\\|\\R\end{subarray} \ ; \ [M]\begin{subarray}{l}\diagup R\\\diagdown H\end{subarray} \longrightarrow [M] + \begin{subarray}{c}R\\|\\H\end{subarray}$$

223

Organotransition metal chemistry

In this elementary step, the electron count at the metal is reduced by two. This mechanism is followed in the decomposition of the octahedral complexes *fac* $[(PhMe_2P)_2PtMe_3I] \rightarrow trans[(PhMe_2P)_2PtMeI] + C_2H_6$. Ethane is the only hydrocarbon product. The reaction is first order in starting material and has an activation enthalpy of $129\,kJ\,mol^{-1}$. If homolysis to radicals were the rate determining step, $\Delta H^{\ddagger}$ would be near to the Pt—Me bond energy of *ca.* $250\,kJ\,mol^{-1}$. The intramolecular nature of the process is verified by decomposing together a mixture of *fac* $[(PhMe_2P)_2Pt(CD_3)_3I]$ and *fac* $[(PhMe_2P)_2Pt(CH_3)_3I]$. C_2H_6 and C_2D_6 only are produced; there is no cross-product CH_3CD_3.

7.1.5 *Carbon—hydrogen bond activation*

Normally intramolecular elimination of alkane from alkyl(hydride) complexes occurs readily and is favoured thermodynamically. There is interest, however, in the possibility of carrying out the reverse reaction, the addition of a C—H bond to an unsaturated transition metal centre. Alkanes are susceptible to electrophilic attack, for example by Lewis or Brønsted acids which convert linear alkanes into their branched isomers *via* carbonium ion intermediates. Linear and cyclic alkanes can be converted into aromatic hydrocarbons and hydrogen over metal surfaces such as platinum. These reactions are carried out on a large scale industrially in the reforming of petroleum.

Intramolecular cyclometallations are special cases of the desired reaction. They are likely to compete successfully on entropy grounds over an intermolecular process involving an alkane. Cyclometallation is favoured by steric congestion at the metal centre. Such congestion should therefore be reduced as far as possible in designing a system which might activate alkanes. Alkanes lack both lone pairs and low lying empty orbitals. Systems have been discovered, however, with which they do react. Interaction with a metal centre may occur initially through M—H—C bridges ('agostic' hydrogens).

In a reaction $R—H + [M] \rightarrow [M]{\overset{R}{\underset{H}{\diagdown}}}$ the alkyl hydride will be stable only if the fragment [M] is of high energy content, that is, is itself unstable. When $CpIr(CO)_2$ or the related compound $(\eta^5\text{-}C_5Me_5)IrH_2(PMe_3)$ is photolysed in the presence of an alkane, loss of CO or H_2 respectively occurs to form very reactive, unsaturated species CpIrL $(L = CO$ or $PMe_3)$. These can add to an alkane forming an alkyl hydride.

224

The activation energy for the addition is very low; methane adds to $(\eta^5\text{-}C_5H_5)Ir(CO)$ in a frozen matrix even at 12 K. (See also reactions of tungstenocene, p. 292). Another approach which involves a reversal of alkene hydrogenation is mentioned on p. 186.

Benzene displaces cyclohexane from $Cp^*Ir(PMe_3)Cy(H)$ at 130°C. Kinetic measurements suggest a two step mechanism:

$$[Ir]Cy(H) \underset{k_{-1}}{\overset{k_1}{\rightleftharpoons}} [Ir] + CyH; \quad [Ir] = (\eta^5\text{-}C_5Me_5)Ir(PMe_3)$$

$$[Ir] + PhH \xrightarrow{k_2} [Ir]Ph(H); \quad Cy = cyclohexyl$$

From the observed position of equilibrium in this and similar reactions it has been established that the relative Ir—C bond energies fall in the usual order Ir—aryl > Ir—primary alkyl > Ir—secondary alkyl > Ir—tertiary alkyl. The Ir—Ph bond must be at least $100\,kJ\,mol^{-1}$ stronger than Ir—Cy. For alkane activation, $M + RH \rightarrow M(R)H$, to occur, $E(M—R) + E(M—H)$ must be greater than $E(R—H)$. As $E(R—H)$ is of the order of 350–$450\,kJ\,mol^{-1}$, the values of $E(Ir—H)$ and $E(Ir—R)$ probably lie between 240 and $320\,kJ\,mol^{-1}$.

7.1.6 Migratory insertion of carbon monoxide

A group of reactions which bear a formal resemblance to each other are the 'insertion' reactions

$$[M]—R + CO \longrightarrow [M]—C(O)R$$

$$[M]—R + SO_2 \longrightarrow [M]—O—S(O)R$$

$$[M]—R + \ {>}C{=}C{<} \longrightarrow [M]—\overset{|}{\underset{|}{C}}—\overset{|}{\underset{|}{C}}—R$$

Methylpentacarbonylmanganese yields an acyl derivative not only when it reacts with carbon monoxide, but also when treated with other ligands. This suggests that it is one of the bound carbonyl groups which forms the acyl C=O, rather than a CO molecule from outside. This has been confirmed by using ^{13}C labelling.

The position of a ^{13}CO group in a complex can be determined by infrared spectroscopy. A ^{13}CO vibration absorbs at a lower frequency than the corresponding ^{12}CO vibration on account of the heavier mass of the former group. In the reaction of ^{13}CO with $MeMn(CO)_5$, the incoming carbon monoxide initially is situated *cis* to the acyl group. In the reverse reaction a CO ligand *cis* to the acyl group is expelled.

In the carbonylation, two pathways can be envisaged. Either (a) the methyl group migrates to a *cis* carbonyl ligand or (b) a *cis* carbonyl inserts into the metal—methyl bond. Using ^{13}CO labels, Calderazzo and Noack were able to show that *in this series* of complexes path (a), methyl migration, operates (Fig. 7.3, see also problem 3, p. 249).

It has usually been assumed that all alkyl–acyl conversions follow path (a), but there is evidence that path (b) may operate in some systems. It should also be noted that it is only because the carbonyl ligands in the manganese complexes exchange very slowly either with gaseous CO, or amongst themselves, that these experiments are possible.

In the migratory insertion of CO into $MeMn(CO)_5$ kinetic studies indicate an initial equilibrium to give a 16-electron intermediate, presumably *via* a cyclic three-centre transition state. The latter resembles the ground state structures of some zirconium acyls (p. 291). The unsaturated intermediate must be formed in

$$MeMn(CO)_5 \underset{k_{-1}}{\overset{k_1}{\rightleftharpoons}} MeCOMn(CO)_4 \xrightarrow[k_2]{L} MeCOMn(CO)_4L$$

very low concentration as it has never been detected spectroscopically reacting rapidly with the incoming ligand L. The migration is aided by Lewis acids. Presumably there is an initial association with a carbonyl ligand. With $AlBr_3$ this is followed by the rapid formation of an adduct of the acyl, in which a bromine atom of the Lewis acid occupies the position vacated by the methyl group. Subsequent treatment with a ligand L (e.g. CO) displaces the bromine from this site.

Pure optical isomers of complexes such as $CpFe(CO)(PPh_3)(COEt)$, which contain a chiral iron atom on account of their four different substituents arranged in a pseudo-tetrahedral fashion, have been obtained. Decarbonylation leads to

226

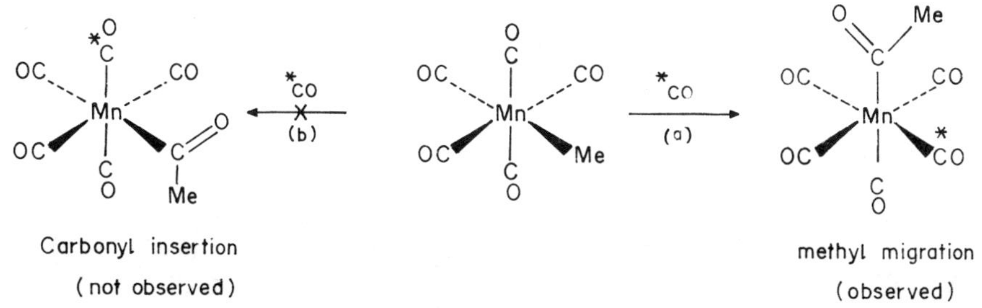

Carbonyl insertion (b) (a) methyl migration

(not observed)   (observed)

Fig. 7.3 Migratory insertion of carbon monoxide.

inversion at iron, showing that the ethyl group migrates to the site vacated by the departing carbonyl ligand.

Alkyl $\rightleftharpoons$ acyl migrations are key steps in many commercial catalytic processes involving carbon monoxide such as the hydroformylation of alkenes (p. 387) and the conversion of methanol into acetic acid (p. 385).

7.1.7 *Some reactions catalysed by palladium complexes*

(a) COUPLING VIA BIS-η^1-INTERMEDIATES.   Attempts to couple an organometallic reagent with an organic halide can give unsatisfactory results ($R^2M + R^1X \rightarrow R^1R^2 + MX$) (p. 88). Metal–halogen exchange and elimination reactions can lead to a mixture of products. Organocopper compounds are sometimes suitable, but with the exception of iodides, the reactions are often sluggish and are accompanied by extensive cross scrambling leading to R^1R^1 and R^2R^2.

One way of overcoming these problems in certain cases is to attach the two groups R^1 and R^2 to a transition metal centre. The following cycle shows how this can be done catalytically. The grouping [M] can be, for example, L_2Pd or L_2Ni (L $= Ph_3P$). The catalyst is conveniently introduced as L_4Pd, L_4Ni or L_2NiCl_2.

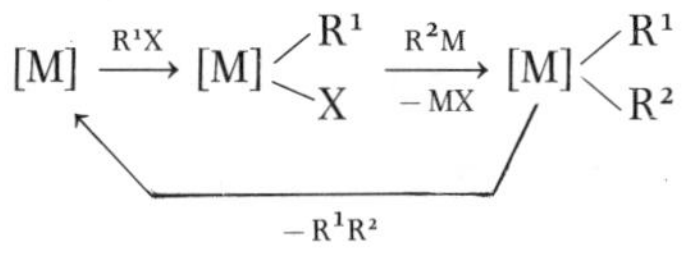

$$[M] \xrightarrow{R^1X} [M]{<}^{R^1}_{X} \xrightarrow[-MX]{R^2M} [M]{<}^{R^1}_{R^2}$$

$$\underset{-R^1R^2}{\underbrace{\qquad\qquad\qquad\qquad}}$$

A serious restriction on the scope of these reactions arises from the property of transition metal alkyls to undergo rapid β-hydrogen transfer with elimination of alkene. Alkyl halides (R^1X) which have a β—C—H(sp^3) function are therefore unsuitable. This limitation does not apply quite so rigidly to the group R^2, probably because loss of the coupled product R^1R^2 from $L_2MR^1R^2$ competes successfully with loss of alkene. Nevertheless the reactions are most generally useful when R^1 and R^2 are aryl, alkenyl, alkynyl, benzyl or allyl. Using nickel or palladium catalysts suitable halides have been coupled with organo derivatives of, for example, M = Li, Mg, Zn, Cd, Hg, Al, Zr and Sn. Retention of stereo- and regiochemistry in alkenyl and allyl groups is usually observed.

227

$RC \equiv CH + ClZr(H)Cp_2 \longrightarrow$ ... $\xrightarrow[L_4Ni]{ArX}$ ...

Organotin reagents are particularly versatile because they are inert to many functional groups including —NO_2, —CN, —OR, —CO_2R, —COR and —CHO. Typically the reactions are carried out under reflux in THF or chloroform. An acyl chloride R^1COCl can replace the halide R^1X, providing a route to ketones R^1COR^2. The rate determining step in the cycle is probably the transfer of the group R^2 from tin to palladium. This apparently involves electrophilic attack by the palladium complex at the carbon atom bonded to tin. The rate of transfer of R^2 from $Bu_3^nSnR^2$ falls in the order $R^2 = PhC \equiv C— > PhCH = CH— > H_2C = CH— > Ph— > PhCH_2 >$ alkyl. It is therefore possible to use the readily accessible tributyltin derivatives $Bu_3^nSnR^2$ as only the one group of interest, R^2, is transferred.

An alternative approach to ketones is to conduct the reaction between R^1X and R^2M under an atmosphere of carbon monoxide. Pressures of at least 3 atm are required and even so some direct product R^1R^2 is sometimes obtained.

geranyl chloride   égomaketone (92%)

228

(b) THE HECK REACTION. Organopalladium species also add across carbon—carbon double bonds; the stereochemistry of the addition is *cis*. If there is a β-hydrogen in the resulting palladium complex which can, by rotation, move into a *syn* position relative to palladium, loss of [Pd]—H occurs and a substituted alkene is generated.

As described above, the initial [Pd]—R complex can be formed by the transfer of R from another organometallic RM. Compounds of mercury, boron or tin have often been used. Alternatively an aryl, heteroaromatic or vinyl halide is added to 'PdL$_2$', generated *in situ* from PdL$_4$ or by reduction of L$_2$PdX$_2$. In the second method a base such as a tertiary amine is also required to react with the hydrogen halide which is produced.

The Heck reaction involves electrophilic attack by palladium on an alkene. In contrast to mercuration (p. 63), however, *cis* addition of [Pd]—R to the double bond occurs. The palladium becomes attached, as expected, to the more negative carbon atom of the alkene. Steric effects are also significant. The less hindered the olefin, the faster the rate. In the absence of a strong electronic effect the new carbon—carbon bond is formed at the less sterically congested centre.

Allyl alcohols couple with aryl halides to give aldehydes or ketones.

229

Organotransition metal chemistry

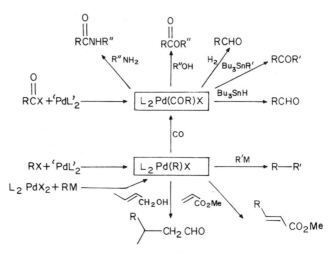

Fig. 7.4 Summary of reactions catalysed by palladium complexes. R,R' = e.g. aryl, vinyl, benzyl, heteroaromatic, allyl.

(c) CARBONYLATION REACTIONS.   The organopalladium species L_2PdRX form acyl intermediates $L_2Pd(COR)X$ with carbon monoxide. These react with alcohols to form esters, while amines afford amides. They are also cleaved by hydrogen yielding aldehydes. All these conversions take place catalytically. They are summarized in Fig. 7.4.

7.1.8 *Some alkyl derivatives of cobalt. Vitamin* B_{12} *coenzyme and the cobaloximes.*

Octahedral complexes of cobalt(III) (d^6) are generally diamagnetic 18-electron compounds which are kinetically inert to substitution. This applies to the group of organometallic compounds which includes Vitamin B_{12} coenzyme, a rare example of a natural product which contains a metal—carbon bond. The human body contains 2–5 mg, mostly in the liver. The coenzyme is implicated in a whole range of biochemical processes of the general type

$$-\underset{|}{\overset{X}{C}}-\underset{|}{\overset{H}{C}}- \rightleftharpoons -\underset{|}{\overset{H}{C}}-\underset{|}{\overset{X}{C}}- \quad X = \text{e.g. } NH_2, \ C(=CH_2)CO_2^-, \ CH(NH_3)CO_2^-$$

in which two groups X and H on adjacent carbon atoms are exchanged. This exchange may be followed by elimination of XH, in which case the process is irreversible.

$$-\underset{|}{\overset{X}{C}}-\underset{|}{\overset{H}{C}}-OH \rightarrow -\underset{|}{\overset{H}{C}}-\underset{|}{\overset{X}{C}}-OH \rightarrow -\underset{|}{\overset{H}{C}}-C\overset{\diagup O}{\diagdown} + XH \quad (X = OH, NH_2)$$

230

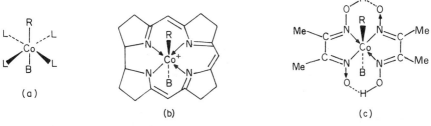

Fig. 7.5 Simplified representation of Vitamin B$_{12}$ and its analogues. (a) General structure. (b) The corrin ring in Vitamin B$_{12}$. (c) Cobaloxime–a bisdimethylglyoxime complex of cobalt.

The general structure of the Co(III) (oxidized) form of these organometallics is shown in Fig. 7.5. In Vitamin B$_{12}$ coenzyme (B$_{12a}$) the four coplanar ligands are provided by a corrin ring, the alkyl group is 5′-deoxyadenosyl and the Lewis base 5,6-dimethylbenzimidazole. Many model compounds have also been prepared, the chemistry of which mirrors to some extent that of the natural product. One of these systems is illustrated, in which the ligands are dimethylglyoxime (the cobaloximes).

The reduction of cyanocobalamin (b, R=CN) occurs in two one-electron steps, through B$_{12r}$, a cobalt(II) species which exhibits radical character, to B$_{12s}$, an anionic strongly nucleophilic species. Reduction of the model cobaloxime [Co(dmgH)$_2$Clpy] proceeds similarly. The best route to alkyl derivatives is to react (Cb)$^-$ with an alkyl halide or tosylate. These (Cb)$^-$ species are very strong nucleophiles to saturated carbon centres. With tosylates substitution occurs with inversion at carbon, indicating an S$_N$2 mechanism.

$$(Cb)^+ \xrightarrow{+e^-} (Cb)\cdot \xrightarrow{+e^-} (Cb)^-$$

$$B_{12} \xrightarrow{+e^-} B_{12r} \xrightarrow{+e^-} B_{12s}$$

$$(Cb) + RX \longrightarrow (Cb)\text{—}R + X^- \qquad (Cb) = (L_4BCo)$$

With some alkyl halides, however, a radical mechanism may operate

$$(Cb)^- + RX \longrightarrow (Cb)\cdot + R\cdot + X^-; \quad (Cb)\cdot + R\cdot \longrightarrow (Cb)\text{—}R$$

In neutral or acid solutions the anions (Cb)$^-$ are protonated to form (Cb)—H, which adds to alkenes (Fig. 7.6).

The Co—C bond energies in cobaloximes and cobalamins are low, falling in the range 85–125 kJ mol^{-1} (cf. $\bar{D}$(Hg—Me) 130 kJ mol^{-1}). Homolytic cleavage to (Cb)$\cdot$ and R$\cdot$ is considered to be the initial step in the action of the coenzyme. Homolysis can be brought about by photolysis or thermally. Nevertheless some of these complexes are chemically rather robust. MeCo(dmgH)$_2$H$_2$O is an air-stable crystalline material which is only slowly attacked by concentrated sulphuric or

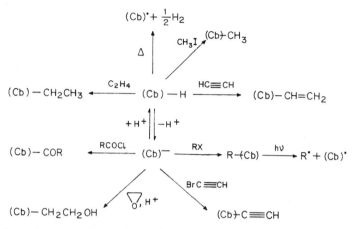

Fig. 7.6 Some reactions of reduced cobalamins and cobaloximes.

hydrochloric acids. The methyl group is cleaved by the electrophile $HgCl_2$ giving
MeHgCl.

7.2 Alkylidene and alkylidyne (carbene and carbyne) complexes

Alkylidene and carbene complexes contain the grouping [M]=CRR'. Free
carbenes CRR' are very reactive and short lived species (p. 61). The isolation of
complexes of these ligands is a good illustration of the stabilization of labile species
by coordination to a transition element (cf. cyclobutadiene, p. 269 and trimethylenemethane p. 270).

Carbene is considered to be a good σ-donor and a weak π-acceptor. In its singlet
state it has a lone pair of electrons in an orbital of sp^2 character which is coplanar
with the two C—H bonds; this acts as the donor orbital. There is also an empty
$C(2p)$ orbital which can accept electron density from metal orbitals of π-
symmetry. A M=CH$_2$ bond would therefore be expected to show significant
double bond character. This is observed in alkylidene complexes such as
$Cp_2Ta(=CH_2)CH_3$ in which the Ta=CH$_2$ bond is considerably shorter than the
single Ta—CH$_3$ bond.

In a large group of complexes (Fischer carbenes) one of the substituents on the
carbene contains a heteroatom X such as nitrogen, oxygen or sulphur which itself

232

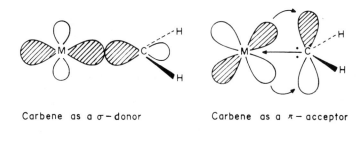

Carbene as a σ–donor   Carbene as a π–acceptor

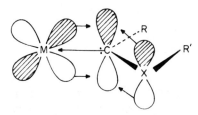

Competition for C_{2p} orbital of carbene
from lone pair on heteroatom X

(X = e.g. NR'', O, S).

Fig. 7.7 Bonding in metal carbenes.

possesses non-bonding electrons. These can be donated to the $C(2p)$ orbital of the carbene. In such cases π-donation from the metal is reduced, the M—C bond order is lowered while the C—X bond order is raised. Electron density is consequently higher at the metal centre so that the M—$C_{carbene}$ bond is polarized $\overset{\delta-}{M}$—$\overset{\delta+}{C}_{carbene}$; in the absence of heteroatomic substituents the polarization of the M—$C_{carbene}$ bond is usually in the opposite sense (Fig. 7.7).

7.2.1 *Fischer carbene complexes*

Most of the work on carbene complexes which bear heteroatomic substituents has been carried out by Professor E.O. Fischer and his group at the Technische Universität, Munich. The first carbene complex was described in 1964 by Fischer and Maasböl. The carbonyl ligands in transition metal carbonyls are polarized M—$\overset{\delta+}{C}$$\equiv$$\overset{\delta-}{O}$ and are susceptible to nucleophilic attack (p.332). Alkyl and aryl-lithium reagents add to a carbonyl group of the hexacarbonyls $M(CO)_6$ (M = Cr, Mo, W) to give an anionic species which can be formulated either as an acyl or a carbene complex. Treatment with a strongly electrophilic trialkylox-onium salt (a source of R^+) or protonation followed by methylation of the resulting hydroxycarbene derivative, affords the product.

233

$$\left[(OC)_5 \, M{-}C{\underset{R}{\overset{O}{<}}} \right]^{-} Li^+$$

$$(OC)_5 M{-}C{\equiv}O + RLi \longrightarrow (OC)_5 M{\leftarrow}C{\underset{R}{\overset{\bar{O}Li^+}{<}}} \xrightarrow{Me_4N^+Cl^-} (OC)_5 M{\leftarrow}C{\underset{R}{\overset{\bar{O}\,\overset{+}{N}Me_4}{<}}}$$

can be isolated,
crystalline

$$Me_3O^+BF_4^-$$
$$-LiBF_4$$

$$H^+$$

$$(OC)_5 M{\leftarrow}C{\underset{R}{\overset{O{\diagdown}Me}{<}}} \xleftarrow[Et_2O]{CH_2N_2} (OC)_5 M{\leftarrow}C{\underset{R}{\overset{OH}{<}}}$$

labile

$(M=Cr,Mo,W)$
$(R=alkyl,aryl)$

The molecular structure of a typical Fischer carbene complex is shown in Fig. 7.8. Particularly notable is the short C—O bond which is intermediate between a typical single (1.46 Å) and a double bond (1.21 Å). This indicates significant O—C π-bonding. Moreover the Cr—C$_{carbene}$ bond is longer than the average Cr—CO distance, which suggests less π-character in the former.

Fischer carbenes behave chemically as if a considerable positive charge resides on the carbene carbon atom. They are, for example, susceptible to nucleophilic substitution by a range of reagents such as amines, amino acid esters, thiols, alcohols, alkoxides and carbon nucleophiles. The reactions with amines are assisted by hydrogen bonding to excess reagent or to the solvent.

$$(OC)_5 \, Cr{\leftarrow}C{\underset{R}{\overset{OMe}{<}}} + {:}Nu^- \longrightarrow (OC)_5 \, Cr{\leftarrow}C{\underset{R}{\overset{Nu}{<}}} + OMe^-$$
(or NuH)   (or MeOH)

$NuH = R_2'NH, \quad R'NH_2, \; H_2NCHR'COOR'', \; R'SH, \; R'OH, \; ArOH.$

The positive carbene carbon atom relays some of its charge to the hydrogen atoms of the adjacent methyl group in $(OC)_5Cr(OCH_3)CH_3$. The pK$_a$ value of this complex is about 8, making it one of the strongest neutral carbon acids. On

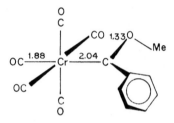

Fig. 7.8 Molecular structure of Ph(MeO)C=Cr(CO)$_5$.

treatment with CD_3O^- in CD_3OD the hydrogen atoms are exchanged for deuterium; nucleophilic substitution of $-OCH_3$ by $-OCD_3$ also occurs giving $(OC)_5CrC(OCD_3)CD_3$.

Can the carbene ligand be released from these complexes and used in synthesis? None of their reactions has yet been shown to involve free carbenes although the group C(OR')R is retained on reaction with various reagents or on thermal decomposition.

Vinyl- and phenyl-methoxycarbene complexes condense with alkynes to form benzene or naphthalene derivatives respectively. One carbonyl ligand is also incorporated into the new ring as a phenolic group. The reaction is a key step in a recent synthesis of Vitamin E.

(+ isomer with Me and
long side chain reversed)

In contrast to Fischer carbenes, the alkylidene complexes of Nb and Ta behave as if the polarization is $\overset{+}{M}-\overset{-}{C}RR'$. In this they resemble phosphorus ylids $R_3\overset{+}{P}-\overset{-}{C}RR'$ (p. 136) and they react with ketones in a Wittig type reaction.

$$(Me_3CCH_2)_3Ta=CHCMe_3 + RCOR' \longrightarrow cis \text{ and } trans \text{ } RR'C=CHCMe_3 + [Ta]=O \text{ product}$$

7.2.2 Carbyne complexes

Treatment of $(OC)_5M=C(OMe)R$ (R = Me, Ph) with Lewis acids $(BX_3, AlX_3, GaX_3; X = Cl, Br, I)$ affords carbyne complexes.

$$(OC)_5 M{=\hspace{-3pt}=}C\underset{R}{\overset{OMe}{<}} \quad \xrightarrow[-45°C]{BX_3} \quad X{-}M{\equiv}C{-}R + X_2BOMe + CO$$

(with the central structure showing M bearing CO ligands: two C≡O above and two C≡O below)

These are rather labile and decompose at 30°–50°C liberating RC≡CR. The *sp* carbon atom has two orthogonal *p*-orbitals which can form two π-bonds with appropriate metal orbitals. The short M—C distance (1.67 Å where X = I, M = Cr, R = Me) indicates a triple bond.

A vinylidene complex is formed when $W(CO)_4$ (dppe) is heated with phenylethyne in THF. This is reversibly protonated, affording a cationic carbyne. The carbonyl ligand *trans* to the triple bond can be substituted by halide or by neutral ligands.

$$W(CO)_6 \xrightarrow{dppe} W(CO)_4\,dppe \xrightarrow[THF]{PhC{\equiv}CH} \Big[W(CO)_3\,dppe\,(PhC{\equiv}CH) \Big] \longrightarrow$$

$$mer\,[(dppe)(CO)_3 W{=}\overset{\alpha}{C}{=}\overset{\beta}{CHPh}]$$

$$\xrightarrow[Et_3N]{HBF_4} mer\Big[(dppe)(OC)_3 W{\equiv}CCH_2Ph\Big]^+ BF_4^- \xrightarrow{Et_4\overset{+}{N}X^-} trans\Big[(dppe)(OC)_3 X\,W{\equiv}CCH_2Ph\Big]$$

While electrophiles (e.g. H^+) attack the β-carbon atom in vinylidenes, nucleophiles add to the α-position yielding η^1-vinyl complexes.

$$\overset{\displaystyle N}{\underset{\displaystyle |}{[M]^-{-}C{=}CHR}} \xleftarrow{N^-} [M]{=}C{=}CHR \xrightarrow{E^+} \overset{\displaystyle E}{\underset{\displaystyle |}{[M]^+{\equiv}C{-}CHR}}$$

Carbyne (alkylidyne) complexes are implicated in the metathesis of alkynes (p. 377).

7.2.3 Bridged methylene complexes

The $CH_2(CR_2)$ group acts not only as a terminal ligand but can, like CO, form a bridge between two metal atoms. These metal atoms can be the same or different. It has been suggested that any μ-CH_2 compound can probably exist if the corresponding μ-carbonyl is known. One preparation involves the addition of Fischer carbenes to coordinatively unsaturated complexes.

$$[M']: + [M]{=}C\underset{R'}{\overset{R}{<}} \longrightarrow [M']{-}[M]; \; cf. \; [M'] + {>}C{=}C{<} \longrightarrow [M']{-}\| \; or \; [M']{<}\|$$

$$Pt(PMe_3)_4 + (OC)_5 W{=}C\begin{smallmatrix} \\ OMe \\ \\ Me \end{smallmatrix} \longrightarrow (Me_3P)_2 Pt{-}W(CO)_5$$

A complex with a CH_2 group bridging a main group and a transition element has been prepared from Cp_2TiCl_2 (p. 290) and $AlMe_3$.

$$Cp_2TiCl_2 + 2\ AlMe_3 \longrightarrow$$

Tebbe's reagent
(reddish−orange crystals)

It reacts with ketones to produce methylene derivatives in higher yields than are obtained by the Wittig reaction. It adds to alkynes and to alkenes forming metallocycles. This is a pivotal step in the accepted mechanism for olefin metathesis (p. 373).

Tebbe's + compound

$$+ Me_2N{-}\langle\bigcirc\rangle{-}N\ AlMe_2Cl$$

The four membered titanacyclobutane ring is nearly planar. Studies of Tebbe's compound are also of interest in connection with Ziegler-Natta polymerization of alkenes. Commercial catalysts incorporate $TiCl_3$ and aluminium alkyls (p. 371).

7.3 Complexes of alkenes

η^2-Alkene complexes have been prepared for nearly all the d-block transition elements. As ethene and its derivatives are good π-acceptors (p. 196), many of the most stable complexes are formed by elements late in the transition series. It is these elements which provide many of the catalysts, both homogeneous and heterogeneous, used to promote alkene reactions such as hydrogenation (p. 182), hydroformylation (p. 387) and oligomerization (p. 365).

7.3.1 *Preparation of η^2-alkene complexes*

(a) LIGAND DISPLACEMENT.   Carbon monoxide can sometimes be displaced from a metal carbonyl by an alkene:

$$Fe_2(CO)_9 + C_2H_4 \xrightarrow[\text{2 days}]{\text{50 atm}} (\eta^2\text{-}C_2H_4)Fe(CO)_4 + Fe(CO)_5$$

The complexes formed by mono-olefins are often rather unstable. Thus tetracarbonyl(ethene)iron, a yellow oil, b.p. $34°C/12$ mm, decomposes slowly at room temperature with loss of ethene to give $Fe_3(CO)_{12}$. Photolysis can be useful in effecting the displacement of CO, for example when the thermal reaction has a high activation energy (p. 169). Often complexes of alkenes which carry electron attracting groups form more readily and are more robust than those of ethene itself.

$$Ni(CO)_4 + H_2C{=}CHCN \rightarrow [Ni(CH_2{=}CHCN)_2]_n \xrightarrow{(RO)_3P} \overset{NC}{\underset{\|}{{}}}{-}Ni\{P(OR)_3\}_2$$

polymeric

Bidentate diolefins such as 1, 5-cyclooctadiene or norbornadiene can chelate to a metal centre forming much more stable products than those obtained from monoolefins.

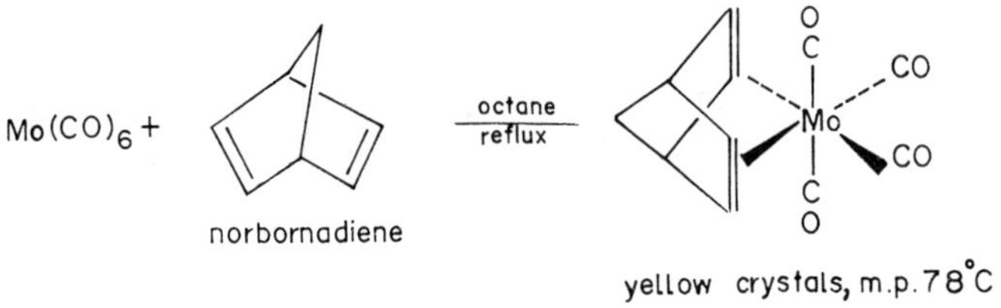

norbornadiene

yellow crystals, m.p. $78°C$

$$\text{cf. } Mo(CO)_6 + C_2H_4 \xrightarrow{h\nu} Mo(CO)_5(C_2H_4) \xrightarrow{h\nu} Mo(CO)_4(C_2H_4)_2$$
$$\text{dec.}$$

Tricarbonyl($\eta^2:\eta^2$-1, 5-cyclooctadiene)iron, prepared by photolysis, rearranges on heating to the η^4-1, 3-cyclooctadiene derivative.

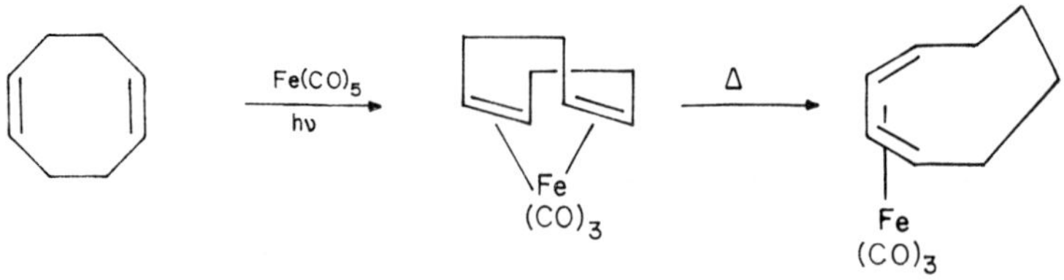

Zeise's salt, $K[PtCl_3(C_2H_4)]H_2O$ (p. 196) can be prepared by passing ethene into an acidic aqueous solution of K_2PtCl_4. The reaction requires several days and elevated pressures. Addition of tin(II) chloride as a catalyst permits the conversion to be carried out in about 4 h at ambient temperature and pressure. Palladium also forms analogous anionic complexes, but the ethene is oxidized in aqueous media to ethanol, forming the basis of the Wacker process (p. 380). Under other conditions halogen bridged dimers or η^3-allyl complexes (p. 261) are formed.

The use of the benzonitrile complex illustrates an approach often adopted in the preparation of labile derivatives. This is to take as starting material a complex from which the ligands can be even more readily displaced. An excess of the incoming ligand can be used, particularly if it is volatile and can be pumped away at the end of the reaction.

Use of reducing agent: The starting materials for the synthesis of organometallic compounds which are readily available are halides, acetylacetonates etc. In many cases it is necessary to include a reducing agent to remove the halide and to provide electrons to the metal centre (p. 166). For rhodium and iridium, the reducing medium is provided by an alcohol.

The dimeric products are useful starting materials for the preparation of other complexes, as the alkenes are readily displaced. The Z-cyclooctene compounds can be stored indefinitely at 0°C under nitrogen.

Stronger reducing agents such as metals, aluminium alkyls or Grignard reagents are often required, especially for elements of the First Transition Series.

$$Ni(acac)_2 + 1,5,9\text{-cyclododecatriene} + \tfrac{2}{3}AlR_3 \longrightarrow (1,5,9\text{-CDT})Ni + \tfrac{2}{3}Al(acac)_3 + 2R^{\bullet}$$

Trigonal planar, 16e
Colourless needles

7.3.2 Some cationic alkene complexes of iron

The chemistry of some cationic alkene derivatives of iron illustrates how several different methods can sometimes be used to prepare the same compound, how the reactions of an organic molecule can be modified on coordination and how advantages of this can be taken in organic synthesis.

The dimer $[CpFe(CO)_2]_2$ (p. 295) is the starting material for these studies. The $\eta^5\text{-}C_5H_5$ and CO ligands are spectators in the reactions to be described, so it is convenient to write $\eta^5\text{-}C_5H_5Fe(CO)_2$ as Fp. The dimer is thus written Fp—Fp or Fp_2. Fp_2 is reduced by sodium amalgam in THF to a nucleophilic anion Fp^- which attacks alkyl halides:

Cationic η^2-alkene complex

Each of the isomeric η^1-alkyl complexes reacts with $Ph_3C^+BF_4^-$ by abstraction of a hydrogen to iron to form the same cationic η^2-propene complex. The conversion of a neutral η^n-complex to a cationic η^{n+1}-species is a property of the Ph_3C^+ reagent (p. 323). It occurs when the complex cation is more stable than the triphenylmethyl cation. The propene complex is also obtained by protonation of the η^1-propenyl derivative. Protonation of an uncoordinated double bond β- to a metal–carbon linkage also provides a general method of converting neutral η^n- into cationic η^{n+1} species. Note the use of the weakly coordinating ion BF_4^-. PF_6^- or ClO_4^- can also be used, although complexes containing the latter are

240

liable to dangerous unpredictable self oxidation. Coordinating anions attack either the metal centre or the ligand. Thus reaction of Fp(alkene)$^+$ with sodium iodide in acetone displaces the alkene giving FpI.

The η^2-alkene complexes can also be obtained directly from Fp$_2$ (by oxidation in presence of the alkene) or from FpX (X = Cl, Br, I).

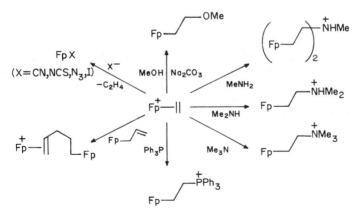

The air-stable 2-methylbutene complex is a useful starting material. The sterically hindered volatile alkene is displaced by many other alkenes at 60°C.

Use in organic synthesis: Free alkenes are susceptible to attack by electrophiles such as acids or halogens, but not normally to attack by nucleophiles. For the cationic iron complexes the reverse is observed. Fp$^+$ therefore provides a convenient group to protect alkenes against electrophilic attack. The Fp group is readily attached to an alkene by exchange with the 2-methylbutene compound. The desired reaction is carried out and the Fp group removed by treatment with sodium iodide. Fp$^+$ can also be used to protect an alkene against hydrogenation or mercuration.

Nucleophilic attack on the 18-electron Fp(alkene)$^+$ cations cannot proceed by coordination of the nucleophile to the metal unless displacement of a ligand also occurs (as with iodide). Generally, however, nucleophiles add to the alkene, yielding substituted alkyl derivatives (Fig. 7.9). Such addition always takes place stereospecifically *trans* to the metal. The Fp group can be removed from Fp—R in

Fig. 7.9 Some reactions of [Fp(C$_2$H$_4$)]$^+$ with nucleophiles.

241

several ways. Oxidation by cerium(IV) probably gives a 17-electron radical cation initially, which then rearranges by alkyl migration to an acyl derivative.

$$FpI + RI \xleftarrow{I_2} CpFe(CO)_2R \xrightarrow[aq]{Ce^{4+}} [CpFe(CO)_2R]^+ \longrightarrow [CpFe(CO)(COR)]^+ \xrightarrow{MeOH} RC\begin{smallmatrix}O\\OMe\end{smallmatrix}$$

$$CpFe(CO)_2R \xrightarrow{HgCl_2} FpCl + RHgCl$$

$$CpFe(CO)_2R \xrightarrow[CS_2]{Br_2/} FpBr + RBr$$

Attack by carbon nucleophiles such as enolates on Fp(alkene)$^+$ is not at all regioselective. The propene complex reacts readily with the dimethyl malonate ion, but a mixture of isomeric products is formed.

$$Fp\overset{}{\underset{CH(CO_2Me)_2}{\diagup}} \longleftarrow \left(Fp \diagup^+ \longleftrightarrow \overset{+}{Fp}-\| \right) \longleftrightarrow \left(\longleftrightarrow Fp\diagdown/ \right) \longrightarrow Fp\diagdown\diagup^{CH(CO_2Me)_2}$$

($1/3$ of product)   ($2/3$ of product)

Lithium alkyls or Grignard reagents generally reduce the alkene complex to Fp$_2$. This problem can sometimes be overcome by using LiCuR$_2$ instead.

The addition of carbon nucleophiles is probably irreversible and kinetically controlled. Amines and alcohols, however, add reversibly and the thermodynamically more stable product is obtained selectively. A novel synthesis of β-lactams involves attack by a primary amine followed by oxidation.

$$R\diagdown\overset{\|}{\underset{Fp^+}{}} \xrightarrow[base]{R'NH_2/} R\diagdown\diagup\overset{}{\underset{R'NH \quad Fp}{}} \xrightarrow[or\ air]{PbO_2} \left[\begin{matrix} R\\ H\diagdown\overset{}{\underset{R'}{N}}\diagup\overset{+/Cp}{\underset{C}{Fe}}\diagdown CO \\ \| \\ O \end{matrix} \right] \xrightarrow{\hat{o}} R\diagdown\overset{}{\underset{R'}{}}\overset{N}{\diagdown}\diagup^O$$

(migratory insertion of CO)   β-lactam

Greatly improved regioselectivity is observed when the alkene ligand has electron-withdrawing substituents. Vinyl ether complexes, prepared from α-bromoacetals, are attacked exclusively at the carbon atom which bears the

$$\overset{RO}{\underset{Br}{}}\diagup\overset{OR}{\underset{R'}{}} \xrightarrow[THF]{NaFp/} \overset{RO}{\underset{Fp}{}}\diagup\overset{OR}{\underset{R'}{}} \xrightarrow[-78°C]{HBF_4Et_2O/\ CH_2Cl_2} \overset{OR}{\underset{Fp^+}{}}\diagup\overset{}{\underset{R'}{}} + ROH$$

$$\overset{OEt}{\underset{Fp^+}{}}/ \xrightarrow[THF,-78°C]{Nu^-} \overset{OEt}{\underset{Fp \quad Nu}{}}/ \xrightarrow[-78°C]{HBF_4Et_2O/\ CH_2Cl_2} \overset{}{\underset{Fp^+}{}}/\overset{Nu}{} \xrightarrow[acetone]{NaI/} \overset{}{}/\overset{Nu}{} + FpI$$

alkoxy group. On protonation the product loses alcohol to give a cationic complex in which the ether group has been replaced by the nucleophile. The new vinyl derivative is released from iron by treatment with NaI in acetone. If $NuH = R'OH$, the alkyl groups in a vinyl ether can be exchanged. Less trivially, the method has been used to make vinyl derivatives of enolates

The molecular structure of $Fp(H_2\overset{\beta}{C}=\overset{\alpha}{C}HOEt)^+$. BF_4^-, determined by X-ray diffraction, shows unsymmetrical bonding of the vinyl group to iron. The distance $Fe-C_\alpha$ (2.32 Å) is significantly longer than $Fe-C_\beta$ (2.09 Å). This, together with the rather long $C-C$ bond (1.42 Å), suggests that the charge distribution may be described in terms of contributions from the canonical structures $Fp-CH_2\overset{+}{C}HOEt$ and $Fp-CH_2CH=\overset{+}{O}Et$.

7.4 Complexes of alkynes

An alkyne possesses two pairs of orthogonal π and π^* molecular orbitals. In many alkyne complexes, the major interactions are analogous to those found in alkene derivatives, viz. σ-donation from a filled bonding $(\pi_\parallel)$ m.o. on the ligand to the metal and π-donation from filled metal orbitals. In addition there can be interaction between the filled $\pi_\perp$ orbital of the ligand and $M(d_{xy})$, so that an alkyne can function as a π-donor as well as a π-acceptor. δ-Overlap between $M(d_{yz})$ and $\pi_\perp^*$ is also permitted on symmetry grounds, although it is thought not to be energetically very significant (Fig. 7.10).

As with alkenes, two extreme representations of the bonding in alkyne complexes can be drawn, in which the alkyne is formally a two-electron donor (p. 197). Arylalkynes form a series of stable complexes with platinum of stoichiometry $(Alkyne)Pt(PPh_3)_2$. The infrared spectrum of the diphenylethyne complex shows a $C-C$ stretching band at $1750\,cm^{-1}$ (cf. $\nu C\equiv C$ in C_2H_2, $2100\,cm^{-1}$ and $\nu C\equiv C$ in C_2H_4, $1623\,cm^{-1}$, Raman). The X-ray crystal structure reveals a planar P_2PtC_2 skeleton. The phenyl groups are bent back considerably from linearity. This observation, together with the fact that the $C-C$ bond length is close to that of a typical double bond (1.33 Å) suggests that the structure can

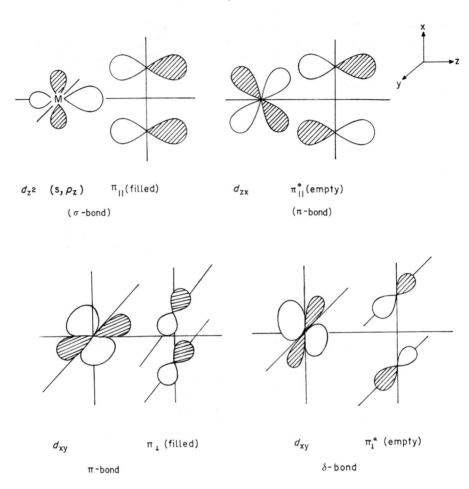

d_{z^2} (s, p_z) $\pi_{\parallel}$ (filled) d_{zx} $\pi_{\parallel}^*$ (empty)

$(\sigma\text{-bond})$ $(\pi\text{-bond})$

d_{xy} $\pi_{\perp}$ (filled) d_{xy} $\pi_{\perp}^*$ (empty)

π-bond δ-bond

Fig. 7.10 Possible bonding interactions in a metal—alkyne bond.

approximately be described as a metallacyclopropene. The resembles the metallocyclopropane structure adopted by the corresponding complex of the electrophilic alkene, C_2F_4. The square planar stereochemistry requires an electron count of 16, making the alkyne formally a two-electron donor. Donation into the $\pi_{\parallel}^*$ orbitals must be important here.

Some platinum complexes in which the alkyne lies perpendicular to the plane defined by P_2PtMe have been prepared as follows:

$$\textit{trans } PtCl(Me)(PPh_3)_2 \xrightarrow[\text{AgPF}_6]{RC\equiv CR} \quad \xrightarrow{MeOH} $$

Here the alkyne is attacked by nucleophiles, like the alkenes in Fp(alkene)$^+$ or the alkynes in the analogous [CpFe(CO)(PPh$_3$)(RC≡CR)]$^+$.

When both π-systems of the alkyne are significantly involved in bonding to a metal, simple electron counting is no longer helpful. Metal $d\pi$ orbitals which interact with strongly π-donating ligands can be raised in energy to such an extent that they remain empty. In some cases e.g. [CpMoL$_2$(RC≡CR)]$^+$, an apparently correct 18-electron count can be obtained by including the π-donating electron pair, making the alkyne a formal four-electron donor. In others e.g. [CpMo(CO)(RC≡CR)$_2$]$^+$ this is confusing, because the π-donor orbitals of both alkyne ligands interact with the same metal d-orbital.

The preparation of the complexes just cited is given below. They undergo some unusual reactions with nucleophiles.

There are also complexes in which an alkyne forms a bridge between two metal atoms. To a first approximation, one of the π-systems interacts with the first metal atom and the other with the second (Fig. 7.11). The cobalt compounds are prepared in high yield by reacting alkynes with Co$_2$(CO)$_8$. The cobalt can be removed cleanly by mild oxidation. This provides a useful procedure for protecting an alkyne during organic synthesis.

Organotransition metal chemistry

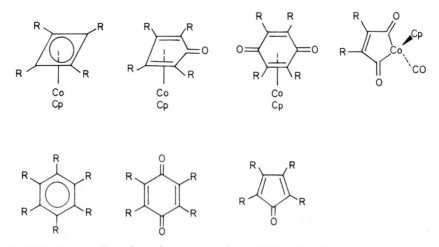

(Semi−bridging CO group present)   cf. Co$_2$(CO)$_8$

Fig. 7.11 Molecular structures of two binuclear alkyne complexes.

7.4.1 *Oligomerization of alkynes*

Alkyne molecules are often linked together in their reactions with transition metal compounds. An early example of this, discovered by Niewland in 1931, was the linear dimerization of ethyne to but-1-en-3-yne, catalysed by copper(I) chloride. The product was for many years an intermediate in the manufacture of chloroprene for synthetic rubber.

$$2HC\equiv CH \xrightarrow[\text{NH}_4\text{Cl}]{\text{CuCl}} HC\equiv C-CH=CH_2$$

The reactions of alkynes with transition metal complexes often lead to bewildering mixtures of products. Substituted alkynes often produce η^4-cyclobutadiene derivatives (p. 269); metal carbonyls also afford products containing cyclopentadienone and quinone ligands, which incorporate two alkyne units together with one or two carbonyl groups. Some of these systems catalyse the cyclotri- or tetra-merization of alkynes to benzenes or cyclooctatetraenes respectively (Fig. 7.12).

Fig. 7.12 Some products from the reaction of CpCo(CO)$_2$ with alkynes, RC≡CR.

246

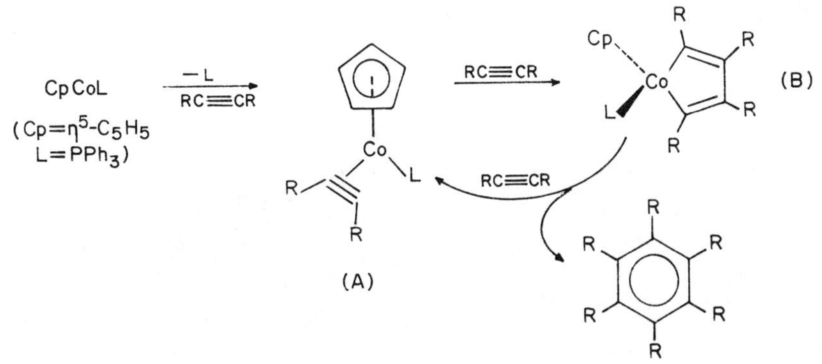

Fig. 7.13 Catalytic cycle for cyclotrimerization of an alkyne to an arene.

In many of these reactions the alkyne complex (A) which is first formed is converted into a metallacyclopentadiene (B), the key intermediate. This has been characterized in the cobalt phosphine system, which catalyses the cyclotrimerization of alkynes to arenes (e.g. 2-butyne to hexamethylbenzene) at 120°C (Fig. 7.13).

The first step is dissociation of L(Ph$_3$P), as addition of more triphenylphosphine inhibits the catalysis. Other cyclopentadienylcobalt complexes CpCoL$_2$ are also effective catalysts, (with L$_2$ = 1, 5-cyclooctadiene, 2C$_2$H$_4$ or, less efficiently, 2CO). By replacing RC≡CR with a nitrile RCN in the final step (i.e. by using a mixture of nitrile and alkyne), pyridines can be synthesized catalytically, although some benzene derivative is always obtained as well. These processes have been developed industrially for making pyridines and in the laboratory for effecting cycloadditions in natural product synthesis:

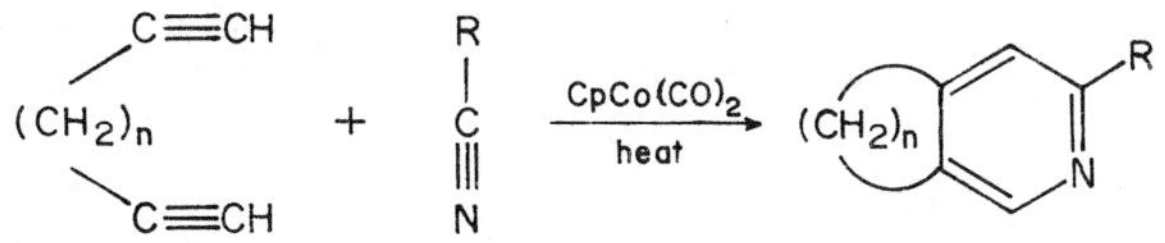

The cobaltacyclopentadiene is converted on heating into a cyclobutadiene derivative. Treatment with carbon monoxide affords a cyclopentadienone complex. It is therefore probably an intermediate in the formation of these products from CpCo(CO)$_2$ (L = CO) and alkynes.

The discovery by Reppe that ethyne is converted catalytically into its tetramer, cyclooctatetraene by nickel(II) cyanide aroused much interest, as it provided a direct route to a compound which had previously been obtained only after a long multi-stage synthesis. Many nickel complexes catalyse the cyclooligomerization of ethyne. Ni(CN)$_2$ and Ni(CH$_2$=CHCN)$_2$ in benzene yield cyclooctatetraene in about 70% yield at 80–120°C/10–25 atm. Other products include benzene and polymer. Tetramer formation is suppressed by coordinating solvents such as pyridine, vinylacetylene (but-1-en-3-yne) and polymer being produced instead.

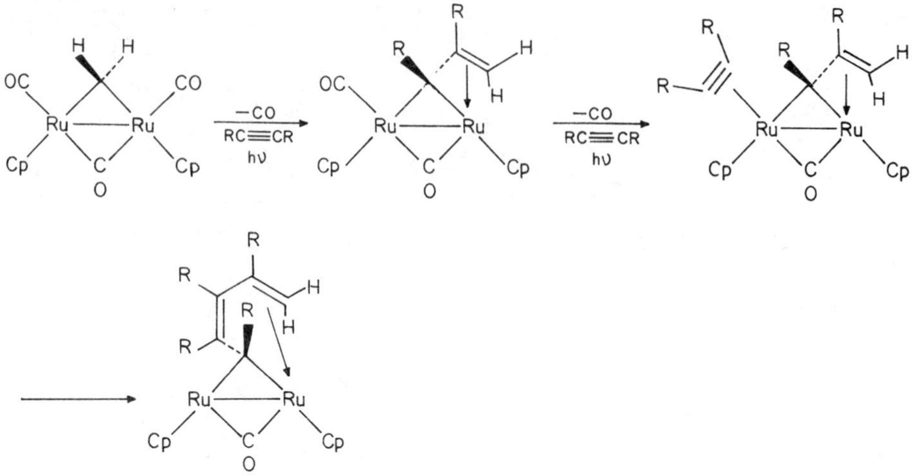

Fig. 7.14 Oligomerization of alkynes at a diruthenium centre.

Addition of phosphine or use of phosphine complexes such as $Ni(CO)_2(PPh_3)_2$ causes trimerization to benzene; the phosphine presumably occupies one coordination site otherwise taken up by ethyne (cf. p. 367). It is envisaged that these oligomerizations take place in a stepwise fashion *via* alkyne and metallacyclopentadiene intermediates rather than in a concerted fashion in which three or four coordinated alkyne molecules link together at once.

Some related reactions of commercial importance, alkyne carbonylation to acrylic acid and acrylic esters, are discussed in Chapter 12.

Oligomerization of alkynes also takes place at dimetal centres bridged by μ-methylene groups. Fig. 7.14 illustrates the types of reactions observed. Similar processes occur at Mo_2, W_2, Fe_2 and Co_2 centres.

7.5 References

Bergman, R.G. (1984) Activation of alkanes with organotransition metal complexes. *Science*, **223**, 902.

Bönnemann, H. (1985) Organocobalt compounds in the synthesis of pyridines. *Angew. Chem. (Int. Edn)*, **24**, 248.

Brookhart, M. and Green, M.L.H. (1983) Agostic hydrogen. *J. Organomet. Chem.*, **250**, 395.

Brown, F.J. (1980) Carbene complexes. *Prog. Inorg. Chem.*, **27**, 1.

Crabtree, R. (1985) Organometallic chemistry of alkanes. *Chem. Revs.*, **85**, 235.

Dötz, K.H. (1984) Carbene complexes in organic synthesis. *Angew. Chem. (Int. Edn)*, **23**, 587.

Fischer, E.O. (1976) Carbene complexes; a personal account. *Adv. Organomet. Chem.*, **14**, 1.

Fischer, H., Kreissl, F.R., Schubert, U., Hoffman, P., Dötz, K.-H. and Weiss, K. (1983) *Transition Metal Carbene Complexes*, Festschrift für E.O. Fischer, Verlag Chemie, Weinheim.

Graham, W.A.G. (1986) From silicon-hydrogen to carbon–hydrogen activation. *J. Organomet. Chem.*, **300**, 81.

Green, M. (1986) The road from alkynes to molybdenum–carbon multiple bonds. *J. Organomet. Chem.*, **300**, 93.

Halpern, J. (1982) Formation of C—H bonds by reductive elimination. *Acc. Chem. Res.*, **15**, 332.

Halpern, J. (1985) Activation of C—H bonds by metal complexes. *Inorg. Chim. Acta*, **100**, 41.

Heck, R.F. (1979) Palladium catalysed reactions of organic halides with olefins. *Acc. Chem. Res.*, **12**, 146.

Heck, R.F. (1985) *Palladium Reagents in Organic Chemistry*, Academic Press, London.

Herrmann, W.A. (1982) Methylene complexes. *Adv. Organomet. Chem.*, **20**, 160.

Negishi, E. (1982) Palladium and nickel catalysed cross coupling. *Acc. Chem. Res.*, **15**, 340.

Rosenblum, M. (1986) The chemistry of dicarbonylcyclopentadienyliron complexes. *J. Organomet. Chem.*, **300**, 191.

Schrock, R.R. (1979) Alkylidene complexes of Nb and Ta. *Acc. Chem. Res.*, **12**, 98.

Toscano, P.J. and Macilli, L.G. (1984) B_{12} and related organocobalt chemistry. *Prog. Inorg. Chem.*, **31**, 106.

Vollhardt, K.P.C., (1984) Cobalt-mediated cycloadditions. *Angew. Chem. (Int. Edn)*, **23**, 539.

Problems

1. Discuss methods of preparation of transition metal compounds containing metal—carbon σ-bonds. Consider factors affecting their stability. Why is $[Ru(C_2H_5)(Cl)(PPh_3)_3]$ stable only under a pressure of ethene?
(*University of Leicester*).

2. Treatment of $MeMn(CO)_5$ with triphenylphosphine gives a product A of molecular formula $C_{24}H_{18}O_5MnP$. The ^1H n.m.r. spectrum shows signals due to phenyl protons (relative intensity 15) and methyl protons (relative intensity 3). The i.r. spectrum shows four bands in the region 2058 to 1927 cm^{-1} and another at 1623 cm^{-1}. Propose, with reasoning, a structure for the product A.
(*University of Leicester*).

3. Two main mechanisms have been proposed for the carbonylation of $CH_3Mn(CO)_5$: the carbonyl insertion (A) and the methyl migration (B).
(a) Illustrate what is meant by these two possible mechanisms, A and B, using CO as the carbonylating agent.
(b) Show that the use of ^{13}C-enriched CO cannot help to distinguish between mechanisms A and B, although it can eliminate the possibility of direct insertion from the gas phase.
(c) Decarbonylation of *cis*-$CH_3COMn(CO)_4(^{13}CO)$, which is formed by the carbonylation reaction in (b), gives a mixture of *trans*- and *cis*-$CH_3Mn(CO)_4(^{13}CO)$ in the ratio 1:2. Show that this result is consistent only with mechanism B.
(d) Explain in detail how the infrared absorption spectra of the reaction products may be used for the preceding analysis.
(*University of York*).

Organotransition metal chemistry

4. The reaction $MeMn(CO)_5 + L$ follows a rate law

$$Rate = \frac{k_1 k_2 [L][MeMn(CO)_5]}{k_{-1} + k_2[L]}.$$

Show that this is consistent with a two step process

$$MeMn(CO)_5 \underset{k_{-1}}{\overset{k_1}{\rightleftharpoons}} MeCOMn(CO)_4 \overset{k_2}{\underset{L}{\longrightarrow}} MeCOMn(CO)_4 L$$

What must be the relative magnitudes of k_1, k_{-1} and k_2 [L] for such a scheme to be valid?

If an excess of L is taken (pseudo first order conditions), sketch a graph to show the variation of the observed rate constant and the concentration of L. Suggest why Lewis acids such as $AlBr_3$ accelerate this reaction.

5. Unstable *cis* hydridoalkyls of platinum can be obtained in solution as follows:

$$\textit{trans}\,[PtHClL_2] \xrightarrow[-78°C]{MeMgX} \textit{cis}[PtHMeL_2] \quad (L = Ar_3P)$$

$$\qquad A \qquad\qquad\qquad B$$

(a) What is the stereochemistry about platinum in A and B?
(b) Decomposition of B according to $B + L \rightarrow PtL_3 + CH_4$ follows the rate law Rate $= k$ [B], independent of the concentration of added L. k^H/k^D for the decomposition of *cis*[PtHMeL_2] and *cis*[PtDMeL_2] is 3.3. Suggest a mechanism.
(c) Suggest why electron withdrawing substituents in $L = (p\text{-}XC_6H_4)_3P$ accelerate the reaction.
(d) How could one show whether the decomposition is intra- or intermolecular?

6. The following scheme illustrates some platinum chemistry:

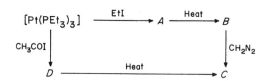

Empirical formulae	Proton n.m.r. data (excluding PEt_3 signals) (δ relative to TMS, relative intensities in parentheses)
A $PtIP_2C_{14}H_{35}$	0.8(2H), 0.3(3H)
B $PtIP_2C_{12}H_{31}$	-12.7(triplet, 1H)
C $PtIP_2C_{13}H_{33}$	0.5(triplet, 3H)
D $PtIP_2OC_{14}H_{33}$	1.8(singlet, 3H)

(a) Propose structures for complexes A to D.
(b) Comment on each of the reaction steps in the scheme.
(c) Deduce the deuterium sites in A, B and C if CD_3CH_2I is used to form A.

(For the interpretation of the n.m.r. data ignore all couplings except those between $^1H\,(I=\frac{1}{2})$ and $^{31}P\,(I=\frac{1}{2})$.)
(*University of Southampton*).

7. Suggest molecular structures for the products of the reactions below, clearly indicating the bonding modes of organometallic ligands and briefly explaining important features where appropriate.

(a) $Fe(CO)_5 \xrightarrow{C_4H_6} Fe(C_4H_6)(CO)_3, \quad C_4H_6 = 1,3\text{-butadiene}$

$Fe(CO)_5 \xrightarrow{C_4F_6} Fe(C_4F_6)(CO)_4 \quad C_4F_6 = \text{hexafluoro-1,3-butadiene}$

(b) *trans*-$PtCl_2(PhNH_2)_2 \xrightarrow[MeC\equiv CMe]{} PtCl_2(C_4H_6)(PhNH_2)$

$Co_2(CO)_8 \xrightarrow[PhC\equiv CPh]{} Co_2(CO)_6(C_{14}H_{10})$

(c) $CrCl_3 \xrightarrow[LiCH_2SiMe_3]{} Li[Cr(CH_2SiMe_3)_4]$

$IrCl(CO)(PEt_3)_2 \xrightarrow[MeI]{} ?$

(d) $Fe(C_5H_5)(CO)_2(CH_3) \xrightarrow[CO]{} Fe(C_5H_5)(CO)_2(COCH_3)$

$Cr(CO)_6 \xrightarrow[Et_2NH]{} (CO)_5Cr{=}C(OH)NEt_2$

(*University of Edinburgh*).

8. Some chemistry of platinum triethylphosphine complexes is shown below:

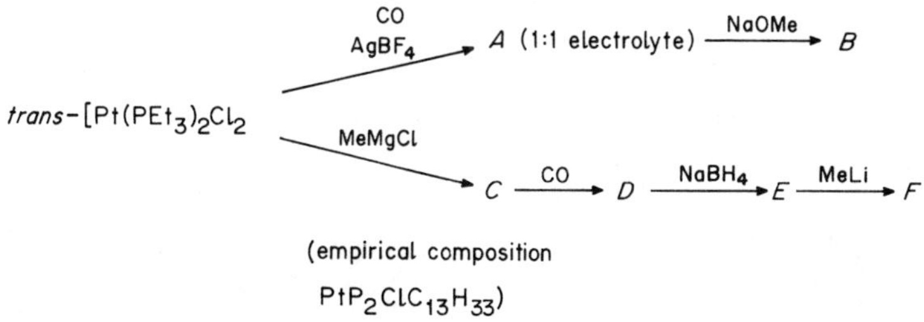

trans$-[Pt(PEt_3)_2Cl_2]$

$\xrightarrow[\text{AgBF}_4]{\text{CO}}$ *A* (1:1 electrolyte) $\xrightarrow{\text{NaOMe}}$ *B*

$\xrightarrow{\text{MeMgCl}}$ *C* $\xrightarrow{\text{CO}}$ *D* $\xrightarrow{\text{NaBH}_4}$ *E* $\xrightarrow{\text{MeLi}}$ *F*

(empirical composition $PtP_2ClC_{13}H_{33}$)

	Infrared data $\nu_{CO}(cm^{-1})$	Decoupled proton n.m.r. data (*excluding* PEt$_3$ signals; δ relative to TMS)
A	2100	—
B	1630	2.8
C	—	0.5
D	1620	2.2
E	—	0.7 and 0.8 (intensities 2:3)
F	—	0.7, 0.8 and 0.5 (intensities 2:3:3)

(a) Propose a structure for each of the complexes *A* to *F*, using and interpreting the spectroscopic data provided.

Organotransition metal chemistry

(b) Discuss each of the reaction steps.

(c) Predict the composition of any gaseous products formed when F is heated, explaining the mechanism of the reactions involved.

(*University of Southampton*).

Allyl and diene complexes of the transition elements

8.1 Allyl (-enyl) complexes

There are two main ways in which the allyl group can link to a transition element. Either it is attached through one of the terminal carbon atoms (η^1) leaving an uncoordinated double bond, or all three carbon atoms of the allyl system are bounded (η^3).

η^1-allyl η^3-allyl

8.1.1 N.m.r. and dynamic processes

A useful way of distinguishing between these bonding modes is by proton n.m.r. spectroscopy. The spectra of one η^1- and one η^3- complex are given in Fig. 8.1. Consider the spectrum of η^1-$C_3H_5Mn(CO)_5$ first. The H^1 ($H^{1'}$) protons are relatively strongly shielded by the adjacent metal atom. They therefore resonate at high field, $1.8\,\delta$. If H^1 and $H^{1'}$ are magnetically equivalent, that is related in space by a symmetry operation on the molecule, this resonance is a doublet of relative intensity 2, $^3H(H^1H^2) \sim 8\,Hz$.*

The remaining protons H^2, H^3, H^4 are substituents on a vinyl (alkenyl) group. Such resonances appear typically at 4.5–$6\,\delta$ in alkenes. The *trans* vicinal coupling constant $^3J(H^2H^3)$ is greater than the *cis* vicinal constant $^3J(H^2H^4)$. The

---

*If the two protons are not so related they are magnetically inequivalent or 'enantiotopic' and two separate signals with different chemical shift values will appear. In this case geminal coupling $^2J(H^1H^{1'})$, the magnitude of which is related to the angle $H^1CH^{1'}$, is observed as well as vicinal coupling $^3J(H^1H^2)$, $^3J(H^{1'}H^2)$.

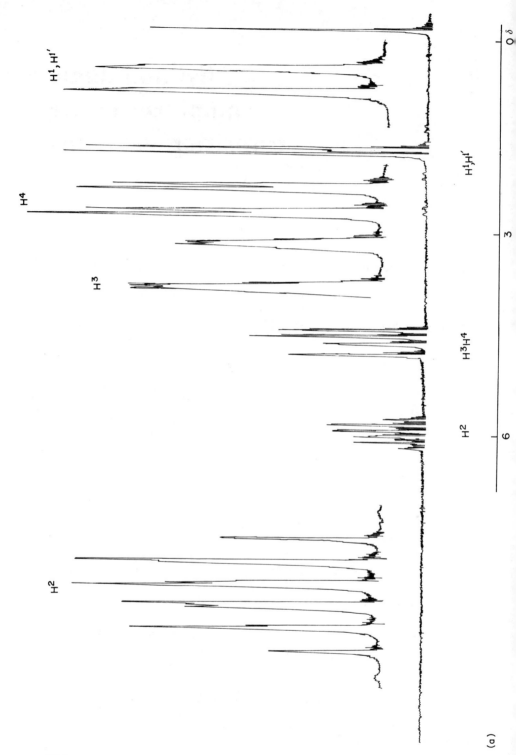

(a)

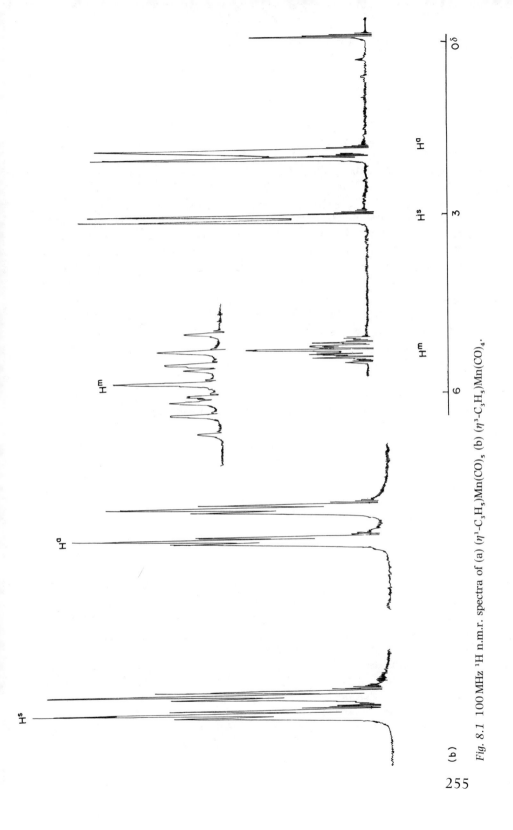

Fig. 8.1 100 MHz ^{1}H n.m.r. spectra of (a) $(\eta^1\text{-}C_3H_5)Mn(CO)_5$ (b) $(\eta^3\text{-}C_3H_5)Mn(CO)_4$.

(b)

255

signals arising from H^3 and H^4 can be assigned on this basis. H^2 gives rise to a multiplet.

The spectrum of $(\eta^3\text{-}C_3H_5)Mn(CO)_4$ can be interpreted in terms of a symmetrical η^3-group. There are three different proton environments; H^m (meso) attached to the central carbon, H^s (syn) and H^a (anti). The anti protons are nearer to the metal centre than the syn protons and hence are more strongly shielded. The signal to highest field is therefore assigned to H^a (doublet). This is confirmed from the coupling constants. The *trans* coupling $^3J(H^aH^m)$ is greater (12.2 Hz) than the *cis* coupling $^3J(H^sH^m)$ (7.4 Hz). Small geminal $^2J(H^aH^s)$ and W $^4J(H^sH^{s'})$ couplings are often resolved in these spectra. H^m appears at low field as a multiplet, a triplet of triplets. (If the difference in chemical shift between two resonances is of the same order of magnitude as the coupling constant, second order spectra, which do not conform to the first order treatment given here, are obtained. This is a general effect which is more likely to be observed using smaller spectrometers operating at low frequency (e.g. 60 or 100 MHz) than with modern high frequency instruments (e.g. 250 or 400 MHz). Chemical shifts, measured in Hz, are proportional to operating frequency, whereas coupling constants are independent of frequency.)

Protons on opposite sides of the allyl group are distinguished in the spectra of some complexes. This can reflect actual asymmetry in bonding as in $(Ph_3P)PdCl(\eta^3\text{-}2\text{-}MeC_3H_4)$. It can, however, merely be caused by slightly different environments of the two syn or anti protons in a molecule of low symmetry. As a general guide to assignment of spectra of η-bonded ligands, the proton resonances normally lie progressively to lower field on going from the front (i.e. H^a, H^s) to the back of the ligand.

The spectrum of $(\eta^3\text{-}C_3H_5)Mn(CO)_4$ conforms to the A_2B_2X pattern. Some allyl complexes show this pattern at low temperatures, but as the temperature is raised the resonances broaden, finally coalescing to an A_4X spectrum in which the syn and anti protons show an averaged signal. Such observations indicate that dynamic processes are proceeding at a rate comparable with the lifetime of the nuclear spin states.

The main mechanism responsible for such changes involves $\eta^3 \rightleftharpoons \eta^1 \rightleftharpoons \eta^3$ interconversions. Formation of the η^1-allyl leaves a vacant coordination site. This can be occupied by a solvent molecule or other ligand. $\eta^3 \rightleftharpoons \eta^1$ interchange is often induced by the addition of ligands or coordinating solvents which stabilize the η^1 intermediate (Fig. 8.2).

Another type of dynamic process, rotation of the allyl group about the metal ligand axis, has also been observed in certain complexes. In this case interchange of syn and anti protons does not occur. The ^1H n.m.r. spectrum of tris(allyl)rhodium shows three separate A_2B_2X patterns at $-74°C$, one from each of three inequivalent allyl groups. On warming to $-10°C$ two of these coalesce. It is suggested that rotation of one of the allyl groups about the allylrhodium axis occurs making the other two allyl ligands equivalent (Fig. 8.3).

Bis(allyl)nickel and the corresponding palladium and platinum compounds

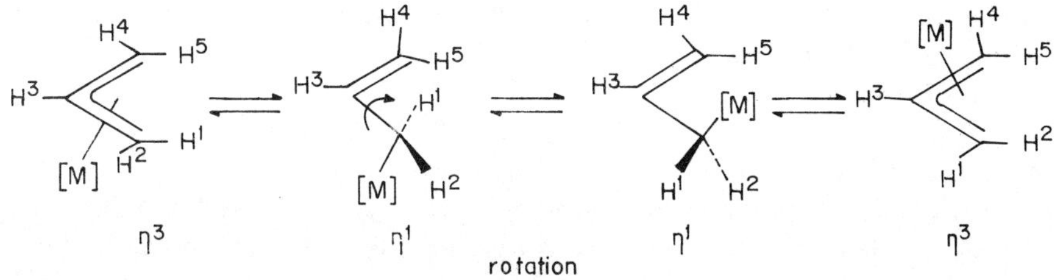

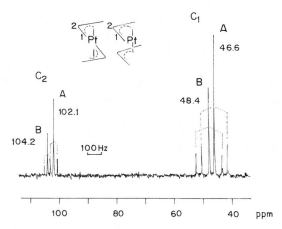

Fig. 8.2 Interchange of syn and anti protons.

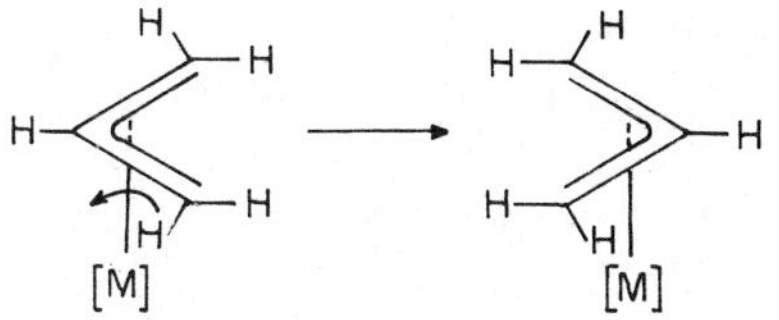

Fig. 8.3 Rotation about M-allyl axis – syn and anti protons not interchanged.

Fig. 8.4 ^{13}C-{^{1}H}-n.m.r. spectrum (25.2 MHz) of (η^3-C$_2$H$_5$)Pt (9°C, toluene-d_8). The satellites are due to coupling with ^{195}Pt (33.7% natural abundance): A (*trans*)J(PtC$_1$) = 225.2, J(PtC$_2$) = 61.1 Hz; B(*cis*)J(PtC$_1$) = 226.0, J(PtC$_2$) = 54.5 Hz. (From Jolly, P.W. and Mynott, R. (1981) *Adv. Organomet. Chem.*, **19**, 257.)

show *cis–trans* geometrical isomerism in solution in hydrocarbon solvents. This is revealed by ^{1}H and ^{13}C n.m.r. studies, when two sets of resonances are observed. At $-24°$C in toluene-d_8 Ni(C$_3$H$_5$)$_2$ consists of 70% *trans* and 30% *cis* isomer. Each isomer gives rise to two signals (C^1, C^3) and C^2 in the ^{13}C spectrum (^{1}H broad band decoupled). In the case of Pt(C$_3$H$_5$)$_2$ (Fig. 8.4) these are accompanied by satellite

257

Allyl and diene complexes

doublets. 33% of the molecules contain ^{195}Pt ($I = \frac{1}{2}$) which couples to the adjacent ^{13}C nucleus. Note that the meso carbon C^2 resonates to low field of the terminal carbon nuclei C^1, C^3.

Trialkylphosphines add to $M(C_3H_5)_2$ (M = Ni, Pd, Pt) giving 1:1 adducts in which the two allyl groups are *cis* to each other. The n.m.r. spectra reveal dynamic $\eta^3 \rightleftharpoons \eta^1$ interconversions between $-100°$ and $0°C$. Addition of excess $L(L = PR_3$, CO) cleaves the allyl groups giving NiL_4. This property has been exploited in catalysis (p. 367).

8.1.2 *Preparation of η^3-allyl complexes*

The methods of preparation of allyl complexes are in general similar to those used for η^1-alkyls. If a free coordination site is available, however, any η^1-allyl which may form initially is converted into the η^3-allyl. Binary allyls of most of the transition elements have been prepared by reacting allyl Grignard reagents with the anhydrous metal halide in ether at low temperatures.

$$MX_n + n\,\text{allMgX} \longrightarrow Mall_n + n\,MgX_2 \quad (\text{all} = C_3H_5 \text{ or substituted allyl})$$

They include $Mall_4$ (M = Zr, Hf, Th, U, Mo, W), $Mall_3$ (M = V, Cr, Fe, Co, Rh, Ir) and $Mall_2$ (M = Ni, Pd, Pt). Binary allyls are generally very air sensitive, have low thermal stability and are labile towards displacement of allyl groups.

Bis(allyl)nickel forms orange yellow crystals, m.p. 0°C after recrystallization from pentane. Its molecular structure has been determined by neutron diffraction at 100 K, which locates the hydrogen atoms precisely, as well as by high resolution X-ray methods. It has a sandwich structure in which the allyl groups are arranged *trans* to each other, that is, with the arrowheads pointing in opposite directions. The syn hydrogen atoms are bent 8.9° out of the allyl plane towards the metal and the anti hydrogens 29.4° away from the metal. This distortion reduces non-bonded interactions between the two anti hydrogens which would arise in a planar allyl group. Moreover the $C(2p)\pi$ orbitals are better directed

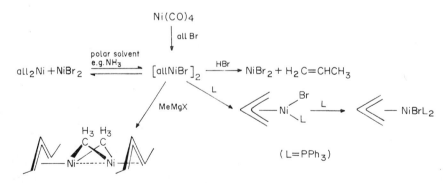

Fig. 8.5 Preparation and reactions of η^3-allylnickel halides.

258

towards the metal; this is thought to be why the meso hydrogen also lies below the allyl plane. Such features are common to many η-bonded ligands (p. 280). Bulky substituents, at the meso position, for example, are often bent away from the metal through steric effects.

Allylnickel halides are catalysts for the stereospecific polymerization of dienes. Some methods of preparation and reactions are summarized in Fig. 8.5.

In solution in polar solvents the complexes couple with primary and secondary alkyl, aryl, vinyl and allyl bromides and iodides at or below room temperature. This was exploited by Corey as early as 1956 in his classic synthesis of α-santalene. Coupling *via* the Grignard reagent gave low yields.

α -santalene

Under conditions of high dilution to prevent intermolecular reactions, two allylic halide groups in the same molecule can be coupled to give a cyclic product. Macrocyclic lactones have been made in this way. Coupling occurs only at the primary centres. Note that the nickel reagents do not attack ester groups. Neither do they react with acid chlorides, ethers, nitriles, olefins or alkyl, aryl or vinyl chlorides (in contrast to bromides or iodides). Aldehydes and cyclic ketones are attacked, however, above 40°C affording homoallylic alcohols.

The palladium analogue $[Pd(\eta^3\text{-}C_3H_5)Cl]_2$ is prepared from Na_2PdCl_4 and allyl chloride in the presence of a reducing agent such as CO, $SnCl_2$ or Zn. It is a yellow crystalline substance, stable in air, which decomposes on heating only at its melting point, 159°C. Its structure, determined by X-ray diffraction, shows symmetrical η^3-allyl groups in which the C—C bond distances (1.36 Å) are equal and the Pd—Cl distances (about 2.41 Å) are similar to those found in other complexes in which an alkene is situated *trans* to Pd—Cl. The plane of the three allyl carbon atoms intersects the plane of the (PdCl)$_2$ system at an angle of

259

Allyl and diene complexes

111.5°. At − 140°C, when thermal motion of the atoms is frozen out, all the metal–carbon distances are equal within experimental error. The CCC angle is close to 120°.

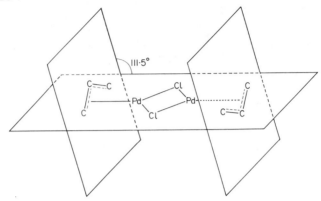

η^3-Allylpalladium halides are formed very readily, for example from alkenes and palladium salts. The reaction is assisted by a base such as sodium acetate, which removes hydrogen chloride. Yields are improved if copper(II) acetate is added to reoxidize any palladium metal which may be formed in side reactions (cf. p. 380). Allylpalladium reagents have found considerable application in organic synthesis, both in stoichiometric and catalytic reactions (p. 261).

Allyl carbonyl complexes can also be prepared by reacting an anionic species with an allyl halide. An η^1-allyl derivative is formed initially. If the formal coordination number of the metal in this intermediate is five it loses carbon monoxide spontaneously to give an η^3-allyl. Thus $Co(CO)_4^-$ and $Fe(CO)_3(NO)^-$ yield $(\eta^3\text{-}C_3H_5)Co(CO)_3$ or $(\eta^3\text{-}C_3H_5)Fe(CO)_2(NO)$ directly. One mole of CO is easily displaced by phosphines (L) from these products to form derivatives e.g. $Co(\eta^3\text{-}C_3H_5)(CO)_2L$ which are much more stable thermally and to oxidation than the parent compounds. When the formal coordination number in the initially formed η^1-complex is six or seven, heating or, better, irradiation is usually required to displace carbon monoxide to form the η^3-allyl. Thus $M(CO)_5^-$ (M = Mn, Re) gives $(\eta^1\text{-}C_3H_5)M(CO)_5$ which can be decomposed to $(\eta^3\text{-}C_3H_5)M(CO)_4$ in a subsequent step:

$$M(CO)_5^- + C_3H_5Cl \xrightarrow{\text{THF}} (\eta^1\text{-}C_3H_5)M(CO)_5 \xrightarrow[-\text{CO}]{h\nu} (\eta^3\text{-}C_3H_5)M(CO)_4$$

Allyl complexes can be formed by the addition of transition metal hydrides to dienes. Two general mechanisms for these additions might be considered. In the first initial dissociation of a ligand from the hydride is followed by coordination of diene.

This is the pathway which is followed by five-coordinate hydrides such as $HCo(CO)_4$ or the isoelectronic $HNiL_4^+$ (L = P(OEt)$_3$), which is formed in solution by protonation of NiL_4 by strong acids. These hydrides are susceptible to substitution of ligand (CO or L) by a dissociative mechanism. Butadiene adds to $HNiL_4^+$ at room temperature to give a mixture of two geometrical isomers (anti-methyl and syn-methyl) in the ratio 88:12 (kinetic control). This suggests that the diene adds preferentially in the *cisoid* conformation which is 12 kJ mol^{-1} higher in energy than the *transoid* conformation (p. 263).

The mixture of isomers is converted into the thermodynamically controlled mixture syn:anti (95:5) by heating to 70°C.

Complexes with anti substituents are generally less stable than their syn isomers. The anti position is the more congested one. Isomerization probably occurs *via* alkene-hydride intermediates. $HCo(CO)_4$ adds to 1,3-butadiene at 25°C to give a mixture of syn (35%) and anti (65%) isomers, but the latter isomerizes to the former on heating to 80°C.

8.1.3 *Allylpalladium complexes – intermediates in organic synthesis*

Allylpalladium complexes find wide application in organic synthesis. With carbon nucleophiles they form new C—C bonds, thus providing a method for building carbon skeletons. Dimeric allylpalladium chlorides are not very susceptible to attack by nucleophiles, but their reactivity can be enhanced by addition of triphenylphosphine, which converts them into monomeric species. These reactions, however, are stoichiometric in palladium, which is undesirable on account of its high cost. Fortunately related catalytic procedures have been discovered. Many allylic compounds react with $Pd(PPh_3)_4$ (p. 181) to form cationic η^3-allyl complexes. These cations react with nucleophiles to form substituted allylic derivatives, while the catalyst, effectively 'PdL$_2$', is regenerated and reenters the cycle.

Allyl and diene complexes

The palladium catalysed substitution of an allylic acetate provides an example.

Addition of the nucleophile can take place by two sterically different pathways. In the first (path (a)), direct attack on the η^3-allyl group occurs *trans* to palladium. This is by far the commoner route and is followed by stabilized carbanions such as $\bar{C}H(CO_2R)_2$, $\bar{C}H(COR)_2$, $Ph\bar{C}HCN$ and $C_5H_5^-$, and normally also with amines. Aryl and alkyl carbanions (e.g. R_2CuLi) or hydride however add initially to the metal centre and then migrate to the allyl group (path (b)). The stereochemistry of the product depends on which mechanism is followed (v.i.).

trans attack, path (a)   *cis* attack, path (b).

Bäckvall has devised a catalytic method for the conversion of 1, 3-dienes into their 1, 4-diacetoxy-derivatives. Less than 3% 1, 2-addition occurs. The reaction is carried out in acetic acid using manganese(IV) oxide, together with a small quantity of benzoquinone, as oxidizing agent. The catalyst is palladium(II) acetate in the presence of lithium acetate. Under these conditions, 1, 3-cyclohexadiene affords *trans*-1,4-diacetoxycyclohexene (path (b)). If a little lithium chloride is also added, coordination of the acetate ion to palladium is blocked by chloride, the better of the two ligands. This suppresses the *cis*-migration pathway (b) so that *cis*-1,4-diacetoxycyclohexene (path (a)) now becomes essentially the only product. If the concentration of lithium chloride is increased still further, chloride replaces acetate as the second nucleophile.

The chloro group in 1-acetoxy-4-chlorocyclohexene is more readily substituted by nucleophiles than the acetoxy group. Direct S_N2 attack of a nucleophile leads to inversion of configuration at carbon. In the presence of palladium catalyst, allylic substitution of chloride *trans* to the metal centre occurs,

262

with retention. (In fact there are two inversion steps, one in the formation of the allylpalladium intermediate and another in its displacement.) Good control of the stereochemical course of these substitutions is therefore possible.

8.2 Diene complexes

The conjugated diolefin 1,3-butadiene exists almost entirely in the planar s-*trans* conformation at 298 K.

The s-*trans* conformer is about $12\,kJ\,mol^{-1}$ more stable than the s-*cis*. Nevertheless it is the *cis* structure which 1,3-diene ligands normally adopt when they are η^4-bonded to transition elements. In this way better overlap between $C(2p)\pi$ orbitals of the ligand and the metal orbitals is achieved.

The largest group of diene complexes which have been investigated are iron carbonyl compounds. When 1,3-butadiene reacts with $Fe_2(CO)_9$, the initial product is a labile η^2-complex, which readily loses carbon monoxide to yield $(\eta^4\text{-}C_4H_6)Fe(CO)_3$. Butadiene(tricarbonyl)iron is a yellow, essentially air stable complex, m.p. 19°C. It was first prepared by Reihlen in 1930 by heating $Fe(CO)_5$ with butadiene in a tube under pressure, but this was an isolated discovery which did not arouse much interest at the time. On account of the ease of preparation and handling of these diene complexes, however, an enormous amount of work has been done in this area since the late 1950s. Only modest precautions are required to protect solutions from oxidation and some operations can be carried out even in air.

While 1,3-butadiene itself is a gas at room temperature, substituted dienes are liquids or solids. In general, therefore, pressure vessels are not required in the preparation of the η^4-diene complexes. Starting materials include $Fe(CO)_5$, $Fe_2(CO)_9$ or $Fe_3(CO)_{12}$. The crucial initial step in each case is the formation of the coordinatively unsaturated 16-electron intermediate $Fe(CO)_4$. This can be induced thermally or, in the case of $Fe(CO)_5$, by irradiation.

263

Allyl and diene complexes

$$Fe(CO)_5 \longrightarrow Fe(CO)_4 + CO$$
$$Fe_2(CO)_9 \longrightarrow Fe(CO)_4 + Fe(CO)_5$$
$$Fe_3(CO)_{12} \longrightarrow 3Fe(CO)_4$$

not isolated

The mildest thermal conditions are achieved with $Fe_2(CO)_9$, which reacts on refluxing in diethyl ether. In this case, however, only half of the iron is converted into the complex. A very gentle method is to photolyse $Fe(CO)_5$ at $-78°C$ with Z-cyclooctene in pentane, yielding the labile intermediate $Fe(\eta^2\text{-}C_8H_{14})_2(CO)_3$. On treatment with a diene at room temperature in a hydrocarbon solvent, cyclooctene is displaced, giving the η^4-diene $Fe(CO)_3$ in high yield. Even styrenes afford η^4-derivatives under these conditions in which one double bond of the aromatic system becomes coordinated.

$$Fe(CO)_5 + 2C_8H_{14} \xrightarrow[-78°C]{h\upsilon} (\eta^2\text{-}C_8H_{14})_2\,Fe(CO)_3 \longrightarrow$$

Prolonged irradiation of $(\eta^4\text{-}C_4H_6)Fe(CO)_3$ with butadiene leads to $(\eta^4\text{-}C_4H_6)_2Fe(CO)$.

The complexes of cyclic dienes, such as 1,3-cyclohexadienes, are precursors to η^5-cyclohexadienyltricarbonyliron salts, which have been applied in origanic synthesis (p. 303). 1,4-Cyclohexadienes, available from the Birch reduction of arenes, are converted into the 1,3-diene complexes. The isomerization occurs through a transient η^3-allyliron hydride (p. 363).

η^4-Derivatives of 1,3-dienes which possess functional groups such as $-CH_2OH$, $-CHO$, $-COR$ and $-COOR$ are prepared by similar methods.

(a) ATTACK OF ELECTROPHILES. (Diene)tricarbonyliron complexes are protonated by strong acids. Addition of HCl gives $[Fe(\eta^3\text{-}MeC_3H_4)(CO)_3Cl]$. With acids containing weakly coordinating anions (e.g. CF_3CO_2H, HSO_3F or HBF_4) the situation is more complicated. ^1H n.m.r. studies indicate that protonation occurs stereospecifically endo, that is from the same side of the diene as the metal,

264

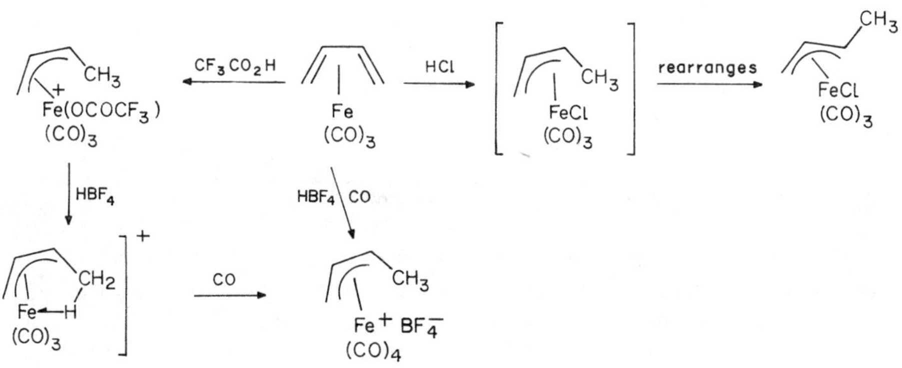

Fig. 8.6 Protonation of butadiene tricarbonyliron.

probably *via* a metal hydride. Intermediates containing an 'agostic' Fe—H—C bridge are likely, although they have not been isolated in the carbonyl series. While Fe(diene)(CO)$_3$ requires a strong acid such as HBF$_4$ in acetic anhydride to protonate it, Fe(diene){P(OMe)$_3$}$_3$ is protonated even by NH$_4^+$PF$_6^-$ in methanol (p$K_a \sim 9$). The resulting salts can be isolated as air-stable crystalline materials. The 'agostic' hydrogen interaction has been confirmed by a neutron diffraction study of the complex [Fe(η^3-C$_8$H$_{13}$){P(OMe)$_3$}$_3$]$^+$ BF$_4^-$, prepared by protonation of the η^4-1,3-cyclooctadiene derivative (Fig. 8.6).

(b) FRIEDEL-CRAFTS ACYLATION. In common with (η^5-C$_5$H$_5$)Mn(CO)$_3$, (p. 294) (η^6-C$_6$H$_6$)Cr(CO)$_3$, (p. 317) and (η^4-C$_4$H$_4$)Fe(CO)$_3$, (p. 269), (butadiene)Fe(CO)$_3$ can be acylated under Friedel-Crafts conditions. The acylation proceeds via an intermediate in which the carbonyl group of the entering acyl is coordinated to the metal:

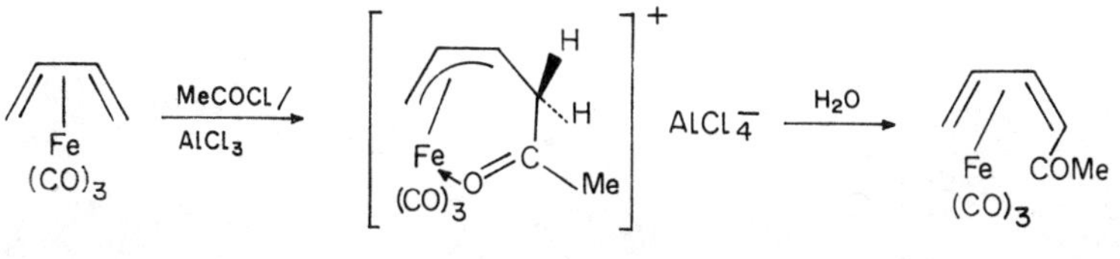

Direct Friedel-Crafts acylation of 1,3-dienes is possible at low temperatures but is accompanied by polymerization of the diene.

(c) HYDRIDE ABSTRACTION BY Ph$_3$C$^+$. Treatment of *cis*-pentadiene Fe(CO)$_3$ with Ph$_3$C$^+$BF$_4^-$ affords an η^5-pentadienyl salt. This is another example of the conversion of a neutral η^n-complex into an η^{n+1}-cation by this reagent (p. 323). The open-chain pentadienyl salts (cf. p. 302) are also obtained by protonation of dienol complexes.

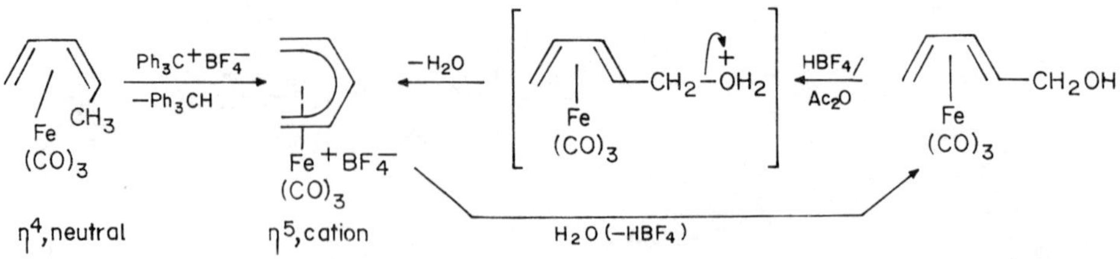

Allyl and diene complexes

η^4, neutral η^5, cation H_2O $(-HBF_4)$

They are yellow crystalline materials, stable in dry air, but quite readily hydrolysed back to dienol derivatives. $Fe(CO)_3$ can be used as a 'protecting group'. We have seen (p. 264) that an $Fe(CO)_3$ substituent can be attached under mild conditions to a diene unit. It can also be readily removed by oxidation, for example using aqueous iron(III) or cerium(IV) or by trimethylamine oxide in a hydrocarbon solvent. A diene unit can therefore be 'protected' using $Fe(CO)_3$ while reactions are carried out on other parts of the ligand molecule. This approach has been employed in terpene and steroid chemistry.

(d) ATTACK OF NUCLEOPHILES. Dienetricarbonyliron complexes are not very reactive towards nucleophiles. Only carbanions stabilized by cyano or ester groups attack the organic ligand at all cleanly. Even with these reagents a mixture of products is often obtained. At $-78°C$ reversible addition to the 2-position in $(\eta^4\text{-}C_4H_6)Fe(CO)_3$ is favoured, but at $0°C$ terminal addition also occurs. Substituted butenes are released on protonation of the anionic intermediates (Fig. 8.7).

The isoelectronic ion $(\eta^4\text{-}C_4H_6)Co(CO)_3^+$ reacts with a wide range of nucleophiles, the terminal position of the diene being attacked exclusively. The proton n.m.r. spectrum of the pyridine adduct is consistent with its having *anti*-stereochemistry. By adding a second nucleophile to the neutral cobalt allyl it is possible, in some cases, to convert a 1,3-diene into a 1,4-bifunctional derivative in a regio- and stereoselective fashion.

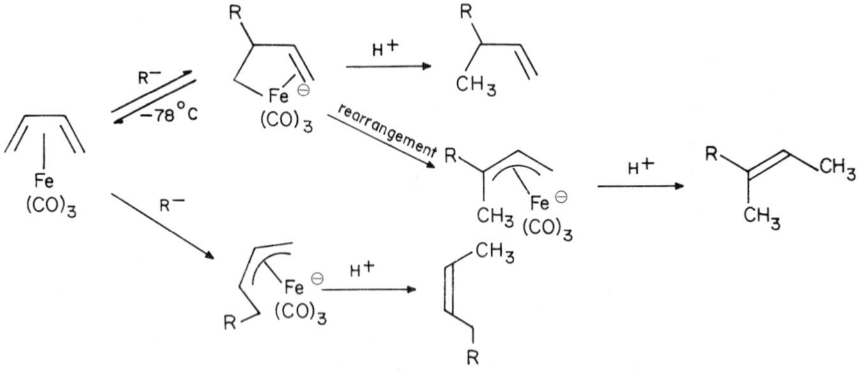

Fig. 8.7 Nucleophilic attack on butadienetricarbonyliron.

266

$$Co_2(CO)_8 \xrightarrow{\quad} [\eta^4\text{-}C_4H_6\,Co\,(CO)_2]_2 \xrightarrow[0°C]{(C_5H_5)_2Fe^+BF_4^-} \underset{\substack{Co^+\,BF_4^-\\(CO)_3}}{} \xrightarrow{(Nu^1)^-}$$

$$\nu\,(CO)\,2150,\,2102\,cm^{-1}$$

$$\underset{\substack{Co\\(CO)_3}}{Nu^1} \xrightarrow{(Nu^2)^-} Nu^2\!\!\diagup\!\!\diagdown\!\!Nu^1$$

$$Z\text{-isomer}$$

Nu^1=H(NaBH$_3$CN); Ph(PhMgBr); pyridine ;–CH(CO$_2$Me)$_2$; allyl (allyl SiMe$_3$); OMe(MeOH).
Nu^1=H, Nu^2=–CH(CO$_2$Me)$_2$

8.2.1 s-trans-Diene complexes

Normally the frontier orbitals of transition metal fragments such as Fe(CO)$_3$ favour bonding to the *cis* conformation of a diene. The fragments $(\eta^5\text{-}C_5H_5)_2$M (M = Zr, Hf) (p. 292), however, are well suited to η^4-bonding to *trans* dienes, as their available orbitals all lie in the xz molecular plane.

Complexes Cp$_2$Zr(η^4-diene) can be prepared from Cp$_2$ZrCl$_2$ and magnesium butadiene (p. 93) in tetrahydrofuran, or by photolysis of Cp$_2$ZrPh$_2$ with butadiene. If the reactions are carried out at low temperature the kinetically controlled *trans*-butadiene complex is produced, *via* and η^2-intermediate. On warming to room temperature an equilibrium between the *cis*-η^4- (55%) and *trans*-η^4- (45%) complexes becomes established.

$$Cp_2ZrCl_2 + \left(Mg\!-\!CH_2\!\!\underset{\substack{\\CH=CH}}{\diagup}\!\!CH_2\right)_n \longrightarrow \left[Cp_2Zr\!-\!\| \right] \rightleftharpoons \underset{\substack{Cp \quad Cp\\s\text{-}trans\text{-}\eta^4}}{Zr}$$

$$s\text{-}cis\text{-}\eta^4$$

When the diene has terminal (1,4-) substituents the equilibrium lies on the side of the *trans* form, while with internal (2,3-) substituents the *cis* structure is favoured. In both structures the zirconium atom adopts pseudo tetrahedral geometry. In the *cis* complex the Zr—C distances to the terminal carbon atoms are shorter than to the internal carbons. This means that the structure approaches the metallacyclopentene limit.

267

Allyl and diene complexes

This indicates a decrease in π-bonding relative to complexes of the late transition elements, in which approximately equal M—C distances to all four carbon atoms are observed. While the free ligands show short–long–short C—C bond alternation, in the s-*cis*-zirconium complexes the order is long–short–long.

The s-*trans* complexes show a very different bonding pattern. Here it is the bonds from the metal to the internal carbon atoms of the diene which are shorter, and the C—C bond alternation is the same (short–long–short) as in the free ligand. The diene unit, however, deviates considerably from planarity, there being a torsion angle of $127°$ in the butadiene complex.

These complexes exhibit very novel chemistry which is only now being developed. Their reactions have considerable synthetic potential. η^1-Alkyl complexes show a propensity to migratory insertion of carbon monoxide. The s-*cis* derivatives are converted into cyclopentenones on reaction with carbon monoxide followed by hydrolysis:

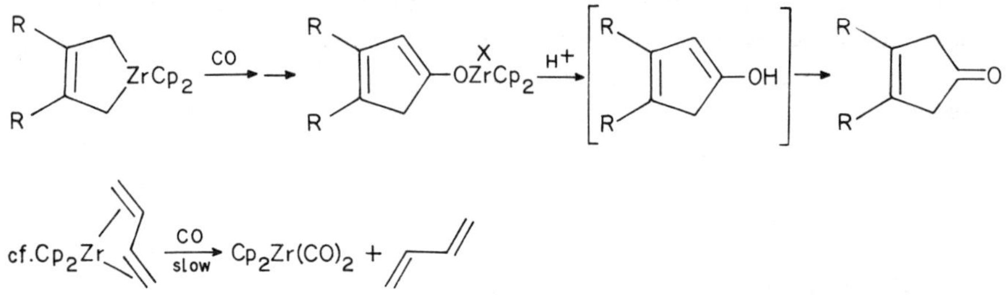

The s-*trans* complexes, however, react with alkenes, alkynes, ketones and aldehydes by insertion into a M—C bond, leading to metallacycles.

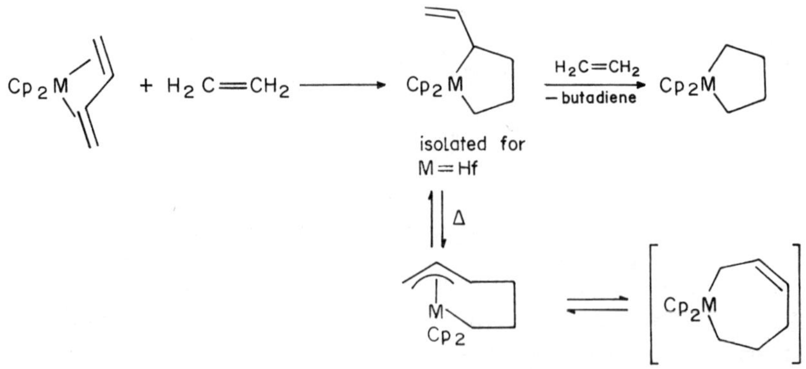

268

8.2.2 Cyclobutadiene complexes

For many years chemists, including the famous German scientist, Richard Willstätter, who was the first to obtain cyclooctatetraene, strove to isolate cyclobutadiene, but it proved to be very elusive. In 1956 Longuet-Higgins and Orgel predicted that its complexes with transition elements should be stable, on account of the favourable symmetry properties of its π-orbitals (p. 200). In 1959 this was confirmed by the isolation of $[(\eta^4\text{-}C_4Me_4)NiCl_2]_2$ by Criegee and Schröder and of $(\eta^4\text{-}C_4Ph_4)Fe(CO)_3$ by Hübel and his group. The first complex of the unsubstituted ligand, $(\eta^4\text{-}C_4H_4)Fe(CO)_3$ was prepared in 1965. Since then the parent hydrocarbon has been generated by irradiation of photo-α-pyrone in a frozen noble gas matrix at 8–20 K and its infrared spectrum recorded. It has a rectangular D_{2h} symmetry in the singlet ground state, rather than the square D_{4h} geometry. This is predicted on the basis of the Jahn-Teller theorem. The reason that free cyclobutadiene cannot be isolated under normal conditions is that it rapidly undergoes Diels-Alder condensation with itself. This can be slowed down by introduction of bulky substituents so that tri-*tert*-butyl-cyclobutadiene survives at room temperature for several hours.

The first cyclobutadiene complexes were obtained either from alkynes (p. 246) or by reaction of *cis*-dichlorocyclobutenes with metal carbonyls. Many more derivatives were then prepared by transfer of the cyclobutadiene ligand from one transition metal to another, especially from palladium in $[(Ph_4C_4)PdCl_2]_2$. In their complexes cyclobutadienes adopt a planar square structure.

(R=H,Me)

Cyclobutadienetricarbonyliron can also be prepared by irradiation of photo-α-pyrone with $Fe(CO)_5$. It is a pale yellow solid, m.p. 26°C.

α-pyrone   'photo-α-pyrone'

The coordinated cyclobutadiene readily undergoes electrophilic substitution. It can be acylated under Friedel-Crafts conditions, sulphonated and mercurated. Like ferrocene (p. 284) it performs the Vilsmeier reaction with $PhN(Me)CHO/POCl_3$ giving the aldehyde $(\eta^4\text{-}C_4H_3CHO)Fe(CO)_3$ and the Mannich condensation with CH_2O/Me_2NH giving $(\eta^4\text{-}C_4H_3CH_2NMe_2)Fe(CO)_3$, which occur

269

Allyl and diene complexes

only with very reactive aromatic compounds such as indole or phenol. Nitro- and halogeno- derivatives cannot be obtained directly by electrophilic substitution as the starting complex is too susceptible to oxidation.

Oxidation of η^4-$C_4H_4Fe(CO)_3$ with aqueous iron(III) or cerium(IV) releases the transient cyclobutadiene which either reacts with itself or which can be trapped by a dienophile in a Diels-Alder reaction:

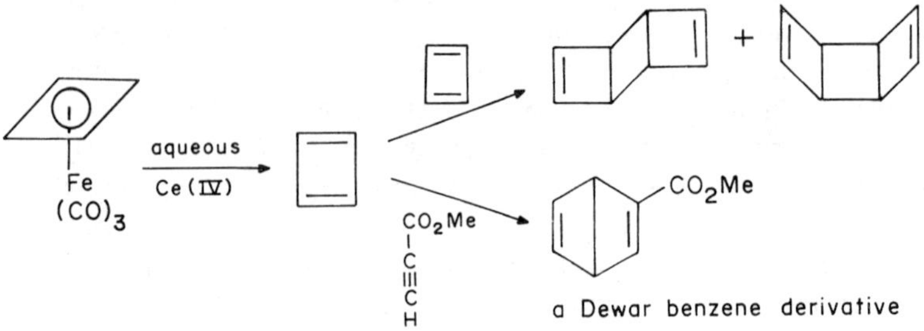

a Dewar benzene derivative

The sandwich complex which would complete the isoelectronic series (η^6-$C_6H_6)_2Cr$, $(\eta^5$-$C_5H_5)_2Fe$.. is bis(cyclobutadiene)nickel, $(\eta^4$-$C_4H_4)_2Ni$. The tetraphenyl derivative has been prepared by the following route.

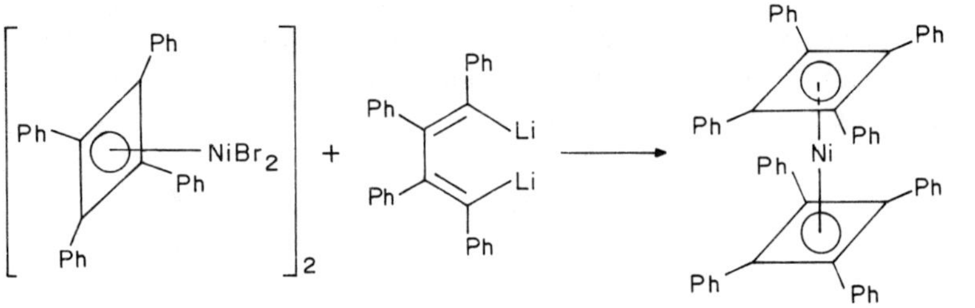

8.2.3 Trimethylenemethane complexes

The reactive species, trimethylenemethane, has been observed only by photolysis of an adduct of allene and diazomethane in a matrix of hexafluorobenzene at $-185°C$. The e.s.r. spectrum shows that it has two unpaired electrons (a triplet ground state) which is consistent with a planar structure with 3-fold symmetry (D_{3h}). This species can, like cyclobutadiene, be stabilized by coordination to tricarbonyliron. The parent compound, a diamagnetic pale yellow solid, m.p. 28°C can be obtained by the two routes shown.

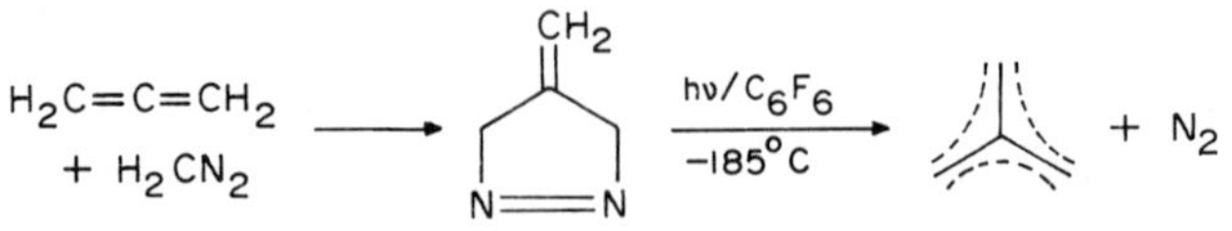

The symmetrical structure of the trimethylenemethane ligand is indicated in solution by the ^{13}C n.m.r. spectrum which shows three resonances: 211.6

270

(singlet, CO), 105.0 (singlet, $\underline{C}(CH_2)_3$ 53.0 (triplet, J(CH), 158 Hz, $C(\underline{C}H_2)_2$), and in the gas phase by electron diffraction.

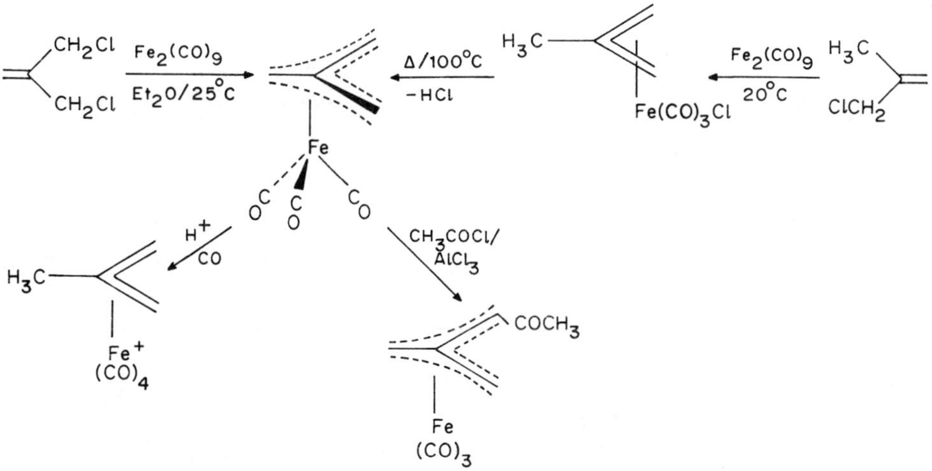

The Diels-Alder reaction is a $[4+2]$ cycloaddition between a diene and a dienophile (a substituted alkene or alkyne). It leads to the formation of six-membered rings. Trost has developed a related procedure for making five-membered rings. A trimethylenemethane, briefly trapped as its palladium complex, adds to an alkene, which is activated by an electron-withdrawing group.

Trimethylenemethane   Diels–Alder
cycloaddition

$(X = \text{electron–withdrawing group})$

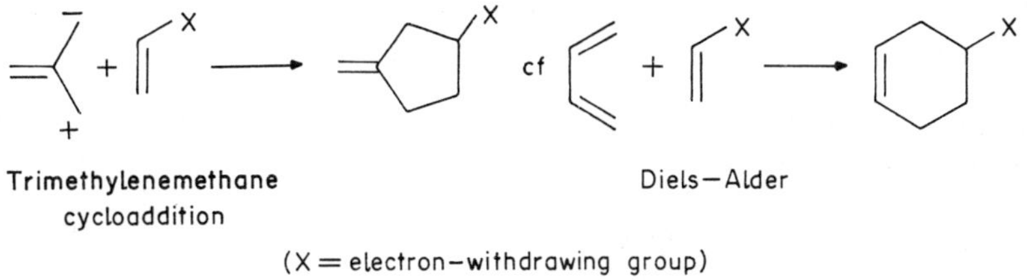

Suitable precursors are allylic acetates with Me_3SiCH_2 substituents at the 2-position.

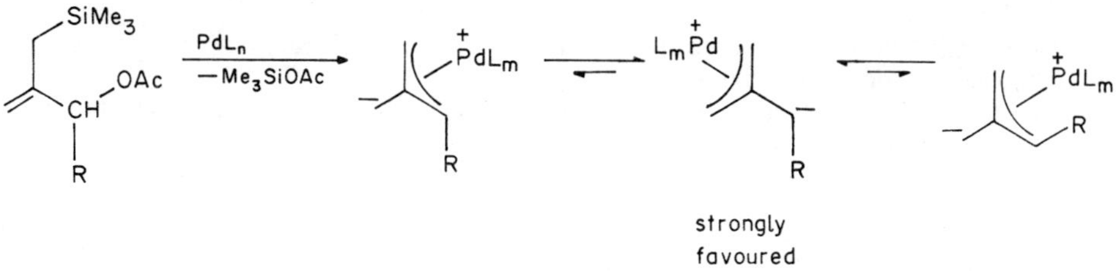

strongly
favoured

In the presence of a palladium(0) complex, which acts as a catalyst, the precursor eliminates Me_3SiOAc to form, fleetingly, a trimethylenemethane palladium

271

derivative. This rapidly adds to the diene affording a methylenecyclopentane. Although there is no direct evidence for the trimethylenemethanepalladium intermediate, a stable osmium complex has been isolated from the reaction of the precursor with $Os(CO)_2(PPh_3)_3$. If cyclic alkenes are used, fused ring systems are formed. The reaction has been used in the synthesis of several complex organic molecules including loganin, a key intermediate in the biosynthesis of alkaloids.

$(Z = CO_2Et, COR, CN, SO_2R)$

Five-membered rings are present, for example, in the physiologically important prostanoids as well as in monoterpenes, the iridoids, in the antibiotic hirsutic acid and in nepetalactone, an essential oil from catmint.

8.3 References

Bäckvall, J.E. (1983) Palladium in some selective oxidation reactions. *Acc. Chem. Res.*, **16**, 335.

Billington, D.C. (1985) η^3-Allyl-nickel halides as selective reagents in organic synthesis. *Chem. Soc. Revs.*, **14**, 93.

Efraty, A. (1977) Cyclobutadiene complexes. *Chem. Rev.*, **77**, 691.

Erker, G., Krüger, C. and Müller, G. (1985) η^4-Conjugated diene zirconocene and hafnocene complexes. *Adv. Organomet. Chem.*, **24**, 1.

Jolly, P.W. (1985) η^3-Allylpalladium complexes. *Angew. Chem. (Int. Edn)*, **24**, 283.

Jolly, P.W. and Mynott, R. (1981) ^{13}C n.m.r. of organotransition metal complexes. *Adv. Organomet. Chem.*, **19**, 257.

Trost, B.M. (1986) 3 + 2 cycloaddition approaches to 5-membered rings. *Angew. Chem. (Int. Edn)*, **25**, 1.

Trost, B.M. (1986) Transition metals and olefins – a promising land; a personal account. *J. Organomet. Chem.*, **300**, 263.

Yasuda, H., Tatsumi, K. and Nakamura, A. (1985) Early transition metal diene complexes. *Acc. Chem. Res.*, **18**, 120.

Problems

1. Comment on each of the following.

(a) Diallylmercury gives an n.m.r. spectrum with ^{1}H resonances giving three groups of peaks at different chemical shifts with an intensity ratio of 2:1:2 at all accessible temperatures. The compound has an infrared spectrum showing the characteristic frequencies of a vinyl group.

(b) Diallylpalladium gives an ^{1}H n.m.r. spectrum at low temperatures with three groups of resonances, intensity ratio $2:2:1$. At room temperature the two stronger resonances merge into a single rather broad peak.

(c) Two isomeric forms can be isolated for the π-allyl compound $Co(CO)_3(CHCH_3CHCH_2)$.

(*University of East Anglia*).

2. M is a First Transition Series element. It forms a carbonyl F of empirical formula $M(CO)_5$ which reacts with sodium amalgam in tetrahydrofuran to give a solution G. Treatment of G with 3-chloro-1-propene gives a compound H of molecular formula $C_8H_5O_5M$. The infrared spectrum of H shows carbonyl stretching bands between 2110 and 2004 cm^{-1}, the ^{1}H n.m.r. spectrum of H indicates protons in *four* chemically distinct environments.

On heating H to 100°C one mole of carbon monoxide is eliminated to give I, $C_7H_5O_4M$ [$\bar{\nu}$(CO) between 2110 and 1950 cm^{-1}]. The ^{1}H n.m.r. spectrum of I indicates protons in *three* chemically distinct environments.

(a) Identify the metal M.

(b) Propose and draw structures for the compounds F, H and I and for the species present in solution G.

(c) Interpret the ^{1}H n.m.r. features of H and I.

Give a qualitative sketch of the spectrum of I, clearly labelling the protons which give rise to each group of resonances. Discuss the spin–spin couplings which are observed. (Coupling through more than three bonds can be ignored).

(d) Discuss the bonding of the organic ligand to M in compound I.

(*Royal Holloway and Bedford New College*).

3. Treatment of an aqueous methanolic solution of Na_2PdCl_4 with 2-methyl-allyl chloride, $CH_2{=}C(CH_3)CH_2Cl$, under a CO atmosphere yields the dimeric complex $[C_4H_7PdCl]_2$. (*C*). The ^{1}H n.m.r. spectrum of C at 298 K shows three singlets at δ 2.2, 2.9 and 3.8 with relative intensities of $3:2:2$ respectively. Addition of 2 moles of AsPh$_3$ to 1 mole of dimer (i.e. molar ratio As:Pd $= 1:1$) yields another complex D.

The ^{1}H n.m.r. spectrum of D at 193 K shows five singlets at δ 2.1, 2.7, 3.2, 3.6 and 4.5 with relative intensities of $3:1:1:1:1$ respectively in addition to the phenyl proton resonances. On warming the solution of D to 323 K the ^{1}H n.m.r. spectrum changes in profile to show two singlets at δ 2.1 and 3.5 relative intensities of $3:4$ respectively in addition to the phenyl proton resonances.

Suggest structures for the compounds C and D, interpret the ^{1}H n.m.r. data, and describe briefly the type of bonding occurring between the organic group and the palladium atom in C.

(*University of Warwick*).

4. The reaction between $H_2C{=}C(CH_2Cl)_2$ and $Fe_2(CO)_9$ was found to yield a volatile, low-melting solid A of empirical formula $C_7H_6O_3Fe$. The mass spectrum of A contained a strong parent ion peak at m/z 194 and there were other strong peaks at m/z 166, 138 and 110.

The ^1H n.m.r. spectrum of A consisted of one sharp singlet. The ^{13}C spectrum consisted of two singlets and a triplet: under conditions of complete proton decoupling the triplet became a singlet, but the spectrum was otherwise unaltered.

When A was dissolved in strong acid, and then Cl^- was added to the solution, a covalent compound B was isolated. B was found to possess the empirical formula $C_7H_7O_3ClFe$. The ^1H n.m.r. spectrum of B contained three singlet resonances of relative area 3:2:2. B could also be prepared by the reaction of $H_2C=CMeCH_2Cl$ with $Fe_2(CO)_9$.

Suggest structures for compounds A and B, showing them to be consistent with the spectroscopic data. Comment as fully as you can on the bonding between metal and organic ligands in A and B. Show that both A and B can be regarded as 'obeying the 18-electron rule'.
(*University of York*).

5. Some chemistry of sodium cyclopentadienyltricarbonyltungstate is shown below:

$$Na[(C_5H_5)W(CO)_3] + CH_2=CH-CH_2Cl \longrightarrow A \xrightarrow{\text{uv}} D$$

$$\downarrow \text{HPF}_6$$

$$B \xrightarrow{\text{NaBH}_4} C$$

Elemental analysis for B gave 35.4% W, 25.4% C, 5.96% P.

Mass spectrum M^+ (based on ^{184}W) amu	ν_{CO}/cm^{-1}	Proton nmr (δ relative to TMS, relative intensities in parentheses)
A: 374	2019, 1950, 1904	6.0(1H), 5.25(5H), 4.6(1H), 4.5(1H), 2.32(2H).
B: —	2107, 2053, 2007	5.99(5H), 4.36(1H), 3.29(1H), 3.20(1H), 2.22(3H).
C: 376	2005, 1947, 1905	4.59(5H), 2.78(1H), 1.55(6H).
D: 346	1960, 1880	4.63(5H), 3.25(1H), 2.48(2H), 1.12(2H).

(a) Propose structures for A to D, showing how you have used the data above.
(b) Comment on the reactions in the scheme.
(c) Describe the bonding of the acyclic hydrocarbon ligand to the metal in D.
(*University of Southampton*).

6. (a) Make a qualitative sketch of the proton n.m.r. spectrum of (η^4-1,3-

274

butadiene)tricarbonyliron, E. Identify the protons which give rise to each group of resonances and discuss the spin–spin couplings which are observed. (Ignore coupling through more than three bonds.)

(b) At ambient temperature the ^{13}C n.m.r. spectrum of E (broad-band proton decoupled) consists of three lines at 213.2, 86.4 and 41.5 ppm. downfield from tetramethylsilane ($\delta = 0$). As the temperature of the sample is lowered the resonance at 213.2δ broadens, collapses and finally resolves into two lines at 217.4δ and 211.3δ, with relative intensities 1:2. The rest of the spectrum does not change with temperature. Assign the spectrum and suggest an explanation for the variation with temperature. How could the assignments be confirmed?

(c) How can compound E be prepared? How does it react with (i) hydrogen chloride, (ii) HBF_4/CO, (iii) $MeCOCl/AlCl_3$, (iv) Ph_3P?

7. The reaction between 1 mmol of $K[Rh(PF_3)_4]$ and 1 mmol of 1-chloro-3-methylbut-2-ene (1,1-dimethylallyl chloride) in diethyl ether gave a yellow solution and a white precipitate. Fractionation of the yellow solution yielded PF_3 (*ca.* 1 mmol) and a yellow volatile liquid A.

Analysis of A: found C, 13.8%; H, 2.1%; P, 21.3%; F, 39.2% and Rh, 23.6%. The mass spectrum of A showed a highest mass peak at m/z = 436. When compound A was heated to 60°C it was converted initially into an isomeric derivative B which subsequently formed a third isomer C. The ^1H n.m.r. spectra of compounds A, B and C, measured at room temperature, are listed in the table. These spectra can be assigned by taking into account proton–proton and proton–rhodium couplings [$1(^{103}Rh) = \frac{1}{2}$; natural abundance of ^{103}Rh isotope is 100%], proton–phosphorus couplings are not observed under these conditions. Interpret these data as fully as you can. Deduce the molecular structures of the complexes A, B and C and suggest a mechanism for the isomerization.

Chemical shift/δ	Relative intensity	Multiplicity	*Coupling constants/Hz
Compound A			
1.7	3	singlet	—
2.1	3	singlet	—
2.7	1	doublet of doublets	10.5, 2.0
3.2	1	doublet of doublets	6.5, 2.0
4.9	1	8 lines of equal intensity	10.5, 6.5, 1.0(Rh)

Chemical shift/δ	Relative intensity	Multiplicity	*Coupling constants/Hz
Compound B			
1.5	3	doublet	6.0
2.0	3	doublet	1.0(Rh)
3.1	1	doublet	2.0
3.6	1	doublet	2.0
4.6	1	quartet (1:3:3:1)	6.0
Compound C			
1.8	3	doublet	6.0
2.0	1	doublet	2.0
2.1	3	doublet	1.0(Rh)
3.2	1	doublet	2.0
3.5	1	quartet (1:3:3:1)	6.0

*Proton–proton coupling constants, unless otherwise indicated.

(*Royal Holloway College*)

8. The adduct formed between $(\eta^3\text{-}C_3H_5)_2Pd$ and PMe_3 has been studied by proton n.m.r. from $-109°$ to $-3°C$. Propose a structure for the 'frozen' adduct

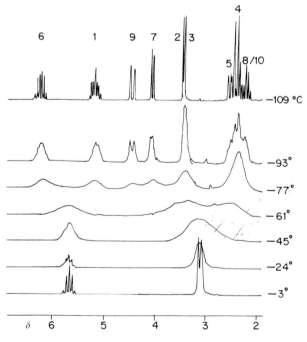

which is present at $-109°C$ and assign the n.m.r. spectrum. Suggest an explanation for the changes in the spectrum which occur as the temperature is raised. (Resonances from PMe_3 are not shown).

9. The dimeric dichloropentamethylcyclopentadienylrhodium complex $[(\eta^5-C_5Me_5)ClRh(\mu-Cl)_2RhCl(\eta^5-C_5Me_5)]$, A, reacts with 1, 3-butadiene in ethanol in the presence of base (sodium carbonate) to give an air stable crystalline product, B.

Elemental analysis of compound B, which contains C, H, Cl and Rh only, gave the following results. Found C, 51.15%; H, 6.75%; Cl, 10.79%. Relative molecular mass, 329. The proton magnetic resonance spectrum of compound B is given below. (Any small couplings to ^{103}Rh $(I = \frac{1}{2})$ have been omitted.)

^1H n.m.r. (Compound B). (Chemical shifts downfield from tetramethylsilane, $\delta = 0$).

Chemical shift	Relative intensity	Multiplicity
3.96	1	Multiplet
3.58	1	Multiplet
3.28	1	Doublet of doublets (J, 6.6 Hz, 1.5 Hz)
2.88	1	Doublet of doublets (J, 11.0 Hz, 1.5 Hz)
1.73	15	Singlet
1.65	3	Doublet, (J, 6.2 Hz)

Interpret the n.m.r. spectrum of compound B as completely as possible, giving reasons for your assignments.

Suggest a mechanism for the reaction of compound A with butadiene in ethanol in the presence of sodium carbonate.

(*Royal Holloway College*).

9 Five electron ligands

9.1 Introduction

The cyclopentadienyl group, C_5H_5, forms complexes with all the d-block transition elements. Its usual mode of bonding is η^5, although there are examples of η^3 and η^1 attachment. In the following series of compounds the 18-electron configuration of the metal is maintained by variation of the number of electrons provided by one of the cyclopentadienyls (Fig. 9.1).

While the η^5-cyclopentadienyl ligand can undergo reactions such as electrophilic substitution at the ring (p. 283), it is often rather inert. It can therefore play the part of a bystander, while changes are going on at the metal centre or at other ligands in the molecule. It effectively occupies three

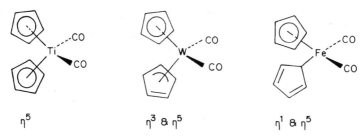

η^5 η^3 & η^5 η^1 & η^5

Fig. 9.1 Various bonding modes in cyclopentadienyl complexes.

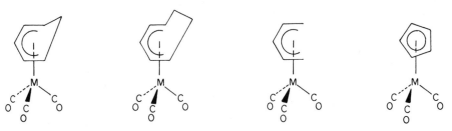

Fig. 9.2 η^5-Cyclohexadienyl, η^5-cycloheptadienyl, η^5-pentadienyl analogues of η^5-cyclopentadienyltricarbonylmanganese.

278

coordination positions and fulfils a role similar to the rather inert alkyl or aryl groups in organic chemistry.

Related to cyclopentadienyl are the open chain pentadienyl group and also the cyclohexadienyl and cycloheptadienyl ligands in which the η^5 system is contained in a larger ring. In general the chemistry of such ligands has not been explored to the same extent as that of cyclopentadienyl, but their complexes usually show greater reactivity and some significant differences (p. 301) (Fig. 9.2).

9.2 Bis(cyclopentadienyl)complexes – the metallocenes

Bis(cyclopentadienyl)complexes have been isolated for the elements of the First Transition Series from vanadium to nickel inclusive and also for ruthenium and osmium. The most general method of preparation is to treat an anhydrous halide of the metal with sodium cyclopentadienide in tetrahydrofuran under an atmosphere of nitrogen or argon.

$$MX_2 + 2NaC_5H_5 \xrightarrow{\text{THF}} M(C_5H_5)_2 + 2NaX \quad (X = Cl, Br)$$

The dihalides of Mn, Fe, Co and Ni are readily available. Trihalides of vanadium and chromium can be used, but an excess of NaC_5H_5 is then required as a reducing agent. Alternatively MX_3 can be reduced *in situ* to MX_2 using zinc or magnesium before NaC_5H_5 is added.

Cyclopentadiene can be deprotonated by dimethylamine in the presence of anhydrous $TiCl_4$, $FeCl_2$ or $NiCl_2$ to yield Cp_2TiCl_2, (p. 290), Cp_2Fe (ferrocene) or Cp_2Ni (nickelocene) respectively. Ferrocene has even been prepared in a rapidly stirred two phase system (benzene/water) from $FeCl_2 \cdot 4H_2O$, cyclopentadiene and potassium hydroxide using a crown ether, 18-crown-6, as a phase-transfer catalyst.

The metallocenes are coloured crystalline solids, all of which melt at about 173°C. They are all thermally robust and sublime readily under reduced pressure. Ferrocene is remarkably resistant to air, moisture and heat, surviving even up to 470°C. All the other metallocenes listed in Table 9.1 are oxidized in air, manganocene and cobaltocene being especially vulnerable. Manganocene is also very easily hydrolysed, making it a rather difficult substance to handle.

9.2.1 Structures

The sandwich structure of ferrocene was suggested in 1952, soon after its discovery, by G. Wilkinson and independently by E.O. Fischer*, and confirmed

---

*These two chemists were responsible for much of the early work on sandwich compounds. They were jointly awarded the Nobel Prize for Chemistry in 1973.

Five electron ligands

Table 9.1 The metallocenes. (MCp$_2$)

Metal	V	Cr	Mn	Fe	Co	Ni
Colour	Purple	Scarlet	Amber (Pink)	Orange	Purple-black	Dark green
m.p. (°C)	167	172	173	173	173	173
Sensitivity to air O$_2$	Medium	High	Very high	Inert	Very high	Medium
Sensitivity to hydrolysis	Low	High	Very high	Inert	Inert	Inert (in neutral solutions)
No. of εs	15	16	17	18	19	20
e^*_{1g}			↑ ↑		↑	↑ ↑
a_{1g}	↑	↑	↑	↑↓	↑↓	↑↓
e_{2g}	↑ ↑	↑↓ ↑	↑ ↑	↑↓ ↑↓	↑↓ ↑↓	↑↓ ↑↓
μ_{eff} (obs)	3.78	3.27	<5.9	0.0	1.70	2.89
μ_{eff} (spin only)	3.87	2.83	5.92	0.0	1.73	2.83
v(M—C) Å	2.28	2.17	2.38 (2.11)[†]	2.06	2.11	2.20
$\Delta H^{\ominus}_f$ (kJ mol^{-1})	145	178	201	168	237	285
D(M—Cp) (kJ mol^{-1})	420	340	266	352	324	301

[†]Mn(C$_5$Me$_5$)$_2$

shortly afterwards by X-ray crystallography, first of ruthenocene and later of ferrocene itself.

The structures of metallocenes of the First Transition Series have also been measured in the gas phase by electron diffraction. The cyclopentadienyl rings have planar C$_5$ skeletons, are parallel and equidistant from the metal atom. In general the C—C bond lengths are essentially equal (between 1.423 and 1.440 Å) and are slightly longer than those found in benzene (1.397 Å). There is only a low barrier to rotation of the C$_5$H$_5$ rings about the ring-metal axis, about 4 kJ mol^{-1} compared with 12 kJ mol^{-1} for the C—C bond in ethane. The eclipsed conformation (D$_{5h}$) is more stable than the staggered one (D$_{5d}$) for ferrocene and this is probably true of the other metallocenes.

There has been some confusion over the favoured conformation of ferrocene in the solid. Crystalline ferrocene undergoes a phase transition at 164 K from a triclinic low temperature modification to a monoclinic form. The molecular structure found below 164 K is close to the eclipsed conformation, the rings being mutually rotated by about 9° away from it. Above 164 K rotation of the ligand rings leads to disorder in the crystal, which caused early workers to suggest incorrectly that the staggered conformation was favoured. It has been found by electron diffraction (gas phase) and also by neutron diffraction (solid state) that the hydrogen atoms in ferrocene are bent out of the C$_5$ plane towards the iron atom by about 3°. This tilting allows the C(2p) orbitals to point more directly towards the metal leading to better overlap with metal orbitals.

9.2.2 Bonding in metallocenes

In the literature bonding in metallocenes has usually been discussed on the basis of D_{5d} symmetry, even though the eclipsed conformation (D_{5h} symmetry) is slightly more stable. This has the advantage that the symmetry labels of related orbitals of $(C_5H_5)_2Fe$ and of $(C_6H_6)_2Cr(D_{6h})$ are the same (Fig. 6.10, p. 205). The main interactions in which the metal d orbitals are perturbed are as follows: Ligand $\pi_1 + d_{z^2}$ (a_{1g}, σ-symmetry), ligand $\pi_3 + d_{x^2-y^2}$, d_{xy} (e_{2g}, δ-symmetry) and ligand $\pi_2 + d_{zx}$, d_{yz} (e_{1g}^*, π-symmetry). These orbitals are filled sequentially, as shown in Table 9.1, as the number of electrons increases from V to Ni. Here is a series of organometallic compounds with electron numbers ranging from 15–20. The eighteen electron compound, ferrocene, is certainly the most robust and inert but with the reasonably strong donor characteristics of C_5H_5 other configurations arise for elements of the First Transition Series. This is in contrast with carbonyls which contain the strongly π-accepting ligand CO. All the metallocenes apart from ferrocene, ruthenocene and osmocene are paramagnetic, the ground state electron configurations being indicated in the Table. While vanadaocene, chromocene and ferrocene are low spin complexes, manganocene has five unpaired electrons in the ground state, which is therefore high spin. The high spin and low spin states (the latter has configuration $(e_{2g})^3(e_{1g})^2$ in manganocene are very close in energy and in 1, 1'-dimethylmanganocene both forms can be observed in equilibrium, for example in the UV photoelectron spectrum, over a wide temperature range.

The metal—carbon bond distances at first sight seem to vary in a rather random fashion. If, however, the value for $Mn(C_5Me_5)_2$, which like its neighbours has a low-spin ground state, is taken there is a general contraction from V to Fe following the contraction in atomic radius. The longer distances in cobaltocene and nickelocene reflect the presence of electrons in antibonding $(e_{1g})^*$ orbitals.

Perhaps the most striking thing about the thermochemical data is the observation that these extremely heat resistant materials, ferrocene in particular, are so strongly endothermic with respect to the elements. The mean bond dissociation energies parallel the binding energies of the metals as discussed in Chapter 1. The iron-ring bond in ferrocene ($352\,kJ\,mol^{-1}$) is about as strong as a carbon—carbon single bond (ca. $360\,kJ\,mol^{-1}$).

9.2.3 Chemistry of ferrocene

It was the remarkable inertness of ferrocene towards air, water and thermal decomposition as well as its novel structure which caught the imagination of many chemists in the 1950s. Nevertheless, ferrocene undergoes many reactions leading to numerous derivatives.

(a) OXIDATION. Despite its 18-electron configuration, ferrocene is oxidized in solution even by fairly mild reagents such as iodine, copper(II) or iron(III) to give

Five electron ligands

the ferricenium ion.

$$Cp_2Fe + 3/2I_2 \rightleftharpoons Cp_2Fe^+I_3^-, \quad K = 175 \pm 3 \text{ in benzene, } 25°C$$

The polarographic half-wave potential for the oxidation is 0.34 V relative to the standard calomel electrode. With iron(III) chloride the usual product is the salt $Cp_2Fe^+ \cdot FeCl_4^-$. Dilute solutions of the ferricenium ion are blue, but appear blood red when concentrated. The assignment of a low spin configuration (a_{1g}^2, e_{2g}^3) to the ground state, in contrast to the high spin configuration of the isoelectronic molecule, manganocene, is also in accord with the magnetic moment of $Cp_2Fe^+PF_6^-$ ($\mu_{eff} = 2.3$ B.M.) at 20 K.

The determination of glucose in blood is important in the control of diabetes mellitus. An electrode which produces a response proportional to glucose concentration over the range 0–40 mM and which functions either in plasma or in whole blood has recently been developed. Glucose is oxidized to gluconolactone by the enzyme glucose oxidase. The oxidized form of the enzyme can be regenerated using ferricenium, or 1,1'-dimethylferricenium ion.

$$D\text{-glucose} + \text{glucose oxidase(Ox)} \rightleftharpoons \text{gluconolactone} + \text{glucose oxidase(Red)}$$
$$\text{Glucose oxidase(Red)} + 2Fc^+ \rightleftharpoons \text{glucose oxidase(Ox)} + 2Fc + 2H^+$$
$$2Fc \rightleftharpoons 2Fc^+ + 2e^-$$

The electrode is made of graphite, coated with 1,1'-dimethylferrocene. Glucose oxidase is immobilized on its surface by entrapment inside a polycarbonate membrane. When the electrode is poised at a potential sufficiently positive to produce the ferricenium ion, a response is obtained. The total current which flows is proportional to the amount of glucose which is present in the sample.

(b) REDUCTION.　Reduction of ferrocene is very difficult. Lithium in ethylamine cleaves the molecule yielding the cyclopentadienyl anion. At $-30°C$ in dimethoxyethane a reversible one electron reduction is observed electrochemically at -2.93 V (vs SCE) to a 19-electron radical anion Cp_2Fe^- which is fairly stable at low temperatures.

(c) PROTONATION.　Strong acids such as sulphuric acid, tetrafluoroboric acid or hexafluorophosphoric acid, which are not also strong oxidizing agents protonate ferrocene at the iron atom. The resulting hydride, Cp_2Fe—H^+ shows a high field signal at -12.1 p.p.m. in the proton n.m.r. spectrum which is broadened by exchange with the protons of the acid. The cyclopentadienyl protons appear as a doublet on account of coupling with the nucleus of this metal-bonded hydrogen atom. There is rapid exchange of hydrogen for deuterium in the C_5H_5 rings when ferrocene is treated with deuterated acids. This exchange could occur by any of three pathways – (a) direct electrophilic attack from above the ring (exo-attack), (b) direct electrophilic attack from below the ring, which might be favoured as this is a region of high electron density near to the metal atom (endo attack) or (c) initial addition of H^+ to the metal followed by intramolecular

282

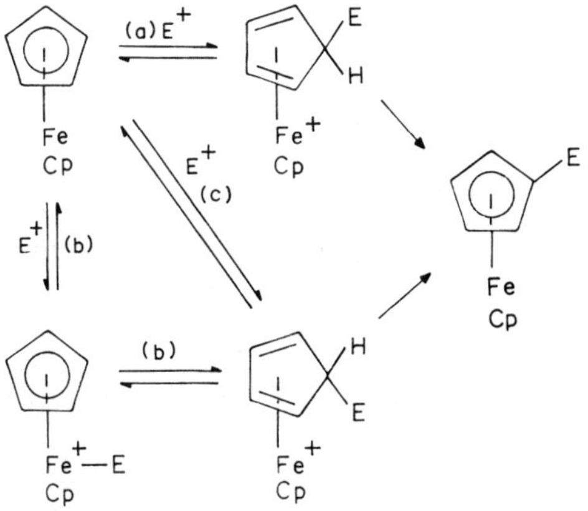

Fig. 9.3 Possible pathways in the electrophilic substitution of ferrocene.

migration to the endo position. Finally elimination of H^+ from the intermediate yields the substituted product (Fig. 9.3).

While there are precedents for all the steps (a), (b), and (c) from analogous reactions on other complexes it has not been possible to establish which mechanism or mechanisms apply in electrophilic substitution of ferrocene or of related molecules.

(d) ELECTROPHILIC SUBSTITUTION.   Many media which effect the electrophilic substitution of benzene are strongly oxidizing (e.g. nitration, halogenation). With ferrocene the result is oxidation to the ferricinium ion or disruption of the complex rather than substitution at a ring carbon atom. Consequently the reagents which can be used to substitute ferrocene are in general restricted to those with only weak oxidizing properties. To counterbalance this, ferrocene is very much more reactive towards electrophiles than benzene or even methoxybenzene, being on a par with phenol.

Friedel-Crafts acylation of ferrocene occurs under mild conditions, acetic anhydride and phosphoric acid yielding monoacetylferrocene, for example. Acetyl chloride and aluminium chloride afford mainly acetylferrocene or 1, 1'-diacetylferrocene depending on the molar proportions taken. In disubstitution, the electron-withdrawing acetyl group deactivates the first ring, directing the second reagent molecule to the other ring.

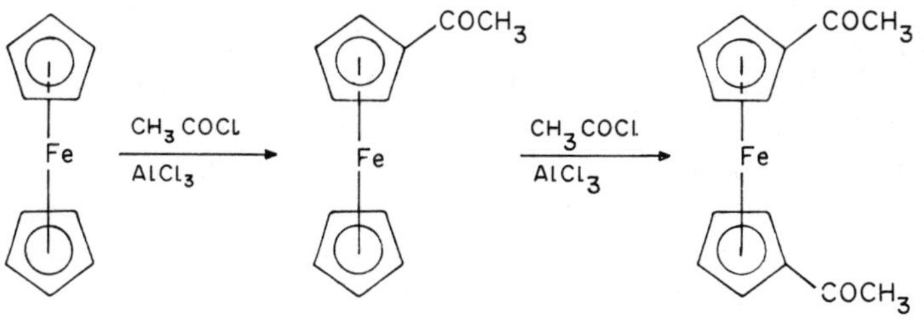

283

Five electron ligands

Ferrocene undergoes the Vilsmeier reaction with $PhNMeCHO/POCl_3$ to yield ferrocenal, FcCHO, a reagent which is useful in the synthesis of numerous derivatives. With formaldehyde and secondary amines (Mannich condensation) the monodialkylamino derivative $FcCH_2NR_2$ is produced. These reactions do not occur with benzene itself but only with very reactive compounds such as phenol. Electrophilic mercuration of ferrocene by mercury(II) acetate to give FcHg(OAc) occurs at a rate 10^9 times that of the corresponding reaction with benzene. It is possible that the initial step involves an intermediate with an iron–mercury bond. A blue adduct $Cp_2Fe \cdot 2HgCl_2$ has been isolated; the structure of a similar ruthenium compound $Cp_2Ru—Hg(\mu\text{-Br})_2Hg—RuCp_2$ has been determined by X-ray diffraction.

MANNICH   VILSMEIER   KNOEVENAGEL

Sulphonation can be achieved with sulphuric acid or, better, with chlorosulphonic acid, $ClSO_3H$, in acetic anhydride.

Friedel-Crafts alkylation of ferrocene often leads to polysubstitution and is accompanied by side reactions, including oxidation to Cp_2Fe^+ which is inert to electrophiles. The reactions with aryldiazonium salts are thought to involve aryl radicals

$$Cp_2Fe + ArN_2^+ \xrightarrow{-N_2} Cp_2Fe^+ + Ar^{\cdot} \longrightarrow CpFe(C_5H_4Ar) + H^+$$

as the ferricinium ion is inert to ArN_2^+.

(e) METALLATION. Ferrocene can readily be metallated. This metallation resembles the hydrogen-exchange reactions typical of aromatic hydrocarbons (p. 40). n-Butyllithium yields mainly lithioferrocene and 1, 1′-dilithioferrocene can be obtained using the Bu^nLi. TMEDA complex. These lithiated derivatives are precursors to a wide range of substituted compounds, some of which, such as halogeno- or nitroferrocene, cannot be obtained by direct electrophilic attack on the parent. Some examples are shown in Fig. 9.4.

284

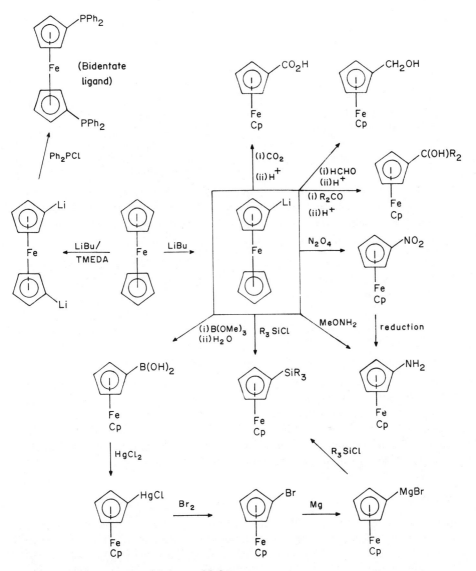

Fig. 9.4 Some reactions of ferrocenyl lithium.

(f) STABILIZATION OF α-CARBONIUM IONS.   A feature of ferrocene chemistry is the ready stabilization of a carbonium ion in which the positive charge is concentrated α- to one ring. Crystalline salts can be isolated from the protonation of ferrocenyl carbinols or vinylferrocenes with acids such as HPF_6, which have weakly coordinating anions.

285

Five electron ligands

The stability of a carbonium ion can be expressed in terms of a pK_{R^+} value, which relates to the equilibrium

$$R^+ + 2H_2O \rightleftharpoons ROH + H_3O^+; \quad pK_{R^+} = \log_{10}\frac{[R^+]}{[ROH]} + H_R$$

H_R is an 'acidity function' which provides a reference point for the pK_{R^+} scale. The more positive the pK_{R^+} value, the more stable is the carbonium ion (Table 9.2). The property of stabilizing α-carbonium ions is also shared by similar organometallic groups. Their effectiveness lies in the order

The molecular structure of $[(C_5H_4CPh_2)FeCp]^+PF_6^-$, determined by X-ray diffraction, shows that the CPh_2 group is bent down through $20°$ out of the plane of the ring towards the iron atom. It is suggested that the vacant p-orbital of the carbonium ion interacts not only with the delocalized π-system of the ring but also with the metal centre. Although the barrier to rotation about the C_{ring}–$\overset{+}{C}R_2$ bond in this and similar compounds is only $80 \pm 10 \, kJ \, mol^{-1}$, this is sufficiently high to prevent racemization during substitution.

9.2.4 Some characteristics of the chemistry of other metallocenes

(a) VANADOCENE AND CHROMOCENE. Vanadocene is a 15-electron complex, so it is not surprising that most of its reactions involve addition to the metal centre to give products in which the electron count has increased to 17 or 18 (Fig. 9.5).

In the adducts Cp_2VX or $Cp_2V(X)Y$, the cyclopentadienyl rings are no longer parallel. Structures of related compounds are described on p. 290.

Chromocene is very easily hydrolysed and oxidized with disruption of the complex. Solutions of a cation $Cp_2Cr^+PF_6^-$, however, can be prepared by

Table 9.2 Relative stabilities of carbonium ions in aqueous acid

Ion	$(C_4H_3\overset{+}{C}Ph_2)CoCp$	$Ph_2\overset{+}{C}H$	$Ph_3\overset{+}{C}$	$(C_5H_4\overset{+}{C}Ph_2)Mn(CO)_3$	$Fc\overset{+}{C}H_2$	$Fc_2\overset{+}{C}H$	$Fc_3\overset{+}{C}$
pK_{R^+}	-15.2	-13.3	-6.6	-6.5	-1.2	4.1	5.8

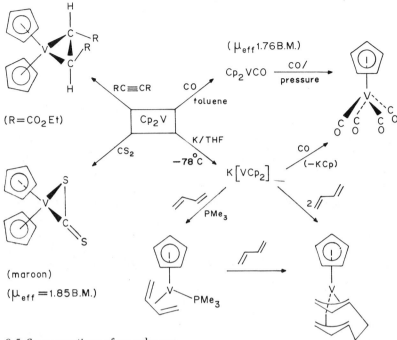

Fig. 9.5 Some reactions of vanadocene.

electrochemical oxidation. At $-78°C$ carbon monoxide forms the adduct $Cp_2Cr.CO$, but further treatment under pressure at higher temperatures leads to a mixture of products.

(b) COBALTOCENE AND NICKELOCENE. Cobaltocene is a 19-electron complex, possessing one unpaired electron in an e_{1g}^* orbital. Its chemistry is dominated by the urge to lose this electron and in so doing to form 18-electron compounds. The simplest way in which this is achieved is by oxidation, which occurs very readily, for example in air or by the action of aqueous hydrogen peroxide. The product is the cobalticinium ion Cp_2Co^+ which forms orange diamagnetic salts with a wide range of anions. This ion is so inert to further oxidation that it is unaffected by boiling concentrated nitric acid or dilute alkaline hydrogen peroxide.

The tendency to form the Cp_2Co^+ ion is further exemplified by the reactions of cobaltocene with acids, when hydrogen is liberated, and with alkyl halides. The latter are thought to proceed by a free radical mechanism

$$RX + Cp_2Co \longrightarrow Cp_2Co^+ + R^{\cdot} + X^-$$

$$Cp_2Co + R^{\cdot} \longrightarrow CpCo{-}\underset{H}{\overset{R}{\diagup}}$$

287

Five electron ligands

Dihalogenomethanes give cyclohexadienyl salts in which the CH_2 group has inserted into one of the cyclopentadienyl rings. A similar reaction with arylboron dihalides affords complexes which contain the borabenzene ligand $C_5H_5B{-}R$. This ligand provides a six-membered delocalized system but only five ligand π-electrons.

Electrochemical reduction of Cp_2Co^+ proceeds in two 1-electron steps at $-0.945\,V$ and $-1.88\,V$ relative to the standard calomel electrode. The final product Cp_2Co^- is so strongly nucleophilic that it attacks even carbon dioxide (cf. a Grignard reagent). This reduction can also be effected by potassium in tetrahydrofuran. The Cp^- anion is quite a good leaving group, so in the presence of suitable ligands such as alkenes or CO, $K^+[CoCp_2^-]$ affords KCp and monocyclopentadienyl complexes $CpCoL_2$. This general strategy has been used to prepare monocyclopentadienyl derivatives from other metallocenes (see Fig. 9.6).

Nickelocene, a 20-electron compound, has two electrons in the antibonding e_{1g}^* orbitals. The monocation Cp_2Ni^+ can be generated chemically with halogens or with dilute nitric acid. Electrochemical oxidation occurs in two steps to Cp_2Ni^{2+} through Cp_2Ni^+. Reduction to Cp_2Ni^- has also been achieved.

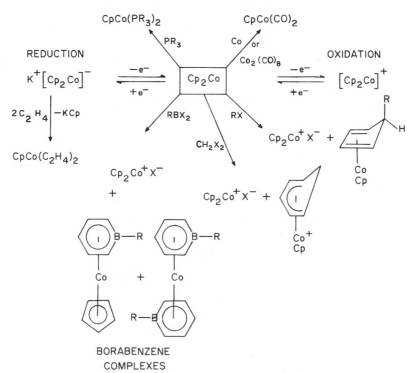

Fig. 9.6 Some reactions of cobaltocene.

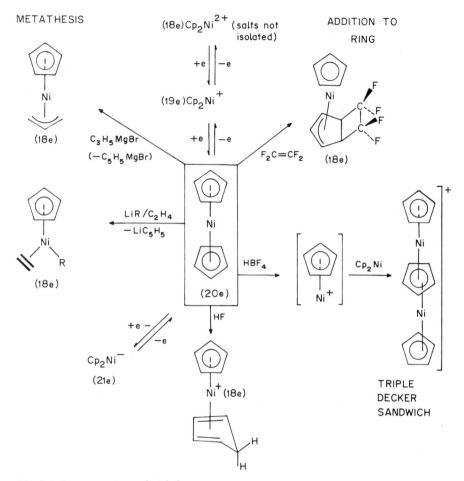

Fig. 9.7 Some reactions of nickelocene.

The drive to attain an 18-electron configuration is illustrated by cycloaddition reactions which occur with tetrafluoroethene or with alkynes (Fig. 9.7). The cyclopentadienyl groups, especially the first one, are fairly easily displaced from nickelocene. Thus phosphines and phosphites yield NiL_4 ($L = R_3P$ or $(RO)_3P$). Nitrogen monoxide the monomeric red liquid $CpNi(NO)$ and nickel tetracarbonyl the dimeric complex $[CpNi(CO)]_2$. Cyclopentadienyl is also displaced by metathesis with allyl magnesium chloride or with LiR/C_2H_4. One fascinating reaction occurs with acids such as HBF_4 in which a 'triple-decker sandwich' is produced. This arises by cleavage of one cyclopentadienyl ligand as cyclopentadiene, and addition of the resulting cation $CpNi^+$ to another molecule of nickelocene.

289

Five electron ligands

(c) TITANOCENE AND DICHLOROBIS(CYCLOPENTADIENYL)TITANIUM. Treatment of titanium(IV) chloride with two equivalents of sodium cyclopentadienide in dimethoxyethane affords the red crystalline complex Cp_2TiCl_2. Its structure is typical of a large number of 16, 17 and 18 electron compounds of general formula Cp_2MX_2 which are formed by the elements of Group IV, V and VI. In these the metal assumes distorted tetrahedral geometry. The chloride ligands in Cp_2TiCl_2

are readily replaced by metathesis in non-aqueous solvents with alkali metal salts MX to yield derivatives Cp_2TiX_2 (X = e.g. F, Br, NCO, NCS, N_3, OR, OCOR, SR). Ammonium pentasulphide gives Cp_2TiS_5 which has been used by Max Schmidt to prepare cyclopolysulphur allotropes such as S_7, S_9 and S_{10}. Partial hydrolysis of the Ti—Cl bonds in Cp_2TiCl_2 occurs in aqueous solutions.

Methyllithium reacts with Cp_2TiCl_2 to yield the dimethyl complex Cp_2TiMe_2, which is an orange yellow solid, stable in air, and unreactive to cold water but not very stable to heat. On photolysis in an atmosphere of hydrogen it affords a violet complex which on heating is converted into one form of the 'titanocene' dimer.

a 'titanocene' dimer

The monomer of titanocene has never been isolated. Decamethyltitanocene, $Ti(C_5Me_5)_2$, however, has been prepared, as has bis(2,4-dimethylpentadienyl)-titanium (p. 302).

(d) HYDROZIRCONATION. Zirconium forms a complex Cp_2ZrCl_2, analogous to Cp_2TiCl_2, which is converted into the polymeric hydride $1/n[Cp_2ZrHCl]_n$ on treatment with $KBHBu_3^s$. This hydride readily adds across the double bond of an alkene in a regiospecific fashion, that is, the zirconium adds to the less sterically hindered end of the double bond. In the case of an internal olefin the initial addition is followed by a series of rapid isomerizations, so that the zirconium finishes at the end of the carbon chain as in hydroboration (p. 66) or hydroalumination (p. 80) (Fig. 9.8). Insertion of carbon monoxide into the Zr—C

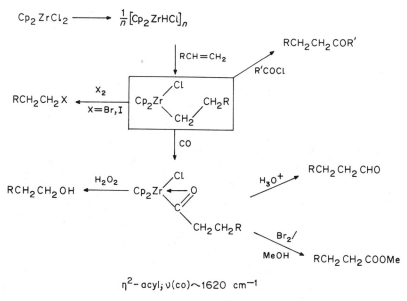

Fig. 9.8 Some applications of hydrozirconation in synthesis.

bond of the initial product occurs readily at room temperature to form an acyl (p. 225). A single crystal X-ray diffraction study of the acetyl derivative $Cp_2ZrCH_3(COCH_3)$ shows that the acyl group is η^2-bonded to the metal forming the three-membered ring illustrated. The Zr—alkyl or Zr—acyl bonds can be cleaved to give a variety of products. As the reagents are readily accessible, cheap and fairly easily handled, hydrozirconation is a viable procedure for the small scale synthesis of organic compounds.

(e) METALLOCENES OF THE ELEMENTS OF THE SECOND AND THIRD TRANSITION SERIES. Apart from the 18-electron compounds ruthenocene and osmocene, no stable monomeric metallocenes of the elements of the Second and Third Transition Series have been isolated. Treatment of halides of these elements with excess sodium cyclopentadienide in tetrahydrofuran leads either to complexes like $(\eta^5\text{-}C_5H_5)_2Nb(\eta^1\text{-}C_5H_5)_2$ which contain both η^5 and η^1 bonded groups, or to low yields of cyclopentadienylmetal hydrides. Better yields of the latter, for example Cp_2TaH_3, Cp_2WH_2 or Cp_2ReH (and the corresponding Nb and Mo compounds) are obtained in the presence of sodium borohydride as a hydride source.

The molecular structure of Cp_2MoH_2 has been determined by neutron diffraction, which is able to locate the hydrogen nuclei precisely. If the metal atom is placed at the origin and the hydrogen atoms in the yz plane there is an orbital along the z axis (approximately d_{z^2}) which in this 18-electron compound contains

Five electron ligands

a pair of non-bonding electrons. (In the 16-electron compound Cp_2TiCl_2 the equivalent orbital is vacant.) Cp_2MoH_2 acts as a base, being protonated by trifluoroacetic acid to give $Cp_2MoH_3^+$ and forming adducts $Cp_2MoH_2.BF_3$ and $Cp_2MoH_2.AlMe_3$ with Lewis acids. The tungsten complex is similar.

Mean bond dissociation energies $\bar{D}(M—H)$ and $\bar{D}(M—Me)$ in the compounds Cp_2MX_2 are as follows: $\bar{D}(M—H)/kJ\,mol^{-1}$: Mo, 252; W, 305. $\bar{D}(M—Me)$: Mo, 159; W, 209. Consider the hypothetical addition of methane to Cp_2M:

$$Cp_2M + H—Me \rightleftharpoons Cp_2M(H)Me$$

in which a C—H bond is broken and M—H and M—Me bonds are formed. Given that $D(C—H)$ in methane is $435\,kJ\,mol^{-1}$, $\Delta H_{reaction}$ is $+24\,kJ\,mol^{-1}$ (M = Mo) and $-79\,kJ\,mol^{-1}$ (M = W). $T\Delta S^{\ominus}$ is estimated to be about $55\,kJ\,mol^{-1}$ at 298 K owing to the loss of one mole of gaseous methane. Although this reaction has not been observed in this system with methane, other hydrocarbons have been shown to react with Cp_2WH_2 when they are irradiated with ultraviolet light.

It is believed that the addition proceeds through a labile tungstenocene species 'Cp_2W'. Spectroscopic investigations of the species produced on photolysis of Cp_2WH_2 in an argon matrix at 20 K are consistent with this interpretation. Cp_2W apparently has a triplet ground state like chromocene, but is much more reactive behaving more like a diradical. This radical character is shared by other bis cyclopentadienyls of the heavy transition elements (apart from Ru, Os). Where they have been observed at all the monomers are very short lived and dimerize rapidly to products rather complicated structures.

9.3 Cyclopentadienylmetal carbonyls

The cyclopentadienyl metal carbonyls form a large group of compounds the chemistry of which has been explored in detail. Consequently only a few general features can be covered here. Most of these compounds obey the 18-electron rule. This is clearly illustrated by the series of mononuclear complexes $CpV(CO)_4$, $CpMn(CO)_3$ and $CpCo(CO)_2$ (Fig. 9.9). The simplest derivatives of the intervening elements Cr, Fe and Ni are binuclear; here the electron count is attained by metal—metal bonding or by electron sharing through bridging carbonyls. The formation of binuclear systems by alternate members of the transition series resembles the similar formation of binuclear carbonyls in the series $Cr(CO)_6$, $Mn_2(CO)_{10}$, $Fe(CO)_5$, $Co_2(CO)_8$ and $Ni(CO)_4$.

292

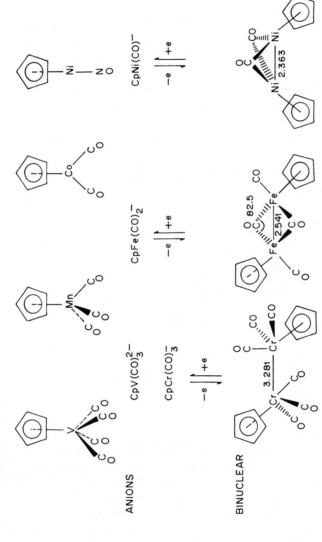

MONONUCLEAR

ANIONS

$CpV(CO)_3^{2-}$ $CpCr(CO)_3^-$

$CpFe(CO)_2^-$ $CpNi(CO)^-$

BINUCLEAR

Fig. 9.9 Structures of some cyclopentadienyl carbonyl complexes.

Five electron ligands

9.3.1 *Mononuclear compounds*

The mononuclear compounds $CpV(CO)_4$, $CpMn(CO)_3$ and $CpCo(CO)_2$ can be prepared by treating the requisite metallocene, formed *in situ* from metal chloride and sodium cyclopentadienide in tetrahydrofuran, with carbon monoxide under pressure. Substitution of a carbonyl group in $CpV(CO)_4$ or in $CpMn(CO)_3$ by donor molecules such as phosphines is rather difficult to effect by heating. The manganese compound especially requires prolonged reaction times and high temperatures. $CpCo(CO)_2$, however, reacts faster. The substitutions of $CpCo(CO)_2$ and of $CpRe(CO)_3$ (and therefore possibly also of $CpMn(CO)_3$) follow an associative mechanism, which is perhaps rather surprising as the complexes are 'saturated', obeying the 18-electron rule. There is good evidence that the bonding of the η^5-Cp ligand changes to η^3 on approach of the incoming ligand. The indenyl complex, for which the η^3 coordination mode should be favoured on account of the gain of aromaticity in the benzene ring, reacts much faster with phosphines than $CpMn(CO)_3$. A rate expression which is second order overall (Rate = k_2[complex][L]) is followed.

L = PR_3, $P(OR)_3$

Because it is so difficult to substitute $CpMn(CO)_3$ thermally, a photochemical method is usually used.

$$CpMn(CO)_3 \longrightarrow CpMn(CO)_2.THF \xrightarrow{\text{L}} CpMn(CO)_2L$$

Often a preliminary photolysis is carried out in tetrahydrofuran which yields a solution of the labile complex $CpMn(CO)_2THF$. This is then converted into the desired product by addition of ligand. A very wide range of compounds can be prepared in this way using, for example, L = R_3P, $(RO)_3P$, R_3As, R_3Sb, R_2S, alkenes, alkynes, sulphur dioxide and even dinitrogen. The $CpMn(CO)_2$ unit is very inert to attack by a wide range of reagents. The photochemical method is also recommended for the substitution of carbonyl groups in $CpV(CO)_4$. Reaction of $CpMn(CO)_3$ or $CpV(CO)_4$ with carbon disulphide and triphenylphosphine under irradiation in tetrahydrofuran yields thiocarbonyl complexes $CpMn(CO)_2CS$ and $CpV(CO)_3CS$ respectively. The CS group shows a band at $1250-1300 \, cm^{-1}$ in the infrared spectrum.

The cobalt complex $CpCo(CO)_2$ is a dark red liquid, b.p. $139°C/710mm$ which is rather sensitive to light, heat and oxygen and which is much more reactive

294

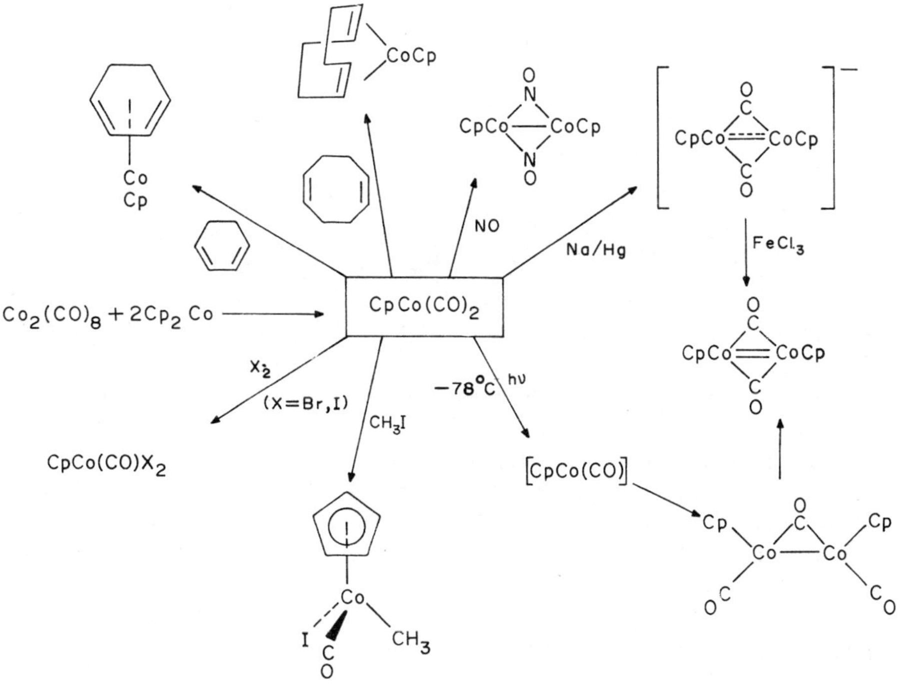

Fig. 9.10 Some reactions of CpCo(CO)$_2$.

than CpMn(CO)$_3$. Carbonyl groups can be replaced by other ligands such as phosphines or dienes. Photolysis at $-78°C$ yields initially CpCo(CO) which can trap another molecule of CpCo(CO)$_2$ to yield Cp$_2$Co$_2$(CO)$_3$ and eventually other species (Fig. 9.10). CpCo(CO)$_2$ is a Lewis base by virtue of a non-bonding electron pair in an orbital centred on the metal atom (p. 210). It forms an adduct with the Lewis acid, mercury(II) chloride, which contains a cobalt—mercury bond. It also adds halogens and alkyl halides with expulsion of one molecule of carbon monoxide. The basicity is greatly enhanced by replacement of the carbonyl groups by phosphines, especially by trimethylphosphine. CpCo(PMe$_3$)$_2$ is a strong base which is very easily protonated, even by NH$_4^+$PF$_6^-$ in methanol. The ^{1}H n.m.r. spectrum of the resulting hydride shows a triplet at $-16.2\,\delta$ (J_{PH} 83 Hz) arising from coupling to two equivalent phosphorus nuclei. CpCo(PMe$_3$)$_2$ also acts as a nucleophile towards alkyl and acyl halides and as a strong Lewis base.

9.3.2 Binuclear compounds

The iron dimer [CpFe(CO)$_2$]$_2$ is easily prepared by heating pentacarbonyliron in dicyclopentadiene at 140°C. It forms maroon-purple crystals which are only slowly oxidized in air, although the solutions are more sensitive. The compound is cleaved by halogens to yield mononuclear derivatives:

$$[CpFe(CO)_2]_2 \xrightarrow{\text{X}_2} CpFe(CO)_2X \qquad (X = Br, I)$$

$$\xrightarrow{\text{HCl/O}_2} CpFe(CO)_2Cl$$

$$\xrightarrow{\text{AgBF}_4/L} [CpFe(CO)_2L]^+ \, BF_4^- \quad (L = CO, PR_3, \text{alkene})$$

Silver tetrafluoroborate in the presence of ligands oxidizes it to cationic species (p. 240).

A typical reaction of many cyclopentadienylmetal carbonyls is reduction to anionic species (Fig. 9.11). Sodium amalgam is often used as the reducing agent. The liquid sodium–potassium alloy can also be used. A solution of the starting material in tetrahydrofuran is stirred with the amalgam under nitrogen and the spent amalgam run off through a stopcock at the base of the flask when the reduction is complete. The solutions of the anions are very air sensitive. The anions themselves are strong nucleophiles which undergo numerous reactions. This chemistry is exemplified with reference to $CpFe(CO)_2^-$. The behaviour of other mononuclear species, especially $CpM(CO)_3^-$ (M = Mo, W) shows many parallels.

Cis and trans geometrical isomers of $[CpFe(CO)_2]_2$ have been prepared. The stable form of the molecule has the trans structure shown in Fig. 9.12, but on crystallization at low temperature a cis isomer can be isolated. In this molecule the Fe_2C_2 group is folded about the Fe—Fe axis, giving an angle of $164°$ between the two Fe—C—Fe planes. In the trans isomer, however, the Fe_2C_2 atoms are coplanar. Observations of the proton and ^{13}C n.m.r. spectra of $[CpFe(CO)_2]_2$ in solution over a range of temperature reveal rapid exchange of carbonyl groups between bridging and terminal positions (Fig. 9.13). In the case of the trans isomer this is sufficient to cause equilibration of the carbonyl groups even at low temperatures (see Newman projection). At $-65°C$ this isomer gives only one averaged carbonyl signal in the ^{13}C spectrum. The activation energy for the exchange is about $30 \, kJ \, mol^{-1}$. The cis isomer, however, affords two signals which arise from CO^b and CO^t respectively. Only two of the carbonyl groups (CO^b) lie in the conformation required to form the two bridges, that is, trans to each other. Rotation about the Fe—Fe bond in the unbridged structure can make the CO^b and CO^t groups equivalent, but this is detected by the n.m.r. experiment only above $-35°C$. The activation energy for the rotation $(47 \, kJ \, mol^{-1})$ is greater than that for the bridge opening. In agreement with this mechanism, the same value has been obtained for the cis–trans isomerization from ^1H n.m.r. measurements.

The iron—iron distance in the cis and also in the trans isomer is quite long, about $2.53 \, \text{Å}$, compared with $2.48 \, \text{Å}$ in metallic iron. X-ray diffraction reveals

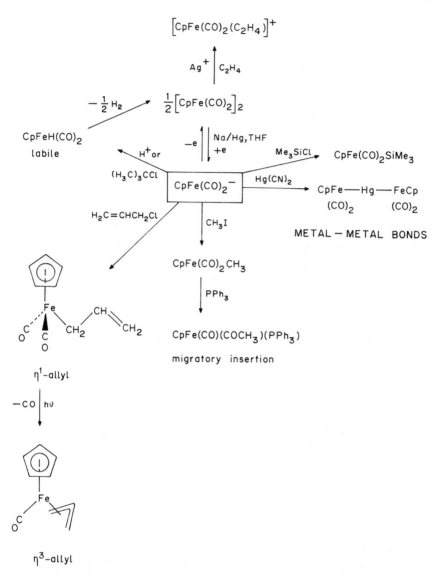

Fig. 9.11 Summary of reactions of the anion $[CpFe(CO)_2]^-$.

that there is little electron density directly between the iron atoms in these structures, but rather a zone of electron density around the Fe_2C_2 system. This suggests that electron sharing between the two $CpFe(CO)_2$ units may occur largely in delocalized multicentre bonds rather than by direct metal—metal interaction. Direct metal—metal bonding, unsupported by carbonyl bridges, occurs however in the osmium analogue $[CpOs(CO)_2]_2$. It is also found in $[CpM(CO)_3]_2$ (M = Cr,

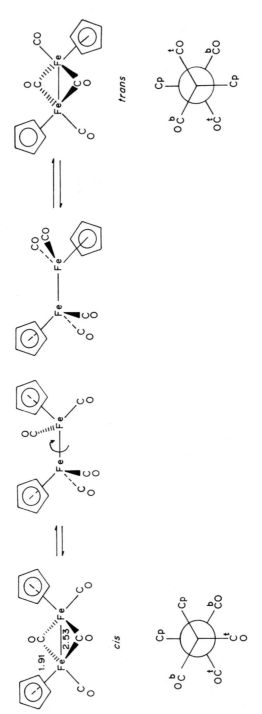

Fig. 9.12 Isomerism in $Cp_2Fe_2(CO)_4$.

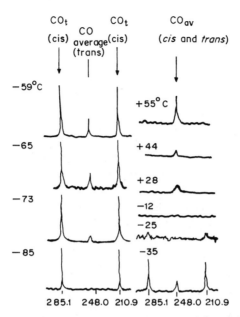

Fig. 9.13 The ^{13}C n.m.r. spectra at various temperatures of the carbonyl-groups of Fe$_2$ $(\eta$-C$_5$H$_5)_2$(CO)$_4$. (Reproduced by permission from O.A. Gansow, A.R. Burke and W.D. Vernon, *J. Am. Chem. Soc.*, 1972, **94**, 2550.)

Mo, W). In the latter complexes, the metal—metal distances are very large, but they decrease down the group (Cr—Cr, 3.28 Å, Mo—Mo, 3.235 Å, W—W, 3.22 Å). The Cr—Cr bond in [CpCr(CO)$_3$]$_2$ is so weak that in solution in phenyl cyanide at $-20°$C the molecule dissociates into radicals CpCr(CO)$_3^{\cdot}$, which have been detected by e.s.r. spectroscopy. Similar thermal dissociation of the molybdenum compound has not been observed, although radicals can be generated by photolysis. In inert solvents a product from this photolysis is the dimer [CpMo(CO)$_2$]$_2$.

$$\left[\text{CpMo(CO)}_3\right]_2 \xrightarrow{\text{h}\nu} 2\,\text{CpMo(CO)}_3^{\cdot} \xrightarrow[\text{toluene}]{-\text{CO}} \left[\text{CpMo(CO)}_2\right]_2$$

$$\downarrow \text{CCl}_4$$

$$2\,\text{CpMo(CO)}_3\text{Cl}$$

The bonding in this dimer poses problems. The 18-electron rule predicts a triple Mo—Mo bond, and this is consistent with the short distance which is observed (2.45 Å). X-ray diffraction, however, reveals rather novel coordination of all four carbonyl groups. The Mo—C—O systems are essentially linear, while the C—O

Five electron ligands

π-system of each ligand interacts with the second molybdenum atom as illustrated.

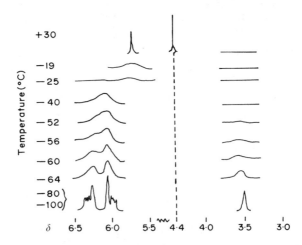

9.3.3 *CpFe(CO)₂(η¹-C₅H₅): Fluxional behaviour*

When $CpFe(CO)_2Br$ is treated with sodium cyclopentadienide, a complex is formed in which two cyclopentadienyl groups are present:

$CpFe(CO)_2\,Br \xrightarrow{\;NaCp\;}$

The proton n.m.r. spectrum, measured at ambient temperature, consists only of two sharp signals of equal intensity (Fig. 9.14). The infrared spectrum, however,

Fig. 9.14 The ¹H n.m.r. spectra at 60 MHz of CpFe(CO)₂(η¹-C₅H₅) in CS₂ at various temperatures. The signal for the Cp protons is shown only on the 30°C spectrum, the amplitude of which is 0.1 times that of the others. (Reproduced by permission from M.J. Bennett, F.A. Cotton, A. Davison, J.W. Faller, S.J. Lippard and S.M. Morehouse, *J. Am. Chem. Soc.*, 1966, **88**, 4371.)

300

shows bands characteristic of a free diene unit. Moreover the 18-electron rule predicts a structure either with one η^5 and one η^1 C_5H_5 group or one containing two η^3-C_5H_5 groups. When the temperature of the n.m.r. probe is lowered one band, assigned to η^5-C_5H_5 does not change. The other, however, broadens, coalesces and finally resolves at low temperature into the pattern A_2B_2X expected from an η^1-C_5H_5 group. These changes are consistent with motion within the molecule (called a 'fluxional' or 'dynamic' process) which is slow at low temperature compared with the lifetime of the nuclear spin state, but which becomes increasingly rapid as the temperature is raised. In the limit all five protons are observed as an averaged signal.

The nature of the 'fluxional' process can be determined by careful examination of the way the spectrum changes with temperature. In a 1, 2-shift the iron atom moves to the adjacent position of the cyclopentadienyl ring. The environment of each proton changes as indicated. Remembering that the positions A and A', B and B' respectively are magnetically equivalent, both H^A and $H^{A'}$ change their environments, but only one of H^B and $H^{B'}$. If this is the mechanism, the signal arising from H^A and $H^{A'}$ will collapse at a lower temperature than that from H^B, $H^{B'}$. If a 1, 3-shift operates the signal from H^B, $H^{B'}$ will collapse at a lower temperature than that from H^A, $H^{A'}$. The observations agree with the former postulate, indicating a 1, 2-shift mechanism.

9.4 Open-chain (acyclic) pentadienyl complexes

1, 4-pentadiene is deprotonated by n-butyllithium in the presence of tetrahydro-furan to yield an orange solution of pentadienyllithium. The potassium derivative can be prepared from either 1, 3- or 1, 4-pentadiene in THF using LiBu^n/KOBu^t or potassium metal in the presence of triethylamine. While the pentadienyl group adopts a W configuration in the lithium derivative, it is U shaped in the potassium salt.

Five electron ligands

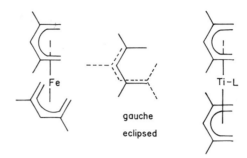

Fig. 9.15 Molecular structures of bis(2,4-dimethylpentadienyl)iron showing the favoured gauche eclipsed conformation of the ligands and of $(\eta^5\text{-Me}_2\text{C}_5\text{H}_5)_2\text{TiL}$ (L = CO, PF$_3$).

Reaction of pentadienyl anions with anhydrous dihalides of the First Transition Series in THF yields a series of acyclic pentadienyl complexes (Fig. 9.15). The 2, 4-dimethylpentadienyl derivatives are generally more robust than the unsubstituted compounds and most work has been done on them.

Bis(pentadienyl)chromium and bis(pentadienyl)dinickel were discovered in 1968. It was not until 1980, however, when R.D. Ernst reported the open chain ferrocenes that systematic study of these compounds began. Bis(2, 4-dimethylpentadienyl)iron is an orange solid, which is essentially stable to air. Like ferrocene it sublimes readily under reduced pressure. In the crystal the rings adopt a gauche eclipsed conformation. N.m.r. studies reveal that this is the most stable conformation in solution also, and that there is a much higher barrier to rotation of the ligands about the vertical axis through the metal ($\sim 60\,\text{kJ}\,\text{mol}^{-1}$) compared with that in ferrocene ($\sim 5\,\text{kJ}\,\text{mol}^{-1}$).

The 14-electron titanium complex is monomeric and thermally robust, unlike titanocene itself (p. 290). It adds only one molecule of CO or PF$_3$ to yield 16-electron adducts Ti(η^5-Me$_2$C$_5$H$_5$)$_2$L in which the two pentadienyl ligands are parallel and eclipsed.

It is to be expected that a wide range of pentadienyl complexes, analogous to cyclopentadienyls, remain to be discovered. Pentadienyltricarbonylmanganese, analogous to cymantrene, CpMn(CO)$_3$ has been prepared from Me$_3$SnC$_5$H$_7$ and Mn(CO)$_5$Br. It adds basic ligands such as PMe$_3$ under mild conditions to give η^3 derivatives. This easy $\eta^5 \rightarrow \eta^3$ interconversion seems to be a feature of pentadienyl chemistry.

Pyrrole forms a few complexes in which it is η^5-bonded as C_4H_4N. These include bis(pyrrolyl)iron, $Fe(C_4H_4N)_2$, which is isoelectronic with ferrocene, and pyrrolyltricarbonylmanganese.

9.5 Cyclohexadienyltricarbonyliron salts

Cyclohexadienyl cations are well known as intermediates in the electrophilic substitution of benzene. Indeed benzene is protonated by superacids to give solutions containing the ion $C_6H_7^+$, which can be characterized spectroscopically. These ions are, however, far too reactive to be isolated as salts, and as they are generated only in acidic media, their reactions with nucleophiles cannot be studied. In contrast, cyclohexadienyltricarbonyliron hexafluorophosphate is a stable yellow solid which can be stored indefinitely at room temperature and can even be recrystallized from water. This is an especially striking example of the stabilization of a reactive organic species by coordination to a transition element.

Birch reduction of arenes affords 1, 4-cyclohexadienes. On heating these with $Fe(CO)_5$ in di-butyl ether, complexes of the isomeric 1, 3-dienes are obtained (see p. 362). Treatment with $Ph_3C^+PF_6^-$ yields the salts. These salts react readily with a wide range of nucleophiles. Attack on $(\eta^5\text{-}C_6H_7)Fe(CO)_3^+$ is regiospecific, occurring exclusively at a terminal carbon atom of the dienyl system. It is also stereospecific, taking place from the *exo* side of the ligand, away from the metal. The $Fe(CO)_3$ group can be removed cleanly by oxidation with trimethylamine oxide in organic media, releasing a substituted cyclohexadiene.

Five electron ligands

Nucleophilic reagents include R_2Cd, $NaCH(CO_2Me)_2$, RO^-, R_2NH, activated aromatics such as 1, 3-dimethoxybenzene, allylSiMe$_3$, BH_4^- and CN^-. Lithium alkyls in dichloromethane at low temperatures give good results, but not in diethyl ether. Tertiary amines and phosphines form quaternary salts.

The 2-methoxycyclohexadienyl complexes are particularly useful in synthesis. The methoxy substituent directs nucleophiles to the remote end of the dienyl system. This directive effect is strong enough to overcome the steric hindrance of a methyl substituent at the 5-position, although a mixture of products is obtained with the 5-ethyl complex.

(only product, R=H, Me)   (R=Et)

Spirocyclic compounds have been prepared by making use of the regioselectivity of these reactions. Dienyl complexes were constructed which contained latent nucleophiles in the side chain which were released on treatment with a mild base. Similar reactions have also been used to introduce angular substituents into bicyclic systems.

Pearson and his coworkers have used η^5-dienyliron salts in approaches to the synthesis of many natural products, including steroids, alkaloids and sesquiterpenes.

9.6 References

Caulton, K.G. (1981) Coordination chemistry of the Mn and Re fragments, $(C_5H_5)M(CO)_2$. *Coord. Chem. Revs.*, **38**, 1.

Ernst, R.D. (1985) Open-chain metallocenes. *Acc. Chem. Res.*, **18**, 56.

Faller, J.W. (1978) Fluxional organometallics. *Adv. Organomet. Chem.*, **16**, 211.

Haaland, A. (1979) 3d-Metallocenes. *Acc. Chem. Res.*, **12**, 415.

Hunt, C.B. (1977) Metallocenes. *Educ. Chem.*, **14**, 110.

Jonas, K. (1985) Reactive organometallic compounds obtained from metallocenes. *Angew. Chem. (Int. Edn)*, **24**, 295.

Mann, B.E. (1986) Fluxionality of polyene and polyenyl complexes. *Chem. Soc. Revs.*, **15**, 125.

Maslowsky, A., Jr. (1978) Metallocenes. *J. Chem. Educ.*, **55**, 276.

Negishi, E. and Takahashi, T. (1985) Organozirconium compounds as new reagents. *Aldrichimica Acta*, **18**, 31.

Pearson, A.J. (1984) Iron-stabilised carbocations as intermediates for organic synthesis. *Science*, **223**, 895.

Werner, H. (1980) The way to novel sandwich complexes. *J. Organomet. Chem.*, **200**, 335. (Includes triple decker and bimetallic sandwiches).

Wilkinson, G. (1975) Ferrocene. A personal account of the exciting early years. *J. Organomet. Chem.*, **100**, 273.

Problems

1. Ferrocene, Cp_2Fe, and cobaltocene, Cp_2Co, can be oxidized to the cations Cp_2Fe^+ and Cp_2Co^+ respectively. Discuss the metal—carbon bond distances and magnetic moments (μ_{eff}) given below in terms of the bonding in these compounds.

Compound	M–C/Å	μ_{eff} (298 K)
Cp_2Fe	2.06	0.0
$Cp_2Fe^+BF_4^-$	2.13	2.2
Cp_2Co	2.12	1.7
$Cp_2Co^+BF_4^-$	2.05	0.0

2. When 1, 3, 5-cycloheptatriene is heated under reflux with $Fe(CO)_5$ the tricarbonyliron complex A, $C_{10}H_8FeO_3$, is obtained. Treatment of A with hydrogen gas affords a second tricarbonyliron complex B, $C_{10}H_{10}FeO_3$. Reaction of A with HBF_4 or B with Ph_3C. BF_4 gives C, $C_{10}H_9BF_4FeO_3$, which is insoluble in non-polar solvents. The ^1H n.m.r. spectrum of C consists of five groups of lines of relative intensity $1:2:2:2:2$.

Propose structures for A, B, and C, giving your reasoning. Draw clear diagrams to illustrate the conformation of the organic ligands in these complexes. Give an interpretation of the ^1H n.m.r. spectrum.

(*University of Leicester*).

3. Suggest mechanisms for the reactions of cobaltocene with (a) CH_2Cl_2 (b) $PhBCl_2$.

4. Consider the following reaction scheme:

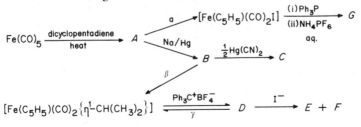

Five electron ligands

(a) Sketch the molecular structures of each of the compounds A to G inclusive.
(b) Name the reagents α, β and γ and state the conditions required to carry out the reactions.
(c) Discuss the isomerism of compound A.
(*Royal Holloway College*).

5. Element A, when heated in air, formed a volatile oxide containing A in its highest oxidation state. On treatment with chlorine A formed a pentachloride as a dark-red solid which, in its simplest structural unit, contained two hexaco-ordinated atoms of A. Fluorides of A in two higher oxidation states are also known.

Reaction of the pentachloride of A with sodium cyclopentadienide resulted in the formation of a yellow solid B.[‡] B was diamagnetic and had a dipole moment. The infrared spectrum of B was similar to that of ferrocene with the addition of a band at $2030 \, cm^{-1}$. The ^1H n.m.r. of B had signals at $\delta = -13.5$ and $\delta = 3.6$ of intensities 1 and 10 respectively. B readily reacted with hydrogen chloride to form an ionic compound C which contained 52.62% A and 10.02% Cl. The ^1H n.m.r. of C had signals at $\delta = -13.6$ and $\delta = 5.6$ of intensities 2 and 10 respectively. On treatment of C with aqueous sodium hydroxide B was regenerated.

Reaction of a complex chloride of A, K_2ACl_6, with copper and carbon monoxide formed D, a carbonyl chloride of A. In the mass spectrometer D showed a molecular ion, the isotopic pattern of which showed its most intense peak at $m/e = 362$. The cracking pattern showed a stepwise loss of five carbon monoxide groups.

Identify A to D and account for all the observations.
[‡]Analysis; C, 37.84%; H, 3.50%; A, 58.66%.
(*University of Exeter*)

6. Some chemistry of ruthenocene derivatives is given below.

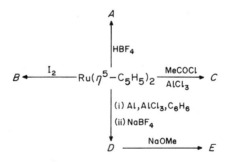

A, B and D are 1:1 electrolytes.
Analytical data: A 31.68%Ru
                  B 13.67%Ru, 68.74%I
                  C 36.97%Ru

D 30.53% Ru
E 36.73% Ru

(a) Describe the metal—ligand bonding in ruthenocene.
(b) Propose a structure for each of the complexes A to E.
(c) Discuss each of the reactions involved.
(*University of Southampton*).

7. Some chemistry of cobaltocene is given in the scheme below, together with the masses of the molecular ions in the mass spectra of the neutral complexes.

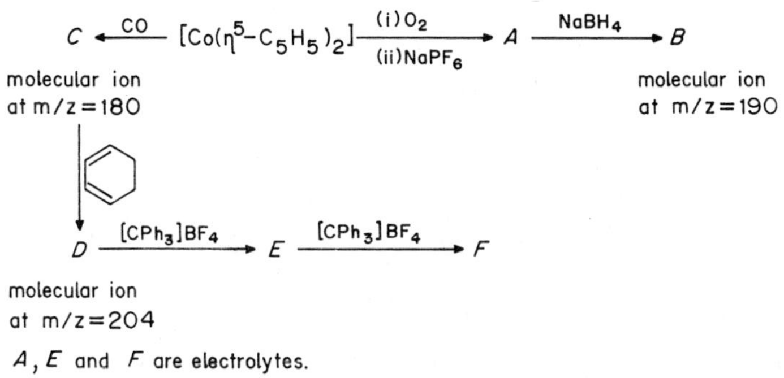

$$C \xleftarrow{CO} [Co(\eta^5\text{-}C_5H_5)_2] \xrightarrow[\text{(ii)NaPF}_6]{\text{(i)O}_2} A \xrightarrow{NaBH_4} B$$

molecular ion
at m/z = 180

molecular ion
at m/z = 190

$$D \xrightarrow{[CPh_3]BF_4} E \xrightarrow{[CPh_3]BF_4} F$$

molecular ion
at m/z = 204

A, E and F are electrolytes.

Proton n.m.r. data (δ relative to TMS, relative intensities in parentheses):

A 3.8
B 5.2(2H), 4.6(5H), 2.7(2H), 2.6(1H), 2.0(1H)
C 5.0
D 4.2(2H), 4.0(5H), 2.4(2H), 0.9(2H), 0.3(2H)
E 7.3(1H), 5.6(5H), 5.4(2H), 4.0(2H), 2.7(1H),1.0(1H)
F 7.9(6H), 7.0(5H)

(a) Identify the complexes A to F, showing how you have used the data given, and comment on the reactions involved, including any stereochemical aspects.
(b) Account for the observation that acidification of B causes evolution of a gas.
(c) Account for the fact that the Co—C bond length in A is shorter than that in cobaltocene.
[Isotopic abundances: ^{1}H, ^{12}C, ^{16}O, ^{19}F, ^{23}Na, ^{31}P, ^{59}Co, 100%; ^{10}B 20%, ^{11}B 80%].
(*University of Southampton*).

8. Suggest a structure for each of the compounds A to E obtained in the following reactions. Show in detail how the n.m.r. data is consistent with your structural assignments. Comment on the mechanism of each reaction and discuss whether C will be more or less thermally stable than B. What would be the effect on the products A to E of replacing H_3PO_4 by D_3PO_4 in the initial reaction to form A?

Five electron ligands

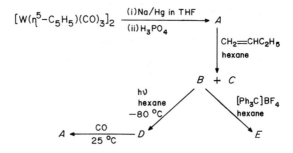

$[W(\eta^5-C_5H_5)(CO)_3]_2 \xrightarrow[\text{(ii) } H_3PO_4]{\text{(i) Na/Hg in THF}} A$

E contains fluorine and is a 1:1 electrolyte.

^{1}H n.m.r. (^{1}H decoupled; δ relative to TMS,
relative intensities in parenthesis)

A 5.70(5H), − 7.40(1H, 1:12:1 pattern)
B 5.61(5H), 1.60(2H), 1.58(2H), 1.55(2H), 0.92(3H)
C 5.61(5H), 1.60(2H), 1.53(1H), 0.92(3H), 0.88(3H)
D 5.46(5H), 2.90(2H), 2.34(1H), 2.31(1H), 2.30(1H), 1.63(3H), − 6.13(1H, 1:12:1 pattern)
E 5.99(5H), 4.36(1H), 3.29(1H), 3.20(1H), 2.80(2H), 2.22(3H)

[^{183}W has nuclear spin I $=\frac{1}{2}$ and is 14% abundant, all other W isotopes have zero spin].
(*University of Southampton*).

9. For the compound $(\eta^1-C_5H_5)Ge(CH_3)_3$, which contains a σ-bonded metal-cyclopentadienyl group labelled as shown in the figure below, the ^{1}H n.m.r. spectrum at − 80°C shows three multiplets at 7.0, 6.0 and 3.5 ppm (relative intensities 2:2:1 respectively) and a singlet at 3.0 ppm (relative intensity 9) downfield from $Si(CH_3)_4$ as reference at 0 ppm. As the temperature is raised to − 40°C the three multiplets start to broaden and collapse but the singlet at 3.0 ppm remains sharp. The resonance for the H_B protons collapses more rapidly than that for the H_C protons. Finally, at 25°C the spectrum shows only two sharp singals, one at 5.6 ppm and the other at 3.0 ppm (relative intensities 5:9 respectively).

Account for the above observations, describe the various exchange processes which could be occurring at 25°C and indicate which process is the most likely on the basis of the variations observed in the spectra.
(*University of Warwick*).

308

10. At 11°C, the proton decoupled ^{13}C n.m.r. spectrum of $(\eta^5\text{-}C_7H_7)Mn(CO)_3$ contains four signals due to the hydrocarbon ligand as shown below. At 35°C, these broaden in the manner shown. Assign these ^{13}C resonances and explain in detail the reason for the differential broadening at 35°C.

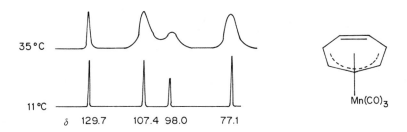

[For the interpretation of the n.m.r. data ignore all nuclei except ^{13}C ($I = \frac{1}{2}$, 1% abundant)].
(*University of Southampton*).

11. Reaction of the pentamethylcyclopentadienyl complex $(C_5Me_5)Rh(CO)_2$ with HBF$_4$ affords a salt A. Treatment of A with NaOMe yields the red solid B, which gives C on heating. Propose structures for A, B and C using the data below, and comment on the bonding in C.

Mass spectrum M^+/amu	ν_{CO}/cm^{-1}	Proton n.m.r. (δ relative to TMS)
A: —	2026, 1739	2.03(singlet, 30H), -10.40 (triplet, $J = 19\,Hz$, 1H).
B: 560	1951, 1778	1.91(singlet).
C: 532	1732	1.65(singlet).

Analysis of A gave 42.6% C, 31.8% Rh and 1.67% B.
Isotopic abundances: ^{1}H, 100%, $I = \frac{1}{2}$; ^{10}B, 19%, $I = 3$, ^{11}B, 81%, $I = \frac{3}{2}$; ^{12}C, 98.9%, $I = 0$; ^{13}C, 1.1%, $I = \frac{1}{2}$, ^{16}O, 100%, $I = 0$; ^{19}F, 100%, $I = \frac{1}{2}$; ^{103}Rh, 100%, $I = \frac{1}{2}$].
(*University of Southampton*).

12. The molar magnetic susceptibility (χ_M^{corr}) of 1,1'-dimethylmanganocene, corrected for diamagnetism, varies with temperature as shown below. Given that the magnetic moment $\mu_{eff} = 797.7 \ (\chi_M^{corr}.T)^{1/2}$, calculate μ_{eff} at each temperature. Plot χ_M^{corr} vs T and μ_{eff} vs T.

T/K	214	235	253	285	310	331	366	371
$10^8\chi_M^{corr}/m^3\,mol^{-1}$.	7.527	8.394	8.771	9.814	9.953	9.840	9.525	9.576

1,1'-Dimethylmanganocene exists as an equilibrium mixture of high spin (hs)

Five electron ligands

$(^6A_{1g})$ and low spin (ls) $(^2E_{2g})$ forms. It can be shown that

$$\mu_{eff}^2 = \mu_{ls}^2 X_{ls} + \mu_{hs}^2(1 - X_{ls})$$

where $\mu_{ls} = 1.98$ B.M. and $\mu_{hs} = 5.92$ B.M. are the magnetic moments of low spin and high spin forms respectively and X_{ls} is the mole fraction of the low spin form. Calculate X_{ls} and the equilibrium constant [low spin]/[high spin] at each temperature. Hence calculate $\Delta H^{\ominus}$ and $\Delta S^{\ominus}$ for the equilibrium. (SI units)

10 Complexes of arenes

10.1 Bis(arene)complexes

The commonest six-electron ligands are benzene and substituted benzenes. Ligands of this type are termed 'arenes'. They form complexes with most of the transition elements, although the variety of derivatives is somewhat smaller than for the cyclopentadienyls. Bis(benzene)complexes $(\eta^6\text{-}C_6H_6)_2M$ are known for M = Ti, V, Nb, Cr, Mo and W and an even wider range of elements including V, Cr, Fe and Co form bis(mesitylene) and bis(hexamethylbenzene)derivatives. Cations such as $Mn(C_6Me_6)_2^+$ and $Fe(C_6H_6)_2^{2+}$ are also common.

Bis(benzene)chromium has the 18-electron configuration and is thus iso-electronic with ferrocene. It forms brown-black crystals, m.p. $281°C$ and is thermally rather stable, although it is readily oxidized in air. In the crystal the molecule has a sandwich structure in which the rings are eclipsed (D_{6h} symmetry) (p. 204). The barrier to rotation of the rings about the ring-metal axis, however, is very low as in ferrocene.

10.1.1 Preparation—Fischer's method

The first general method for the preparation of bis(arene)complexes was reported by Fischer and Hafner in 1954. It involves heating the anhydrous metal halide with the arene and aluminium powder, which acts as a reducing agent, and aluminium trichloride as halide ion acceptor. Because an arene and $AlCl_3$ are involved, the procedure is sometimes called Fischer's 'reducing Friedel-Crafts' reaction, but the name is somewhat misleading as no electrophilic substitution of the arene occurs. The reaction exemplifies well the general strategy for preparing complexes of metals in low formal oxidation states—Metal salt + reducing agent + ligand—noted in Chapter 5 for metal carbonyls. The addition of a catalytic quantity of mesitylene, which readily forms a complex and also readily exchanges with benzene, enables lower temperatures and shorter reaction times to be used.

The first product in these reductions is the tetrachloroaluminate of the complex cation:

$$3CrCl_3 + Al + 2AlCl_3 + 6C_6H_6 \rightarrow 3(C_6H_6)_2Cr^+AlCl_4^-$$

Complexes of arenes

For chromium this can be reduced by alkaline aqueous dithionite or, with loss of half the product, allowed to disproportionate in aqueous solution.

$$2(C_6H_6)_2Cr^+ + S_2O_4^{2-} + 2OH^- \rightarrow 2(C_6H_6)_2Cr + 2HSO_3^-$$

or

$$2(C_6H_6)_2Cr^+ \rightarrow (C_6H_6)_2Cr + Cr^{2+} + 2C_6H_6$$

The green complex bis(benzene)molybdenum can be prepared similarly in up to 70% yields

$$3MoCl_5 + AlCl_3 + 4Al + 6C_6H_6 \rightarrow 3(C_6H_6)_2Mo^+AlCl_4^-$$
$$6(C_6H_6)_2Mo^+ + 8OH^- \rightarrow 5(C_6H_6)_2Mo + MoO_4^{2-} + 4H_2O + 2C_6H_6$$

The method has been applied to many of the transition elements. The cations first produced can in some cases be converted into the neutral complexes by reduction or disproportionation. It suffers from the disadvantage that arenes with functional groups such as halogen, acyl etc. cannot be used as they interact with the aluminium trichloride.

10.1.2 Metal atom synthesis

Bis(arene)complexes are, like the metallocenes, endothermic compounds. The standard enthalpy of formation of bis(benzene)chromium from the elements is $141.4 \, kJ \, mol^{-1}$. Nevertheless they do often exhibit a fair degree of thermal stability. The mean bond dissociation energies $\bar{D}(M-arene)$ in $Cr(C_6H_6)_2$, $Mo(C_6H_6)_2$ and $W(C_6H_5Me)_2$ defined according to

$$M(arene)_2(g) \rightarrow M(g) + 2arene(g); \quad \Delta H^{\ominus} = 2\bar{D}(M-arene)$$

have been estimated as 165, 247 and $304 \, kJ \, mol^{-1}$ respectively. The trend of increasing bond strength down a group, previously noted for transition elements (p. 5) is thus maintained here.

Can $M(arene)_2$ be prepared from the metal and arene? At first sight this might appear to be ludicrous, bearing in mind the unreactive nature both of bulk metals and also of arenes. Ferrocene, however, was prepared by Miller, Treboth and Tremaine by passing cyclopentadiene over hot metallic iron. The reaction $Cr(s) + 2C_6H_6(g) \rightarrow Cr(C_6H_6)_2(s)$ has $\Delta H^{\ominus} = -11.6 \, kJ \, mol^{-1}$ ($+69.5 \, kJ \, mol^{-1}$ for Mo), although the entropy term is unfavourable. This reaction does not occur at room temperature, however, and becomes increasingly disfavoured thermodynamically as the temperature is raised.

Transition metals have high energies of atomization (Cr, 397; Mo, 658; W, $860 \, kJ \, mol^{-1}$). In other words, the gaseous atoms possess high energy content. A reaction such as $Cr(g) + 2C_6H_6(g) \rightarrow Cr(C_6H_6)_2(s)$ is strongly favoured thermodynamically $\Delta H^{\ominus} = -408 \, kJ \, mol^{-1}$ ($-589 \, kJ \, mol^{-1}$ for Mo). The 'metal atom' synthesis takes advantage of this stored energy. While compounds such as $Cr(C_6H_6)_2$ cannot survive at the temperatures required to vaporize chromium metal, it is possible by rapid chilling of the vapour to trap the atoms (or small clusters of atoms) at low temperature. The metal vapour is cocondensed with a large excess of a reactant such as an arene on the inner surface of a vessel cooled in liquid nitrogen. When the mixture is allowed to warm up slowly, reaction occurs under very mild conditions. An apparatus designed for these reactions is illustrated in Fig. 10.1. For relatively volatile elements electrical heating is sufficient but for very high boiling elements such as Ti, Nb and W an electron gun is employed. The method is suitable for the laboratory preparation of small quantities (1–50 g). The experiment is carried out under reduced pressure (ca. 10^{-3} mm) to increase the mean free path of the metal atoms after vaporization so that they are trapped in a matrix of coreactant before they can meet each other to reform bulk metal.

The metal atom synthesis was first applied to the preparation of bis(arene)chromium derivatives by Timms. A wide range of complexes containing functional groups such as F, Cl, OMe and COOR were obtained. By introducing the arene in a solvent, compounds of relatively involatile ligands such as Cr(naphthalene)$_2$ have been prepared. Green has extended the method to the synthesis of compounds such as Ti(C$_6$H$_6$)$_2$, Zr(C$_6$H$_6$)$_2$PMe$_3$, Nb(C$_6$H$_6$)$_2$ and

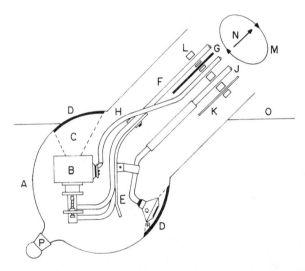

Fig. 10.1 Metal atom reactor. A – Glass reaction vessel. B – Electron beam furnace, model EBS1, G.V. Planer Ltd. C – Vapour beam of metal atoms. D – Co-condensate of metal and substrate vapours. E – Heat shield. F – Furnace cooling water pipes. G – Electrical lead for substrate solution dispersion device. H – Furnace electrical leads. J – Substrate inlet pipe (vapour). K – Substrate inlet pipes (solution). M – Rotation of reaction vessel. N – To vacuum rotating seal, service vacuum lead troughs and pumping systems. O – Level of coolant (usually liquid nitrogen). P – Capped joint for product extraction. Q – Substrate vapour dispersion device. R – Substrate vapour beam. (From Green, M.L.H., 1980, *J. Organomet. Chem.*, **300**, 119.)

Complexes of arenes

$W(C_6H_6)_2$. In these cases the Fischer-Hafner method gives either very low yields (ca. 2% of $W(C_6H_6)_2$) or other derivatives.

10.1.3 Hein's 'polyphenyl complexes'

The investigations by Hein and later by Zeiss on the reactions between anhydrous chromium(III) chloride and phenylmagnesium bromide in diethyl ether provide a fascinating story. The formation of unstable organochromium complexes was first observed in 1903, and in 1919 Hein was able to isolate a polyphenylchromium complex from this reaction. At the time it was not possible to predict the structure of this complex. Only in 1954 Zeiss showed that sandwich compounds containing benzene and biphenyl ligands are produced. The formation of these arene complexes proceeds through η^1-aryl derivatives. When anhydrous $CrCl_3$ in tetrahydrofuran at $-20°C$ is treated with phenylmagnesium bromide in the molar ratio 1:3, deep red crystals of $Ph_3Cr(THF)_3$ can be isolated. When this is washed with ether, tetrahydrofuran is displaced, leaving a black paramagnetic solid. On hydrolysis this affords a mixture of η^6-arenechromium complexes, produced probably through free radical intermediates (Fig. 10.2).

10.1.4 Chemistry of bis(benzene)chromium

The chemistry of bis(benzene)chromium is dominated by the ease of oxidation to the paramagnetic 17-electron cation $Cr(C_6H_6)_2^+$ (μ_{eff} 1.70–1.80 for salts). This

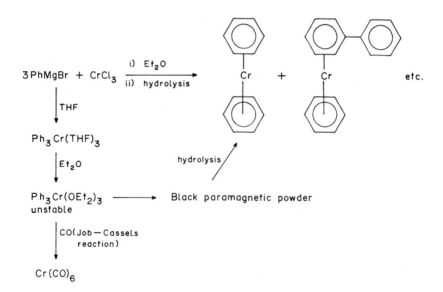

Fig. 10.2 Possible mode of formation of η-arene chromium compounds from η^1-phenylchromium derivatives.

314

tendency is much more pronounced than in ferrocene. $Cr(C_6H_6)_2$ is very air sensitive, while the salts $(C_6H_6)_2Cr^+X^-$ are air-stable. Consequently bis(benzene)chromium does not undergo electrophilic substitution at the arene rings but becomes oxidized instead.

The arene ligands are displaced only with difficulty; CO or PF_3 yield $Cr(CO)_6$ or $Cr(PF_3)_6$ respectively at 200°C/300 atm. pressure. Bis(naphthalene)chromium, made by metal atom synthesis, is much more labile. Two naphthalene ligands are readily displaced by 6RNC or 3bipy, and one by 3CO, $3PF_3$ or $3PR_3$.

Metallation of the rings in $Cr(C_6H_6)_2$ can be achieved using n-pentylsodium or the n-butyllithium.TMEDA complex. The resulting derivatives react with carbon dioxide or with aldehydes or ketones. It is difficult, however, to control the number of metal atoms which enter, so that mixtures of products are formed.

10.1.5 Chemistry of bis(benzene)molybdenum

Bis(benzene)molybdenum, in contrast to the chromium analogue, shows a very rich and varied chemistry. It has a very low molecular ionization energy in the

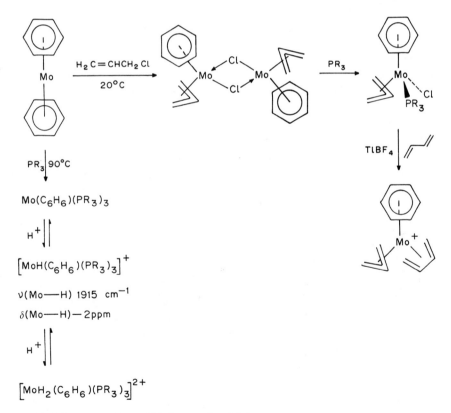

Fig. 10.3 Some reactions of bis(benzene)molybdenum.

315

gas phase (5.01 eV cf. Na 5.14 eV). In solution it is readily oxidized (polarographic half-wave potential $E_{1/2} = -0.71$ V vs. standard calomel electrode). Consequently it has been described as an 'electron-rich' compound. This arises by virtue of the essentially donor character of the arene ligands.

One of the arene ligands in $Mo(C_6H_6)_2$ is easily displaced, for example by phosphines. The products $Mo(C_6H_6)(PR_3)_3$ are strong bases which can be protonated once or twice by acids. The addition of allyl chloride (3-chloropropene) with elimination of one benzene ligand gives a chloro-bridged allyl dimer. These and related reactions provide an entry into mono-arene molybdenum chemistry (Fig. 10.3).

10.2 Arene(tricarbonyl)chromium complexes

The arene(tricarbonyl)chromium complexes exemplified by $(C_6H_6)Cr(CO)_3$ belong to the series of compounds based on the $M(CO)_3$ group, $(C_7H_7)V(CO)_3$ through to $(C_3H_5)Co(CO)_3$. They were first prepared in 1958 by Nicholls and Whiting who heated $Cr(CO)_6$ with the arene under reflux, if necessary in a high boiling solvent. One difficulty with the preparation is that the hexacarbonyl sublimes out of the mixture and hence is removed from the reaction vessel into the condenser. An ingenious apparatus was devised by Strohmeier to overcome this, in which the sublimed carbonyl is washed back using condensed solvent. Recently, however, it has been shown that simple heating of $Cr(CO)_6$ and arene in a mixture of di-n-butyl ether and tetrahydrofuran gives excellent yields, the volatile tetrahydrofuran serving to return any sublimed carbonyl.

Benzenetricarbonylchromium forms yellow crystals, m.p. 163°C which are stable in air, although its solutions are slowly oxidized, especially in light. In fact a good method of removing the $Cr(CO)_3$ group from the arene is to photolyse the complex in air or oxygen. In the solid state the molecules adopt a staggered conformation. On the basis of X-ray and neutron diffraction it has been suggested that the bond lengths in the benzene ring alternate slightly. The ring is planar but the hydrogen atoms are bent out of the plane towards the metal by 1.6°. In this way the C(2p) orbitals point more towards the metal, giving better overlap.

The related complexes $C_5H_5V(CO)_4$, $C_5H_5Mn(CO)_3$ and $C_4H_4Fe(CO)_3$ (p.269) all undergo Friedel-Crafts acylation. Attempts to acetylate $(C_6H_6)Cr(CO)_3$ under normal conditions (excess $AlCl_3$, CH_2Cl_2, 0°C) lead to extensive decomposition. The Perrier method, however, which employs preformed $CH_3CO^+.AlCl_4^-$ and hence avoids excess aluminium(III) chloride, is successful. Benzenetricarbonylchromium is deactivated to electrophilic substitution compared with benzene itself. Evidence from a variety of experiments, for example comparison of the pK_a values of benzoic acid (5.48 in 50% aqueous ethanol) and its $Cr(CO)_3$ complex (4.52) suggests that the $Cr(CO)_3$ group is about as electron-withdrawing as a p-nitro substituent. Some consequences of this property which have applications in organic synthesis are discussed below.

10.2.1 *Nucleophilic attack on arenetricarbonylchromium complexes*

While arenetricarbonylchromium complexes are deactivated to electrophilic substitution with respect to the free arene, their susceptibility to nucleophilic substitution is enhanced. The chlorobenzene complex, for example, undergoes ready substitution by sodium methoxide. In this way it resembles 1-chloro-4-nitrobenzene.

Carbanions such as $\bar{C}H(CO_2Et)_2$ also react; in the early stages of the reaction ring adducts *ortho* or *meta* to the chloro substituent are observed, but these are gradually replaced by the substitution product. The reactivity of $Cr(PhX)(CO)_3$ towards nucleophiles decreases in the order $X = F > Cl > Br > I$.

These results led Semmelhack and his colleagues to study the addition of carbanions to benzenetricarbonylchromium itself. Carbanions derived from species with $pK_a \leqslant 25$ in tetrahydrofuran form cyclohexadienyl anions which are stable below 0°C in the absence of air. Oxidation with iodine affords substituted benzenes. The sequence

$$ArH \longrightarrow (ArH)Cr(CO)_3 \xrightarrow{R^-} [(ArHR)Cr(CO)_3]^- \xrightarrow{I_2} ArR$$

| COMPLEX | ADDITION OF | REMOVAL OF |
| FORMATION | NUCLEOPHILE | METAL |

provides a mild procedure for the nucleophilic substitution of an arene, a reaction which is normally not possible on the free ligand.

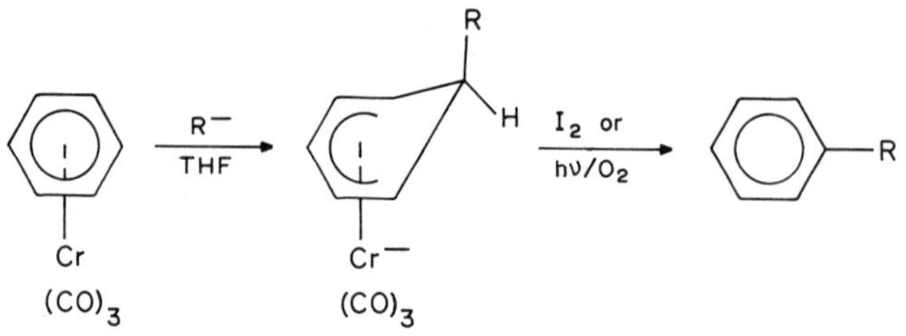

Complexes of arenes

$$R = CMe_2CN, CH_2CN, CH_2CO_2Bu^t$$

Very strong nucleophiles such as phenyllithium, however, attack a carbonyl ligand to form carbene complexes, as described on p. 234.

10.2.2 Other reactions

Coordination of an arene to $Cr(CO)_3$ has further consequences which are valuable in synthesis. On account of its electron withdrawing nature the acidity of both ring (H^r) and side chain (H^α, H^β) hydrogens is enhanced relative to the free arene.

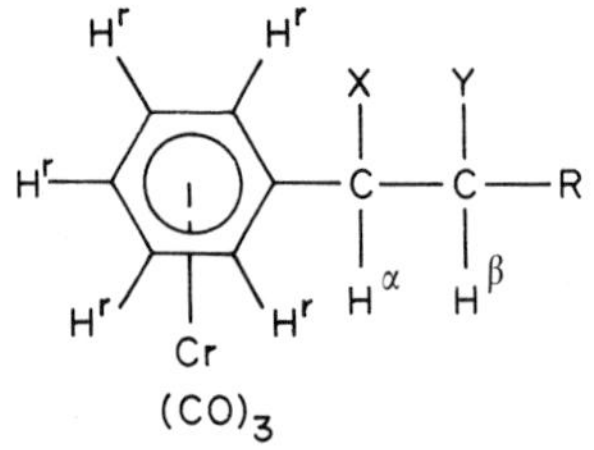

$(C_6H_6)Cr(CO)_3$, for example, is metallated by n-butyllithium in THF at $-40°C$, whereas benzene itself is not under similar conditions. The stabilization of side chain anion sites is illustrated by the following reactions:

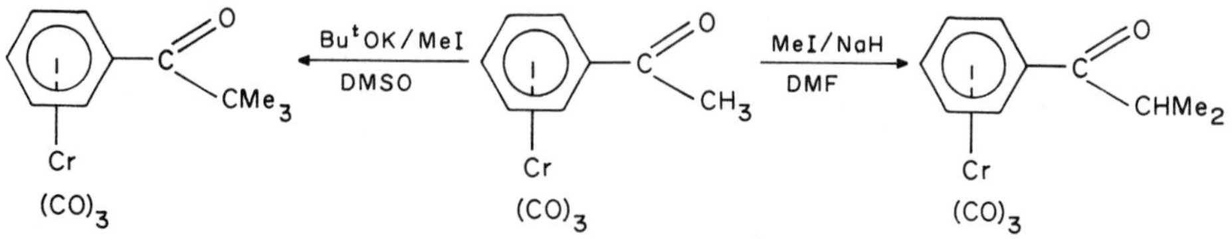

The uncomplexed ketone is inert to NaH/MeI in dimethylformamide.

10.2.3 Stabilization of α-carbonium Ions

The stabilization of carbonium ions adjacent to a ferrocenyl group was noted on p. 285. Compared with the α-ferrocenyl group the stabilization arising from an α-

318

$C_6H_5Cr(CO)_3$ group is relatively small, but it is sufficient to produce noticeable changes in reactivity. Benzyl chloride and its complex $(PhCH_2Cl)Cr(CO)_3$ are both solvolysed by an S_N1 mechanism in ethanol in which the carbonium ion is an intermediate. It has been found that the complex reacts about 10^5 faster than the free halide. These effects have been exploited in organic synthesis. Complexed benzyl alcohols can be converted into ethers or amines by reaction with hexafluorophosphoric acid followed by an alcohol or an amine. The Ritter reaction, which involves electrophilic addition to a nitrile, is also greatly accelerated. While an uncomplexed benzyl alcohol requires over a day for complete reaction, the complex reacts within a few minutes.

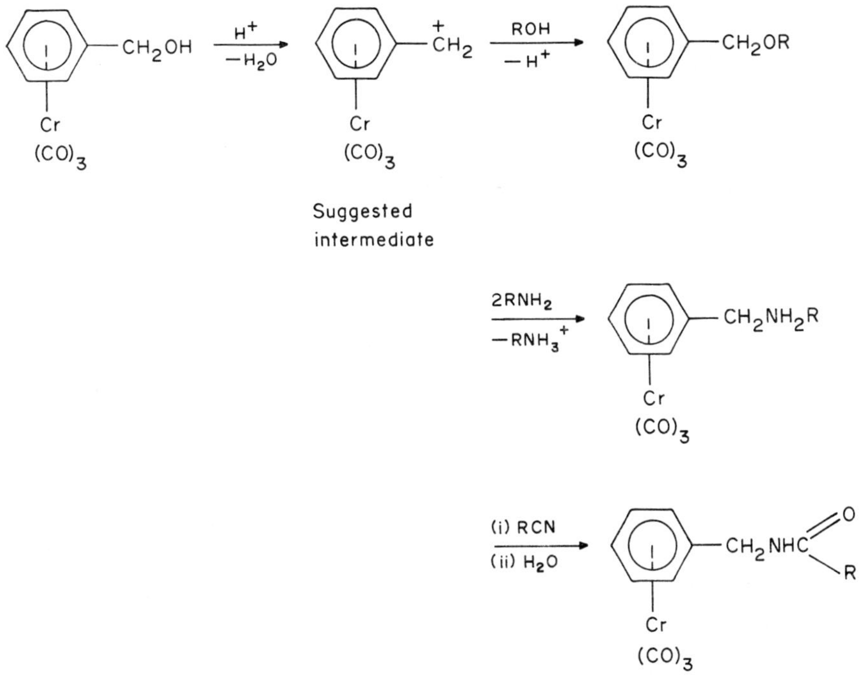

Suggested
intermediate

The $Cr(CO)_3$ group can finally be removed cleanly by photolysis of a solution of the product in air.

10.2.4 *Steric effects*

The steric effect of the bulky $Cr(CO)_3$ group induces reagents such as nucleophiles to attack from the side away from the metal centre (*exo* attack). The reactions are very stereospecific. Thus reduction of the indanone complex occurs almost exclusively from above the ring to give a racemic mixture of the indanol complexes shown.

319

10.3 Chemistry of some arene complexes of iron

Salts of bis(arene)iron cations can be prepared by a rather remarkable reaction in which anhydrous iron(II) chloride, aluminium(III) chloride and the arene are heated under reflux in cyclohexane

$$FeCl_2 + 2AlCl_3 + 2arene \longrightarrow Fe(arene)_2^{2+} AlCl_4^-$$

The tetrachloroaluminate is then converted into the more tractable hexafluorophosphate by treatment with aqueous NH_4PF_6. The dications are susceptible to attack by nucleophiles including cyanide and various carbanions, which add *exo* to one of the rings.

Alkyl or aryllithium reagents can add either to one ring or to both, depending on the ratio of reactants. The addition of aryllithium followed by oxidation of the bis(cyclohexadienyl)iron adduct provides a route to unsymmetrical biphenyls.

When ferrocene is heated with an arene in the presence of aluminium and aluminium(III)chloride, arene(cyclopentadienyl)iron cations are produced together with 'ferrocenophane' biproducts. Following hydrolysis of the reaction mixture the cations are precipitated as hexafluorophosphates, leaving the

320

ferrocenophanes in the organic phase. Reduction by sodium amalgam or electrochemically affords neutral 19-electron complexes such as $Fe(C_5H_5)(C_6H_6)$ which are very easily oxidized and which dimerize as shown. The dimerization can be suppressed by complete alkylation of the arene ring as in $Fe(C_5H_5)(C_6Et_6)$.

10.4 Nucleophilic attack on organotransition metal complexes

Several examples have been given of the reactions of nucleophiles with coordinated organic ligands. Unsaturated hydrocarbons such as ethene, butadiene and benzene are not normally attacked by nucleophiles, but on coordination to a transition metal their susceptibility to such attack is often greatly enhanced, particularly when the complex is cationic. On the other hand many otherwise reactive species such as the cyclohexadienyl cation, 'protonated benzene' (p. 303) can be stabilized by coordination, so that controlled reactions can be performed at ligand sites. In a complex there may be several potential sites at which a nucleophile might attack. Often different nucleophiles prefer different sites. This is illustrated with reference to η^6-arenetricarbonylmanganese salts.

These salts are usually prepared by the action of an arene on $ClMn(CO)_5$ in the presence of $AlCl_3$. They can also be obtained from arenes and $(\eta^5\text{-}MeC_5H_4)Mn(CO)_3$ together with $AlBr_3$.

$$ClMn(CO)_5 + C_6H_6 + AlCl_3 \longrightarrow [(\eta^6\text{-}C_6H_6)Mn(CO)_3]^+ AlCl_4^-$$

$$\xrightarrow{\text{NH}_4^+ \text{PF}_6^-} [(\eta^6\text{-}C_6H_6)Mn(CO)_3]^+ PF_6^-$$

Attack by nucleophiles has been observed at three sites in $(\eta^6\text{-}C_6H_6)Mn(CO)_3^+$, viz. at the benzene ring, at a carbonyl ligand and at the metal centre (Fig. 10.4).

Hydride ion donors or alkyllithium reagents add to the benzene ring, forming air-stable η^5-cyclohexadienyl complexes. The alkyl substituent enters stereospecifically *exo*, that is, away from the metal. Generally nucleophiles attack exclusively from this side, as approach to the coordinated ligand is relatively unrestricted compared with significant steric hindrance on the side of the metal. This stereospecificity can be valuable in organic synthesis. (There are a few cases, particularly involving hydride and large rings where *endo* addition has been observed.)

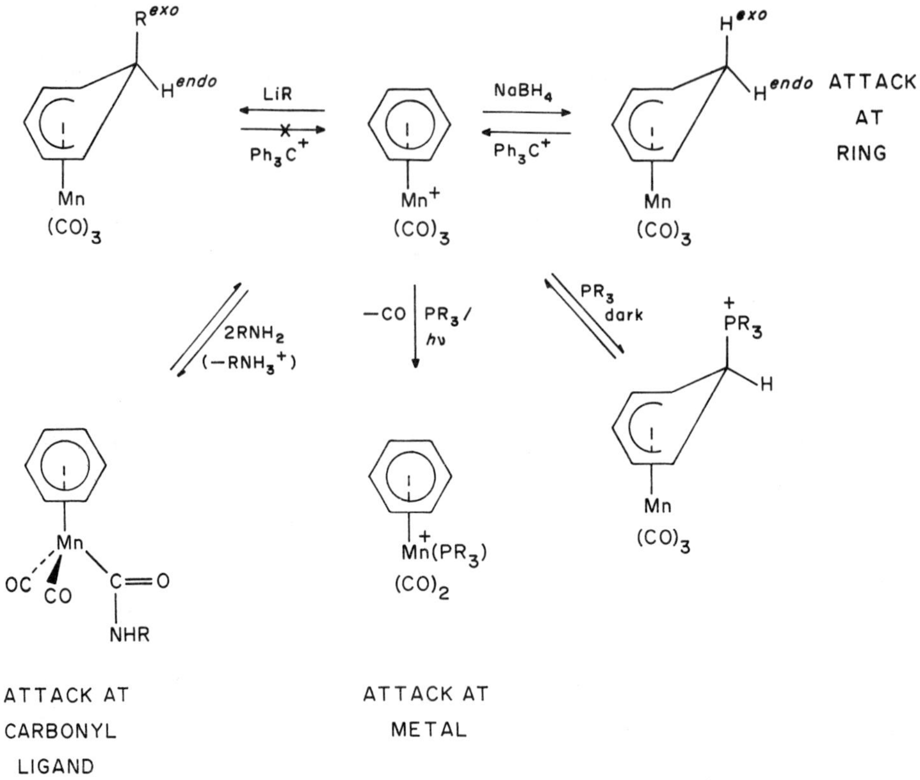

Fig. 10.4 Various sites for nucleophilic attack in $(\eta^6\text{-}C_6H_6)Mn(CO)_3$.

While trialkyl(aryl)phosphines in the dark add to the ring of $(C_6H_6)Mn(CO)_3^+$, in the presence of light substitution of a carbonyl ligand occurs, probably by dissociative photolysis. A third possible site of attack is at a carbonyl ligand. This occurs with hard nucleophiles such as amines and alkoxides and is reversible. The reactivity of carbonyl groups correlates with the $v(CO)$ stretching frequency. The higher this frequency the less back donation into π^* orbitals has occurred and hence the greater the susceptibility to nucleophilic attack at carbon.

η^5-Cyclohexadienyltricarbonylmanganese complexes are reconverted into the η^6-arene cation by the triphenylmethyl cation only if they have an exo-H substituent. The bulky reagent is unable to remove an endo-H on account of steric hindrance. Advantage may be taken, however, of a rearrangement which occurs, probably via an η^6-arene(hydride) intermediate. The new areneMn(CO)$_3^+$ cation can now be attacked by a second nucleophile to give a mixture of disubstituted cyclohexadienyl complexes.

322

A more direct way of activating the cyclohexadienyl complex (η^5-RC_6H_6)Mn(CO)$_3$ is by treatment with NO$^+$PF$_6^-$. One carbonyl ligand is displaced by nitrosyl. As CO is isoelectronic with NO$^+$ the product is a salt [(η^5-RC_6H_6)Mn(CO)$_2$(NO)]$^+$PF$_6^-$. Once again several different sites in this cation are attacked by nucleophiles. Tertiary phosphines (R$_3$P) add *exo* to the ring, forming phosphonium salts, while phosphites (RO)$_3$P displace CO from the metal. Addition of D$^-$ (as NaBD$_4$) is very unusual as it occurs exclusively *endo*. This probably takes place through a transient metal deuteride, followed by transfer of D$^-$ to the underside of the ring.

To summarize, two methods of converting neutral complexes to cationic species are thus available. In the first, *exo* hydride abstraction by Ph$_3$C$^+$, the hapto number of the ligand increases by one ($\eta^n \rightarrow \eta^{n+1}$). In the second, replacement of coordinated CO by NO$^+$, the hapto number is unchanged (Fig. 10.5). These procedures are useful because cationic complexes are much more readily attacked by nucleophiles than neutral species. Moreover the regiospecificity of the reactions is often better (cf. (η^4-C$_4$H$_6$)Fe(CO)$_3$ and (η^4-C$_4$H$_6$)Co(CO)$_3^+$, p. 266). The reagents Ph$_3$C$^+$PF$_6^-$ and NO$^+$PF$_6^-$ are thermally stable crystalline materials, although they are very easily hydrolysed. They dissolve in dichloromethane, a polar but weakly nucleophilic medium, and their reactions are conveniently carried out in this solvent. The salt of the organometallic cation is precipitated on addition of diethyl ether to the reaction mixture.

Cyclohexadienyltricarbonylmanganese can be reduced further by hydride donors to the dienetricarbonylmanganese anion, which is isoelectronic with (η^4-C$_6$H$_8$)Fe(CO)$_3$ (Fig. 10.6). This anion is protonated even by water, forming an η^3-cyclohexenyl complex. N.m.r. studies over a range of temperature show that this species rearranges in solution by two independent processes which link a (diene)metal hydride, an η^3-enyl in which there is an 'agostic' interaction and a 16-electron η^3-enyl complex. The anion undergoes *endo* ring methylation with

η^7	η^6	η^5	η^4
trienyl $Cr(CO)_3$ ⇌	triene $Cr(CO)_3$		
	arene $Cr(CO)_3$		
	arene $Mn(CO)_3^+$		
	or triene $Mn(CO)_3^+$ →	dienyl $Cr(CO)_3^-$	
		dienyl $Mn(CO)_3$ ⇌	→ diene $Mn(CO)_3^-$
		dienyl $Mn(CO)_2(NO)^+$ →	→ diene $Mn(CO)_2(NO)$
		dienyl $Fe(CO)_3^+$ ⇌	⇌ diene $Fe(CO)_3$

η^4	η^3	η^2	η^1
$CpMo(CO)_2$diene$^+$ ⇌	$CpMo(CO)_2$enyl		
	$CpMo(CO)(NO)$enyl$^+$		
	$CpMo(CO)(NO)$enyl$^+$ →	$CpMo(CO)(NO)$alkene	
diene $Fe(CO)_3$ →	enyl $Fe(CO)_3^-$ →	$CpFe(CO)_2$alkene$^+$ ⇌	⇌ $CpFe(CO)_2$alkyl
		alkene $Fe(CO)_4$ →	→ alkyl $Fe(CO)_4$
diene $Co(CO)_3^+$ →	enyl $Fe(CO)_4^+$		
	enyl $Co(CO)_3$ →	alkene $Co(CO)_3^-$	

Fig. 10.5 Summary of some cationic and neutral complexes and of their reactions with nucleophiles. Arrows → link reactant and product in nucleophilic addition to organic ligand. Arrows ← link reactant and product in hydride abstraction by $Ph_3C^+PF_6^-$. Vertical arrows ↓ link reactant and product in reaction with $NO^+.PF_6^-$.

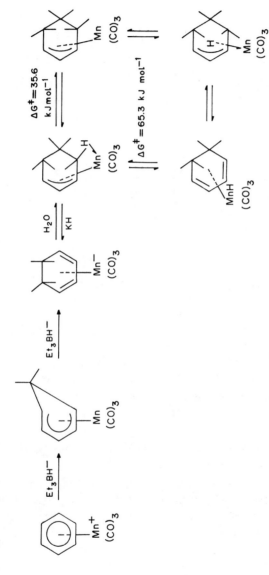

Fig. 10.6 Products from the reduction of $(\eta^6\text{-}C_6H_6)Mn(CO)_3^+ \cdot PF_6^-$.

methyl iodide. The cyclohexadiene ligands are released from the anionic complexes by oxidation in air.

The reaction of a nucleophile with a coordinated ligand may be viewed either as an addition (as above) or as a substitution at carbon. In principle mechanisms ranging from dissociative (D or I_d, S_N1) to interchange associative (I_a, S_N2) are possible. Kinetic measurements on reactions of this type (especially the formation of phosphonium salts by phosphines) reveal the rate law, rate $=$ k_2[complex][PR_3] for the forward reaction. This probably suggests a synchronous I_a (S_N2) process with inversion of stereochemistry at carbon. A mechanism of this type involving slippage of the metal atom to one end of the double bond on approach of a nucleophile has been proposed for alkene complexes, on the basis of molecular orbital calculations.

10.4.1 Choice of organic ligand for nucleophilic attack (Green's rules)

Three rules have been proposed, based on much experimental data, which enable the preferred site of attack to be predicted in a cationic complex which contains more than one organic ligand. It is assumed that the nucleophile seeks out the position of highest positive charge, and that no subsequent rearrangement of the products occurs. That is to say, the reactions are assumed to be under kinetic control. This is not always true; while additions of carbanions are often irreversible, phosphines, amines or alkoxides may add reversibly.

Rule 1: Nucleophilic attack occurs preferentially at *even* coordinated polyenes (i.e. η^2, η^4, η^6 ···) rather than at *odd* coordinated polyenyls (i.e. η^3, η^5, η^7 ···)

Rule 2: Nucleophilic addition to *open* coordinated polyenes/polyenyls (e.g. η^5-pentadienyl, η^5-cyclohexadienyl) is preferred to addition to *closed* (cyclically conjugated, e.g. η^5-cyclopentadienyl) polyenes/polyenyls.

Rule 3: For *even open* polyenes (e.g. η^4-butadiene) nucleophilic attack at the terminal carbon atom is always preferred, for *odd open* polyenyls attack at the terminal carbon atom occurs only if the attached metal containing group is strongly electron-withdrawing.

In brief Rule 1. *even* before *odd*

Rule 2. *open* before *closed*

Rule 3. *even open* attacked at ends

Rule 1 can be understood in terms of a simple molecular orbital description. The HOMO in an even polyene contains an electron pair. On interaction with a metal-centred frontier orbital some formal transfer of electrons from ligand to metal will normally occur. The extent of this transfer depends on the relative energies of the ligand HOMO and metal-centred frontier LUMO. From Fig. 10.7(a) it can be seen that the total charge on an even polyene in an 18-electron complex lies between 0 and +2. The HOMO in an odd polyene, however, contains only one electron. In the bonding orbital formed by interaction of this HOMO with the metal-centred LUMO this electron must be paired with a second electron originally associated with the metal. The total charge on the odd polyene thus lies between −1 and +1 (Fig. 10.7(b)). In a complex which contains both an even and an odd polyene it is reasonable to suppose that the greater positive charge resides on the former.

Rule 2 is perhaps most easily understood by noting that all positions in a cyclically conjugated (closed) ligand are equivalent and hence carry the same charge density. In an open ligand, however, some positions will have a greater positive charge than others. These positions are also the most positive ligand sites

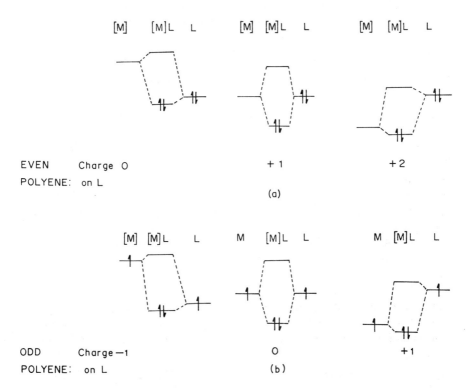

Fig. 10.7 Interaction diagrams for metal based frontier LUMO [M] and ligand HOMO to show ranges of total charge on L in a complex for (a) an even polyene and (b) an odd polyene. The extent of formal charge transfer depends on the relative energies of metal and ligand orbitals. Two extremes and an intermediate case are shown for each of (a) and (b).

327

in the whole molecule, assuming that the same total charge resides on the open as on the closed ligand.

While nucleophiles nearly always attack terminal positions of even open polyenes in cationic complexes, this is not always so for odd open systems. Addition to the end of the η^3-allyl group in $[CpMo(CO)(NO)(\eta^3\text{-}C_3H_5)]^+$ is observed, an η^2-alkene complex being formed. The stereochemistry of this reaction is rather complicated because the cation exists as *exo* and *endo* geometrical isomers. These isomers interconvert in solution to an equilibrium mixture, but sufficiently slowly for their reactions to be studied independently. It should also be noted that the molybdenum centre is chiral, as it carries four different ligands in a roughly tetrahedral arrangement. It happens that nucleophiles such as methoxide or $NaBH_3CN$ (effectively H^-) attack each isomer regiospecifically in such a way that only one of the possible diastereoisomers of the alkene complex is formed.* This diastereoisomer is in fact the same one starting either from the *exo* or from the *endo* isomer of the cation. It is also the more stable one in which the non-bonded interactions between alkene and other ligands are less (Fig. 10.8).

While nucleophiles generally add to terminal positions in allyl complexes, there are a few examples where attack at the central carbon atom occurs. When CO and NO in $[CpMo(CO)(NO)(\eta^3\text{-}C_3H_5)]^+$ are replaced by $\eta^5\text{-}C_5H_5$, the metal centre becomes more electron-rich and the energy of the frontier LUMO is raised. The allyl group may then behave more like an allyl anion than an allyl cation (Fig. 10.8). If this is so, it is the central carbon atom which is most positive (or least negative).

The three rules successfully correlate a large body of experimental data on the site of nucleophilic attack on cationic 18-electron complexes. They do not always work; rule 3 in particular is not very well defined in relation to odd open polyenes. Rule 2 seldom fails, but several exceptions to Rule 1 are now known. Rule 1, for example breaks down in relation to the cation $[Fe(\eta^5\text{-}C_6H_7)(\eta^6\text{-}C_6H_6)]^+$. Sodium borohydride and several carbon nucleophiles including KCN, $NaCH(CO_2Me)_2$ and $PhCH_2MgBr$ react with the cyclohexadienyl group (odd open) in preference to the benzene ligand (even closed). With borohydride the product is formed under kinetic control and not as a result of rearrangement *via* the predicted complex $(\eta^5\text{-}C_6H_7)_2Fe$. Some reactions, in particular those which involve soft nucleophiles, may not operate under charge control but rather are directed by the extent of overlap between the frontier orbitals of the complex and those of the incoming nucleophile. It has also been suggested that the preferred direction of attack from the *exo* side remote from the metal centre is not caused mainly by steric effects as

---

*Diastereoisomers require two chiral centres. On coordination a monosubstituted alkene

loses its plane of symmetry. Alternatively the bonding can be viewed as

showing a chiral centre at carbon. Racemic mixtures (RS) of the molybdenum cations were used.

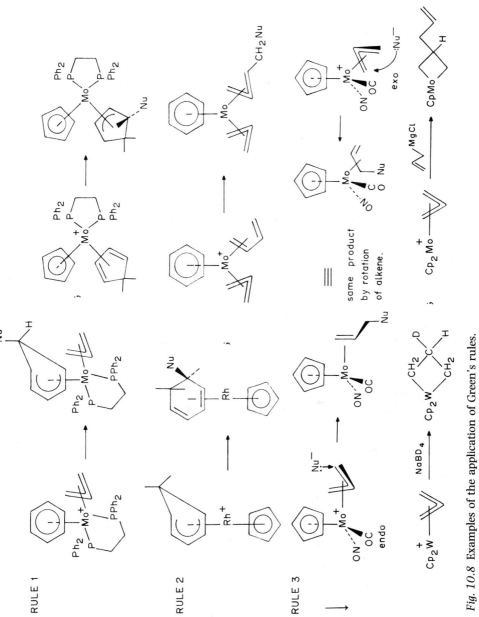

Fig. 10.8 Examples of the application of Green's rules.

proposed above, but because more favourable overlap results in this way. Probably both factors are significant.

10.5 η^6-Cycloheptatriene and η^7-tropylium complexes

Cycloheptatriene complexes of chromium, molybdenum and tungsten are prepared in a similar way to the arene tricarbonyls, by heating the ligand with the hexacarbonyl in a high boiling solvent. The preparation of the tungsten compound can be difficult, giving low yields, so it has been recommended to convert the hexacarbonyl into $W(CO)_3(MeCN)_3$ first. Abstraction of hydride from

the η^6-triene ligand affords the orange air-stable tropylium complexes (η^7-$C_7H_7)M(CO)_3^+ PF_6^-$. These salts are attacked by nucleophiles, although they are much less reactive than salts of the free tropylium ion, e.g. $C_7H_7^+ ClO_4^-$. In many cases ($Nu^- = OMe^-$, $\bar{C}H(CO_2Me)_2$ and H^-) the isolated product is the 7-exo substituted complex. With methoxide ion in methanol low temperature infrared studies indicate that two faster but reversible reactions, addition to the metal centre and also to a carbonyl ligand precede the final reaction. The isomeric 7-endo methoxycycloheptatriene complex is obtained by reacting methoxycyclo-heptatriene directly with $Cr(CO)_6$. The trityl cation abstracts the exo substituent from each isomer, OMe^- from the 7-exo and H^- from the 7-exo compound.

The major product from the reaction of $V(CO)_6$ with cycloheptatriene is $(\eta^7\text{-}C_7H_7)V(CO)_3$, isoelectronic with $(\eta^7\text{-}C_7H_7)Cr(CO)_3^+$. Reduction of VCl_4 with Pr^iMgBr in the presence of cycloheptatriene affords $V(\eta^6\text{-}C_7H_8)_2$, which is converted in two steps into $[V(\eta^6\text{-}C_7H_8)(\eta^7\text{-}C_7H_7)]^+$ and then into $[V(\eta^7\text{-}C_7H_7)]^{2+}$ using $Ph_3C^+.PF_6^-$.

X-ray diffraction has shown that the C_7H_7 ring in $V(\eta^5\text{-}C_5H_5)(\eta^7\text{-}C_7H_7)$ is planar and the carbon—carbon bond distances equal within experimental error.

It might be expected that azulenes would be able to act as 7-electron ligands. In those complexes whose crystal structures have been determined this is not so. In the binuclear iron complex, $(azulene)Fe_2(CO)_5$, for example, azulene acts as a $5 + 3$ electron ligand while in the similar molybdenum compound, $(azulene)Mo_2(CO)_6$, it is a 2×5 electron ligand.

10.6 References

Metal atom synthesis

Blackborow, J.R. and Young, D. (1979) *Metal Vapour Synthesis in Organometallic Chemistry*, Springer Verlag, Berlin.
Green, M.L.H. (1980) The use of atoms of the Group IV, V and VI transition metals for the synthesis of zerovalent arene compounds. *J. Organomet. Chem.*, **200**, 119.
Timms, P.L. and Turney, T.W. (1977) Metal vapour synthesis in organometallic chemistry. *Adv. Organomet. Chem.*, **15**, 53.

Nucleophilic attack on organotransition metal complexes

Davies, S.G., Green, M.L.H. and Mingos, D.M.P. (1978) A survey and interpretation. Tetrahedron Report No. 57, *Tetrahedron*, **34**, 3047.
Kane-Maguire, L.A.P., Honig, E.P. and Sweigart, D.A. (1984) Nucleophilic addition to coordinated hydrocarbons. *Chem. Rev.*, **84**, 52.
Pauson, P.L. (1980) *J. Organomet. Chem.*, **200**, 207.
Semmelhack, M.F., Clark, G.R., Garcia, J.L., Harrison, J.J., Thebtanaronth, Y., Wulf, W. and Yamashita, A. (1981) *Tetrahedron*, **37**, 3956.

Complexes of polyolefins

Deganello, G. (1979) *Transition Metal Complexes of Cyclic Polyolefins*, Academic Press, New York.

Complexes of arenes

Problems

1. Ferrocene was heated under reflux with mesitylene (1,3,5-trimethylbenzene) in the presence of aluminium powder and aluminium(III) chloride for several hours. The aqueous layer obtained by hydrolysis of the reaction mixture was treated with aqueous NH_4PF_6. A yellow solid A ($C_{14}H_{17}F_6FeP$) was precipitated. The ^1H n.m.r. spectrum of A consisted of singlets (δ 6.30, 5.07 and 2.52 ppm relative to Me_4Si). Compound A gave a bright red product B ($C_{14}H_{18}Fe$) on treatment with $NaBH_4$. B with $Ph_3C^+BF_4^-$ in dichloromethane afforded C ($C_{14}H_{17}BF_4Fe$).

Suggest structures for compounds A, B and C.

2.

$$Cr(CO)_6 \xrightarrow[\text{reflux}]{CH_3CN} C_9H_9CrN_3O_3 \ (F)$$

$\nu(CO)$ 1915,1790 cm^{-1}

$\xrightarrow{\quad\quad} C_{13}H_{14}CrO_3 \ (G)$

$\nu(CO)$ 1991,1921,1893 cm^{-1}

$$\xrightarrow[CH_2Cl_2]{Ph_3C^+BF_4^-} [J^+][PF_6^-] \xrightarrow[HPF_6]{\text{1,8-bis(dimethylaminonaphthalene)}} K$$

$\nu(CO)$ 2066,2021 cm^{-1}

$\nu(CO)$ 1985,1931,1905 cm^{-1}

Proton n.m.r. (δ relative to TMS, relative intensity, multiplicity).

$J^+PF_6^-$: 6.57, 6H, multiplet; 3.23, 1H, septet ($J = 7\,Hz$); 1.46, 6H, doublet ($J = 7\,Hz$)

K : 5.78, 2H, triplet ($J = 9\,Hz$); 4.64, 2H, triplet ($J = 9\,Hz$); 4.08, 2H, doublet ($J = 9\,Hz$); 1.56, 6H, singlet

Propose structures for F, G, J^+ and K. Discuss how the spectroscopic data are compatible with the proposed structures.

(*Royal Holloway College*).

3. Suggest how the following conversions could be effected using arenetricarbonylchromium complexes as intermediates.

(a) $PhF \longrightarrow PhNMe_2$

(b) $PhH \longrightarrow$ Ph—⟨ ⟩ $\left(\xrightarrow{(p.47)} PhCHO \right)$

(c) $Ph(CH_2)_4CN \longrightarrow$

(d) $PhCH_2CO_2Me \longrightarrow PhCMe_2CO_2Me$

332

4. The following scheme describes some of the chemistry of (cyclohexa-1,3-diene)(cyclopentadienyl)rhodium, *A*:

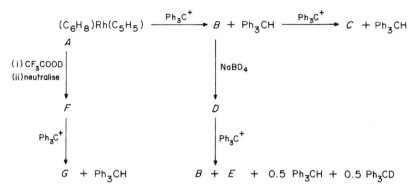

Proton n.m.r. spectra (chemical shifts in ppm on the δ scale; relative intensities in parenthesis)

A 5.23(5), 4.90(2), 3.73(2), 1.50(2), 1.22(2)
D 5.23(5), 4.90(2), 3.73(2), 1.50(2), 1.22(1)
F 5.23(5), 4.90(2), 3.73(2), 1.22(2)

(a) Deduce the structures and stereochemistries of the compounds *A* to *G*.
(b) Comment on the nature of the bonding between the metal and the six-membered ring in *C*.
(c) What species are likely to be present in a solution of *A* in CF_3COOH?
(*University of Southampton*).

5. Predict the products *X* and *Y* of the following reactions.

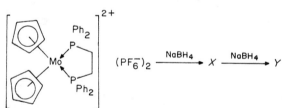

6. Using Green's rules, predict the position of nucleophilic attack in each of the following complexes.

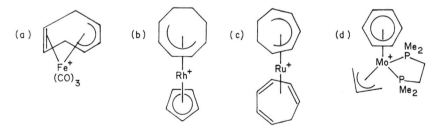

Complexes of arenes

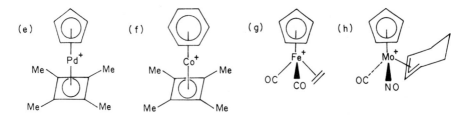

7. A method for the stereo- and regio-controlled substitution of cycloheptene using organomolybdenum compounds has been devised. Suggest reagents for each of the steps (a) to (g) inclusive. Discuss the stereochemistry of each step. (Pearson, A.J. and Khan, M.N.I. (1985) *Tetrahedron Lett.*, **26**, 1407.)

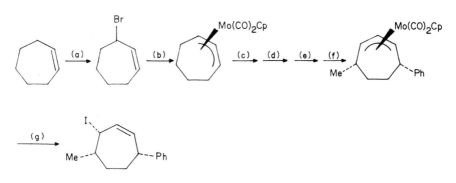

11   Cluster compounds

11.1 Introduction

The study of compounds which contain clusters of atoms, often of high symmetry, is making significant contributions to theories of bonding. Part of the fascination of this area is aesthetic, deriving from the beautiful order which these structures, in their unanticipated variety, reveal. Impetus for this work is enhanced by the realization that clusters, especially of transition metal atoms, may have potential as homogeneous catalysts, where two or more sites can act in conjunction. Moreover clusters sometimes have the structures expected from fragments of the extended close packed arrays found in bulk metals. It has therefore been suggested that they could provide a link between homogeneous and heterogeneous catalyst systems. Study of cluster chemistry could possibly lead to a deeper understanding of the behaviour of metal surfaces in catalysing reactions of industrial importance such as the reforming of hydrocarbons or conversions of 'synthesis gas'.

11.2 Structure and bonding

11.2.1 *Electron-precise structures*

Carbon forms a wide range of catenated compounds linked together by strong C—C bonds. This property is shared to a lesser extent by other elements of the same Group (Si, Ge) and of adjacent Groups to the right in the Periodic Table (P, As, S). The fragments CR_2, PR and S have 2 available orbitals which contain two electrons. They can thus link to form rings $(CR_2)_n$, $(PR)_n$, S_n or chains with suitable end groups. The bonding in these 'clusters' can be described in terms of classical two-electron two-centre bonds. These ring molecules are characterized by a total of $6n$ valence electrons (e.g. 4 from C and 2 from 2R) (Table 11.1).

The fragment CH has three available orbitals and three electrons for forming skeletal C—C bonds. Polyhedral three dimensional structures can arise, some of which are illustrated in Fig. 11.1. The bonding in these rather strained molecules can again be described in classical terms. In all each CH group contributes five valence electrons (four from C and one from H) so that a cluster $(CH)_n$ is characterized by $5n$ valence electrons. Note that square and pentagonal as well as

Table 11.1. Electron counting in clusters

Description	Vertices	Faces	Edges	Restriction on n	Main Group	Transition Metal	Mixed
							Polyhedral electron count
					Main Group	*Transition Metal*	*Mixed*
Rings	n	2	n	$n \geqslant 3$	6m	16t	6m + 16t
Polyhedra	n	(n + 4)/2	3n/2	$n \geqslant 4$ n even	5m	15t	5m + 15t
Deltahedra							
Closo	n	2n − 4	3(n − 2)	$n \geqslant 5$	4m + 2	14t + 2	4m + 14t + 2
Nido	M_n cluster occupies n corners of an (n + 1) cornered deltahedron				4m + 4	14t + 4	4m + 14t + 4
Arachno	M_n cluster occupies n corners of an (n + 2) cornered deltahedron				4m + 6	14t + 6	4m + 14t + 6
Hypho	M_n cluster occupies n corners of an (n + 3) cornered deltahedron				4m + 8	14t + 8	4m + 14t + 8

m = number of Main Group atoms in cluster; t = number of transition metal atoms in cluster. m + t = n = number of vertices.

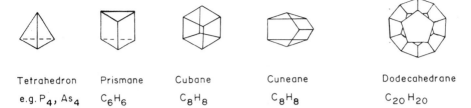

Tetrahedron	Prismane	Cubane	Cuneane	Dodecahedrane
e.g. P_4, As_4	C_6H_6	C_8H_8	C_8H_8	$C_{20}H_{20}$

Fig. 11.1 'Electron precise' clusters.

triangular faces are included in these polyhedra. Euler's theorem for polyhedra states that *the number of vertices + the number of faces = the number of edges + 2* (Table 11.1). As a consequence $(CH)_n$ clusters can arise only with even values of n.

11.2.2 Deltahedral clusters

The largest group of polyhedral clusters formed by Main Group elements are based on figures having triangular faces only (deltahedra) (Fig. 11.2). Most of them are derived formally from the borane anions $B_nH_n^{2-}$, but they also include some metal clusters such as Sn_5^{2-} and Ge_9^{2-}. The existence of the eicosahedral anion $B_{12}H_{12}^{2-}$ was predicted by Longuet-Higgins and Roberts from molecular orbital calculations a few years before its preparation. The B—H fragment possesses four valence electrons in all (three from B and one from H) of which two are used to form the B—H bond. It therefore has three orbitals but only two electrons left for skeletal bonding. Adoption of deltahedral structures maximizes the number of nearest neighbours and permits the most efficient electron delocalization within the cluster. As the grouping CH is isoelectronic with BH^-, the treatment covers the carboranes $B_{n-2}C_2H_n$ as well as the borane anions $B_nH_n^{2-}$. BH_2 can also be formally replaced by CH, giving rise to further carboranes.

The frontier orbitals of a B—H unit consist of a sp_z type hybrid directed towards the centre of the polyhedron and two orbitals (p_x, p_y) which lie tangential to its surface. The n radial orbitals of the former set combine together to give m.o.s of

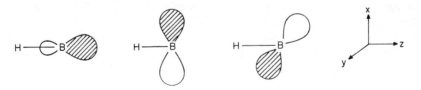

which the lowest lying in-phase combination is strongly bonding in character. The approximate spherical symmetry of the cluster and of the interacting orbitals leads to the result that (as in an atom) the other radial m.o.s are spherical harmonics of this lowest orbital (S^σ, one orbital) and have symmetries resembling

337

Cluster compounds

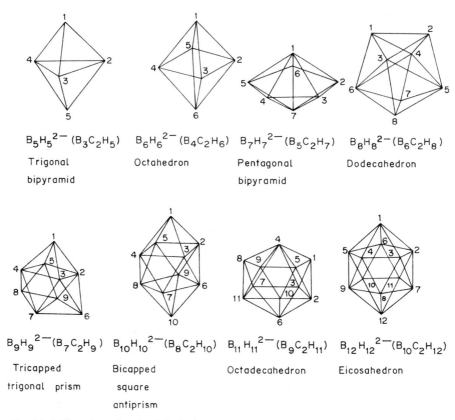

Fig. 11.2 The triangulated-polyhedral structures of $B_nH_n^{2-}$ and $B_{n-2}C_2H_n$ species with the conventional numbering schemes also indicated.

p orbitals (P^σ, three orbitals), d orbitals (D^σ, up to a maximum of five orbitals, depending on cluster size), f-orbitals (F^σ, up to seven orbitals) etc. This is shown for the eicosahedron $B_{12}H_{12}^{2-}$ (Fig. 11.3). Calculations predict that in this case the three $P^\sigma(t_{1u})$ orbitals should be weakly bonding, but they are destabilized by mixing with tangential bonding orbitals of the same symmetry. In general, therefore, the n radial frontier orbitals of the n B—H groups in $B_nH_n^{2-}$ lead to one bonding m.o. and $(n-1)$ antibonding m.o.s.

The $2n$ p orbitals tangential to the cluster give rise to n bonding and n antibonding m.o.s. The $5n$ m.o.s associated with $B_nH_n^{2-}$ ($4n$ B a.o.s and n H a.o.s) are therefore divided up as follows:

Bonding m.o.s	Antibonding m.o.s
n B—H m.o.s	n B—H m.o.s
1 radial m.o.	$(n-1)$ radial m.o.s
n tangential m.o.s	n tangential m.o.s
Total: $2n+1$ bonding m.o.s	Total: $3n-1$ antibonding m.o.s

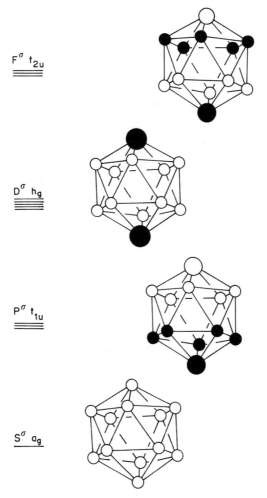

$F^\sigma\ t_{2u}$

$D^\sigma\ h_g$

$P^\sigma\ t_{1u}$

$S^\sigma\ a_g$

Fig. 11.3 Radial molecular orbitals in an eicosahedral cluster. (After Hall and Mingos (1984).)

Thus the *closo* boranes (Greek κλωβόσ, a bird cage), require a total of $4n + 2$ valence electrons (sometimes termed the 'polyhedral electron count', p.e.c.) to fill all the bonding m.o.s in the cluster.

It has been recognized that less symmetrical borane structures can be derived formally from $B_nH_n^{2-}$ by removal of one or more BH vertices (Fig. 11.4), or by capping one triangular face. Carboranes (strictly carbaboranes) formally arise by isoelectronic replacements e.g. BH_2 or BH^- by CH. The p.e.c.s listed in Table 11.2 follow from the result that an open cluster has the same number of bonding m.o.s as the closed figure from which it is derived. The same applies to a mono-capped deltahedral structure.

339

Cluster compounds

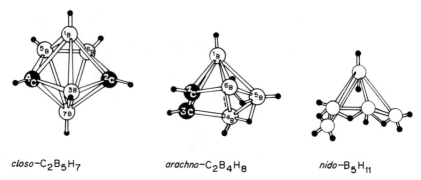

$closo\text{-}C_2B_5H_7$ $arachno\text{-}C_2B_4H_8$ $nido\text{-}B_5H_{11}$

Fig. 11.4 Closo, arachno and *nido* structures derived from a deltahedron with seven vertices, a pentagonal bipyramid.

Table 11.2 Polyhedral electron counts and borane structures

		P.e.c.
Capped *closo* deltahedra	Closed $(n-1)$ cornered polyhedron, one triangular face capped	$4n$
Closo-boranes	Closed polyhedron of n boron atoms	$4n + 2$
Nido-boranes	B_n cluster occupies n corners of an $(n + 1)$ cornered polyhedron	$4n + 4$
Arachno-boranes	B_n cluster occupies n corners of an $(n + 2)$ cornered polyhedron	$4n + 6$
Hypho-boranes	B_n cluster occupies n corners of an $(n + 3)$ cornered polyhedron	$4n + 8$

Nido from Latin *nidus*, a nest; *arachno* from Greek ἀράχνη, a spider's web; *hypho* from Greek ὑφή a web or a weaving.

11.3 Carboranes

Carboranes were first reported in 1962. Much of the early work was done in the USA as part of a classified project on borane chemistry, seeking novel rocket fuels, an object which proved abortive. Since then thousands of compounds of this type have been reported. Carboranes have structures which are related to those of the boranes by isoelectronic replacement of groups (e.g. B^- or BH by C; BH_2 or BH^- by CH). They may belong to *closo, nido* or *arachno* series, although the last are relatively uncommon. The usual entry to carborane chemistry is to heat a borane with an alkyne or to subject the mixture to a silent electric discharge. Thus *nido*-B_5H_9 and ethyne at 200°C give mainly *nido*-$2,3\text{-}C_2B_4H_8$, while at 500°C a

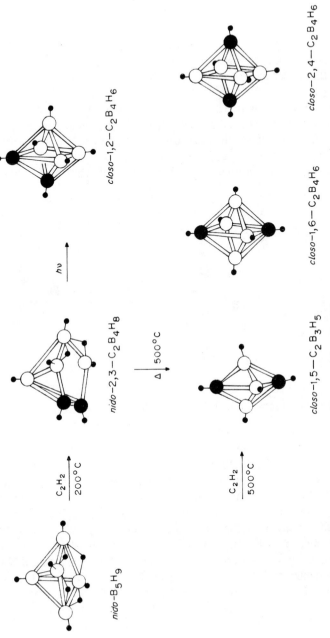

Fig. 11.5 Preparation of some carboranes from pentaborane (9).

$nido$-B_5H_9 $\xrightarrow[200°C]{C_2H_2}$ $nido$-2,3-$C_2B_4H_8$ $\xrightarrow{h\nu}$ $closo$-1,2-$C_2B_4H_6$

$nido$-2,3-$C_2B_4H_8$ $\xrightarrow[500°C]{\Delta}$

$nido$-B_5H_9 $\xrightarrow[500°C]{C_2H_2}$ $closo$-1,5-$C_2B_3H_5$

$closo$-1,6-$C_2B_4H_6$

$closo$-2,4-$C_2B_4H_6$

Cluster compounds

mixture of the more stable *closo*-boranes is produced (Fig. 11.5). *Nido* carboranes can also be pyrolysed or photolysed to *closo* species. Carboranes often exhibit high thermal stability, the *closo* compounds being more stable than the *nido*. Most *closo* boranes are stable to at least $400°C$ under static conditions. At or above this temperature rearrangements to geometrical isomers may be observed. It is found that those *closo* compounds in which the two carbon atoms are separated as far as possible in the structure are the most stable isomers and are often the final products of pyrolysis.

Carboranes may contain one, two, three or four carbon atoms, although those with two have been most thoroughly studied, being readily accessible by the alkyne route. Insertion of ethyne into the *arachno* borane B_4H_{10} gives a mixture of *nido*-carboranes including $1,2\text{-}C_2B_3H_9$ and alkyl derivatives of *nido*-$2,3,4$-$C_3B_3H_7$. Treatment of the former product with more ethyne affords *closo*-$2,3,4,5\text{-}C_4B_2H_6$ (Fig. 11.6).

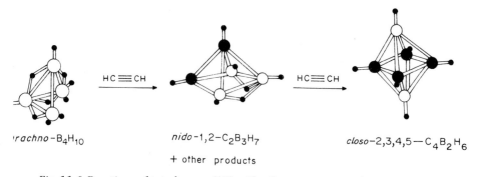

$$\text{'rachno-}B_4H_{10} \qquad \text{HC}\equiv\text{CH} \qquad \text{nido-}1,2\text{-}C_2B_3H_7 \qquad \text{HC}\equiv\text{CH} \qquad \text{closo-}2,3,4,5\text{-}C_4B_2H_6$$

+ other products

Fig. 11.6 Reactions of tetraborane (10) with ethyne

Carboranes show characteristic absorptions in the infrared at $2900\text{-}3160\,\text{cm}^{-1}$ $v(\text{C}-\text{H})$, $2500\text{-}2660\,\text{cm}^{-1}$ $v(\text{B}-\text{H}_{\text{terminal}})$ and $1800\text{-}2050\,\text{cm}^{-1}$, $1460\text{-}1500\,\text{cm}^{-1}$ $v(\text{B}-\text{H}_{\text{bridging}})$, which are useful in characterization. N.m.r. spectroscopy is also valuable, information being available not only from ^1H and ^{13}C spectra, but also from ^{11}B $(I=\frac{3}{2})$ (80% abundance). Interpretation of ^{11}B spectra is simplified as $^{11}\text{B}-^{11}\text{B}$ coupling is not detected. The other naturally occurring isotope of boron (^{10}B) has $I=3$. Both ^{11}B and ^{10}B nuclei couple to attached protons, but as proton signals from $^{10}\text{B}-^1\text{H}$ are of low relative intensity on account of the isotopic abundance, tend to be rather broad and are split into seven components their contribution is usually ignored. The resonance from a proton attached to ^{11}B consists of four lines of equal intensity.

The eicosahedral *closo*-carborane $1,2\text{-}C_2B_{10}H_{14}$ is prepared by the action of ethyne with *nido*-$B_{10}H_{14}$ in the presence of a Lewis base:

$$\text{nido-}B_{10}H_{14} + 2SEt_2 \longrightarrow B_{10}H_{12}(SEt_2)_2 + H_2$$
$$B_{10}H_{12}(SEt_2)_2 + C_2H_2 \longrightarrow \text{closo-}1,2\text{-}C_2B_{10}H_{12} + 2SEt_2 + H_2$$

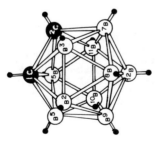

1,2—C$_2$B$_{10}$H$_{12}$ 1,7—C$_2$B$_{10}$H$_{12}$ 1,12—C$_2$B$_{10}$H$_{12}$

Fig. 11.7 Isomers of the *closo*-carboranes, C$_2$B$_{10}$H$_{12}$.

Cluster compounds

It is extremely stable both thermally (m.p. $320°C$) and to oxidation in air. On heating to $470°C$ in the gas phase for several hours or by flash pyrolysis to $600°C$ for $30\,s$, $1,2\text{-}C_2B_{10}H_{12}$ isomerizes to the $1,7$-isomer (m.p. $265°C$). Further isomerization to the $1,12$-isomer (m.p. $261°C$) occurs on heating $1,7\text{-}C_2B_{10}H_{12}$ for a few seconds to $700°C$.

The C—H bonds in *closo*-carboranes are appreciably acidic. Reaction of $1,2\text{-}C_2B_{10}H_{12}$ with butyllithium yields lithio derivatives which provide a starting point for an extensive chemistry of organo-substituted carboranes. Heat resistant silicone polymers have been developed commercially which include carboranes in their backbones. The equilibrium lies to the left in benzene, so the mono-

$$
\underset{B_{10}H_{10}}{HC\text{---}CLi} \;\rightleftharpoons\; \underset{B_{10}H_{10}}{HC\text{---}CH} \;+\; \underset{B_{10}H_{10}}{LiC\text{---}CLi}
$$

lithiated compound is prepared in that solvent, whereas tetrahydrofuran is used to make the dilithio derivative.

Degradation of *closo* carboranes such as *closo*-1, 2- or -1, 7-$C_2B_{10}H_{12}$ with base affords corresponding isomers of the *nido* anion

$$C_2B_nH_{n+2} + 2EtOH + EtO^- \longrightarrow C_2B_{n-1}H_{n+2}^- + B(OEt)_3 + H_2$$

$$C_2B_{10}H_{12} \xrightarrow[\text{EtO}^-]{\text{EtOH}} C_2B_9H_{12}^- \xrightarrow{\text{NaH}} C_2B_9H_{11}^{2-}$$

| p.e.c. 50 | p.e.c. 48 | p.e.c. 48 |
| $(4n+2)$ | $(4n+4)$ | $(4n+4)$ |

11.4 Metallocarboranes and metalloboranes

In these reactions the BH vertex adjacent to the CH vertices is removed. In the $C_2B_9H_{11}^{2-}$ ion five orbitals are directed from the five atoms of the pentagonal face towards the site vacated by this BH group. Six electrons can be assigned to these orbitals, so that the ion strongly resembles the cyclopentadienyl anion $C_5H_5^-$. ($C_2B_9H_{11}^-$ is a five-electron ligand). Transition metal complexes have been prepared which have sandwich structures analogous to that of ferrocene. Reaction between $C_2B_9H_{11}^{2-}$ and $FeCl_2$ in tetrahydrofuran yields the pink diamagnetic complex $[Fe(C_2B_9H_{11})_2]^{2-}$ which is reversibly oxidized to the red paramagnetic anion $[Fe(C_2B_9H_{11})_2]^-$, which corresponds to $Fe(C_5H_5)_2^+$. Carborane complexes of a wide range of transition elements are now known. They include mixed sandwiches e.g. $[CpFe(\eta^5\text{-}C_2B_9H_{11})]$ and $[CpCo(\eta^5\text{-}C_2B_9H_{11})]$ as well as half sandwiches e.g.

$$[M(CO)_3(\eta^5\text{-}C_2B_9H_{11})]^{2-} \quad (M = Cr, Mo, W)$$

and

$$[M'(CO)_3(\eta^5\text{-}C_2B_9H_{11})]^-. \quad (M' = Mn, Re)$$

$$Co^{2+} + 2C_2B_9H_{12}^- \xrightarrow[\text{NaOH}]{\text{hot aq.}} [Co(C_2B_9H_{11})_2]^{2-} \xrightarrow{\text{air}} [Co(C_2B_9H_{11})_2]^-$$

$$Mo(CO)_6 + C_2B_9H_{11}^{2-} \xrightarrow{h\nu} [(OC)_3Mo(C_2B_9H_{11})]^{2-} + 3CO$$

$$BrRe(CO)_5 + C_2B_9H_{11}^{2-} \longrightarrow [(OC)_5Re(\eta^1\text{-}C_2B_9H_{11})]^-$$

$$\xrightarrow[\text{THF}]{\text{reflux}} [(OC)_3Re(\eta^5\text{-}C_2B_9H_{11})]$$

The nickel compound is found in three oxidation states

$$[Ni(C_2B_9H_{11})_2]^{2-} \underset{\text{Na/Hg}}{\overset{O_2}{\rightleftharpoons}} [Ni(C_2B_9H_{11})_2]^- \underset{\text{Cd}}{\overset{Fe^{3+}}{\rightleftharpoons}} Ni(C_2B_9H_{11})_2$$

$[Cu(C_2B_9H_{11})_2]^-$ as well as $[Cu(C_2B_9H_{11})_2]^{2-}$ have also been prepared, even though cuprocene $Cu(C_5H_5)_2$ is unknown. In those compounds in which the valence shell of the metal would have to be assigned 20 or more electrons, if $C_2B_9H_{11}^-$ is assumed to be a five-electron donor, slipped sandwich structures are observed. This is illustrated by $[Cu(C_2B_9H_{11})]^{2-}$ (Fig. 11.8) in which the three boron atoms of each carborane face are closer to the metal atom than the two carbon atoms. Even more pronounced distortion occurs in the mercury complex $Ph_3PHg(C_2B_9H_{11})$.

Closo cages have a p.e.c. of $4n + 2$ electrons, whereas *nido*-carboranes require $4n + 4$ electrons. The larger *closo*-carboranes can be reduced by alkali metals to *nido* anions. The latter can then often be converted into *closo*-metallocarboranes on treatment with a metal halide:

$$C_2B_nH_{n+2} \xrightarrow{2e^-} C_2B_nH_{n+2}^{2-}$$

e.g.

$$closo\text{-}1,7\text{-}C_2B_6H_8 \xrightarrow[\text{THF}]{2e^-/} nido\text{-}C_2B_6H_8^{2-} \xrightarrow[\text{NaCp/THF}]{CoCl_2 +} CpCo(C_2B_6H_8)$$

Metalloboranes have also been prepared in which a metal atom has been incorporated into a borane structure. While the lower boranes such as B_4H_{10} and B_5H_9 are extremely air sensitive as well as being thermally rather unstable, the metalloboranes are often robust. Cobalt(II) chloride with a mixture of $B_5H_8^-$ and Cp^- affords a 5% yield of the red air-stable crystalline $2\text{-}CpCoB_4H_8$, which isomerizes at $200°C$ to the yellow 1-isomer (Fig. 11.9). The isoelectronic complex $1\text{-}(OC)_3FeB_4H_8$ is one product from the reaction between $Fe(CO)_5$ and B_5H_9. The borane ligand in these complexes, B_4H_8, is isoelectronic with C_4H_4 and hence forms complexes of similar structure to C_4H_4CoCp and $C_4H_4Fe(CO)_3$, (p. 269).

The polyhedral electron count usually correlates with the structure in the case of metalloboranes and metallocarboranes, as well as for the parent compounds.

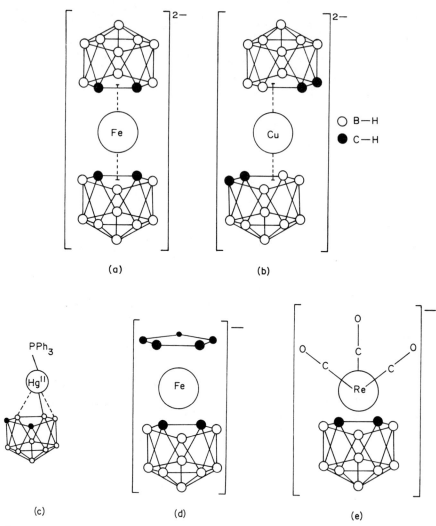

Fig. 11.8 The structures of some metallocarboranes. (a) The symmetrical structure in $[Fe(C_2B_9H_{11})_2]$ (b) The slipped structure. (c) $Ph_3PHg(C_2B_9H_{11})$. (d) A mixed sandwich. (e) A half sandwich.

To calculate the p.e.c. the valence electrons of all the skeletal atoms are summed, together with the number of electrons contributed by the ligands (p. 192). This sum is characteristic of the structure of the cluster (Table 11.1). In order to meet cluster bonding requirements, transition metal atoms require ten more valence electrons (to fill the $(n-1)$d orbitals) than Main Group atoms.

Isoelectronic relationships can also be very helpful in classifying known compounds and predicting new ones. This is illustrated by the series of complexes

346

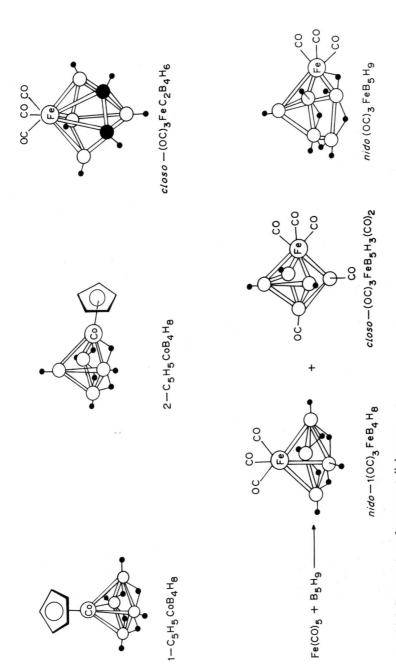

Fig. 11.9 Preparation of some metalloboranes.

$(\eta^5\text{-}C_3B_3H_6)Mn(CO)_3$, $(\eta^5\text{-}C_2B_4H_6)Fe(CO)_3$ and $(\eta^5\text{-}CB_5H_6)Co(CO)_3$, all of which have the *closo*-pentagonal bipyramidal structure with a p.e.c. of 40.

The frontier orbitals of $Fe(CO)_3$ and of CoCp are isolobal with those of B—H (p. 207). They consist of a radial hybrid directed to the centre of the polyhedron and two tangential orbitals at right angles to each other. In the $Fe(CO)_3$ fragment (or CoCp fragment) these orbitals, as in B—H, contain two valence electrons. $Fe(CO)_3$ can therefore replace B—H in deltahedral clusters. $Mn(CO)_3$ provides one less valence electron than $Fe(CO)_3$, so that one extra electron is required from the remaining cluster atoms. With $Co(CO)_3$ one electron less is needed for an isoelectronic arrangement.

11.5 Transition metal clusters

Many cluster molecules are formed by transition metal atoms which are coordinated to π-acid ligands, in particular carbon monoxide, cyclopentadienyl and phosphines. Under these circumstances the 18-electron rule should be obeyed so that a transition metal fragment has ten more valence electrons than the corresponding isolobal Main Group fragment. This relationship is exemplified in Table 11.3.

The 17-electron fragments such as $Co(CO)_4$, $Mn(CO)_5$ or $CpNi(CO)$ form the dimers $Co_2(CO)_8$, $Mn_2(CO)_{10}$ or $CpNi(CO)_2$ in which the total p.e.c. is 34. A carbonyl group normally contributes two electrons to the p.e.c., irrespective of its mode of bonding–bridging or terminal. Rings arise from 16-electron fragments e.g. $Fe_3(CO)_{12}$, $Cp_3Co_3(CO)_3$ (p.e.c. 48) and $Cp_2Mn_2(CO)_4(\mu\text{-}CH_2)$ (p.e.c. = 38; $16t + 6m = 38$). Electron-precise tetrahedral clusters (p.e.c. 60) are exemplified by $Co_4(CO)_{12}$ and $[CpFe(CO)]_4$, as well as by mixed clusters such as $Co_3CH(CO)_9$ (p.e.c. 50; $15t + 5m$, $t = 3$, $m = 1$). The 15t p.e.c. is also consistent with the structure of $Rh_6C(CO)_{15}^{2-}$ (p.e.c. 90) which has a carbon atom encapsulated in the centre of a prismatic cluster. The incorporation of transition metal fragments into borane and carborane deltahedra is discussed above. *Closo*, *nido* and *arachno* deltahedral clusters made up entirely from transition metal atoms are also common. The octahedral cobalt carbonyl $Co_6(CO)_{16}$ belongs to this class. Some carbido–iron carbonyls illustrate some of these structural relationships as well as providing novel chemistry (Fig. 11.10). Reduction of $Fe(CO)_5$ by strong reducing agents such as $Fe(CO)_4^{2-}$ or $Mn(CO)_5^-$ affords the cluster anion $[Fe_6(CO)_{13}(\mu\text{-}CO)_3(\mu_6\text{-}C)]^{2-}$, in the middle of which a carbide carbon, derived from a carbonyl ligand of $Fe(CO)_5$, is encapsulated. Subsequent oxidation with $C_7H_7^+BF_4^-$ removes a vertex $Fe(CO)_3$ group to give a *nido* octahedral cluster (p.e.c. 74) which on further reduction and protonation gives a butterfly *arachno* octahedral derivative (p.e.c. 62) in which there is a bridging CH group with an 'agostic' hydrogen. This reaction sequence shows the conversion, albeit by a very involved route, or coordinated CO to methane.

The structures of large transition metal clusters can be derived by bringing together smaller triangular, tetrahedral and octahedral fragments. These 'con-

Table 11.3. Isolobal fragments and cluster formation

Main Group fragments	B	BH C	BH₂ CH P	CH₂ PH S	CH₃ PH₂ SH	CH₄ PH₃ SH₂
No. of electrons contributed to p.e.c.	3	4	5	6	7	(8)
Transition metal fragments Carbonyls	Co(CO)₂ Mn(CO)₃	Ni(CO)₂ Fe(CO)₃	CuL₂ Co(CO)₃ Mn(CO)₄	Ni(CO)₃ Fe(CO)₄	Co(CO)₄ Mn(CO)₅	Ni(CO)₄ Fe(CO)₅
Cyclopentadienyls Mixed		CoCp CpMn(CO)	NiCp CpFe(CO)	CuCp CpCo(CO) CpMn(CO)₂	CpNi(CO) CpFe(CO)₂ CpCr(CO)₃	CpCuL CpCo(CO)₂ CpMn(CO)₃
No. of electrons contributed to p.e.c.	13	14	15	16	17	(18)
Type of cluster	†	4m + 2* 14t + 2* Deltahedra	5m 15t Electron precise	6m 16t Rings	7m 17t Dimer	8m 18t Closed shell molecules

*Closo.
†Isoelectronic replacement e.g. BH by CH or additional electron required.

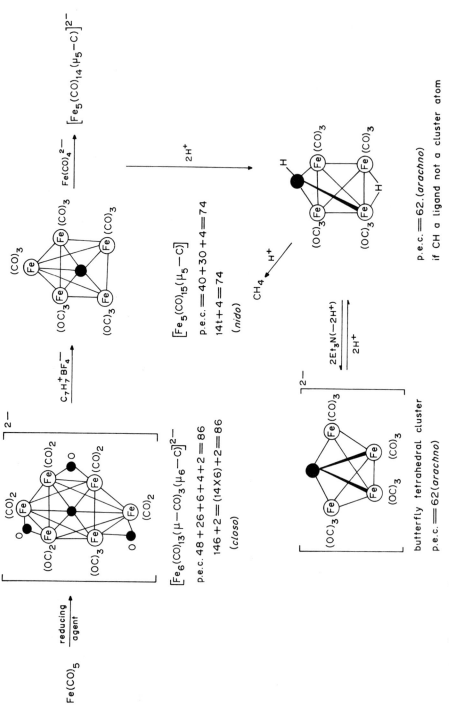

Fig. 11.10 Chemistry of some deltahedral ironcarbidocarbonyl clusters.

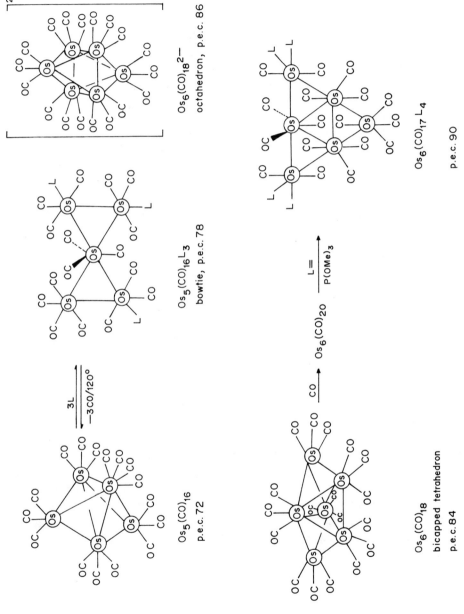

Fig. 11.11 Structures of osmium carbonyl complexes, showing the polyhedral electron counts.

351

Cluster compounds

densations' do not normally occur in practice but are useful on paper as an aid to correlating and sometimes to predicting structures. The principle is supported by molecular orbital calculations which take into account interactions between the frontier orbitals of the two combining fragments. The following rule results from these considerations:

'The total electron count in a condensed polyhedron is equal to the sum of the electron counts for the parent polyhedra *minus* the electron count characteristic of the atom, pair of atoms, of face of atoms common to both polyhedra'. For transition metal clusters which adhere to the 18-electron rule the characteristic electron counts to be substracted are as follows: atom M (vertex sharing) 18; M—M (edge sharing) 34; and M_3 (triangular face) 48. This rule is now applied in the course of a brief account of some chemistry of osmium carbonyls. Osmium forms an especially large number of neutral and anionic cluster carbonyl complexes. The triangular cluster $Os_3(CO)_{12}$ is cleaved by halogens to give linear dihalides

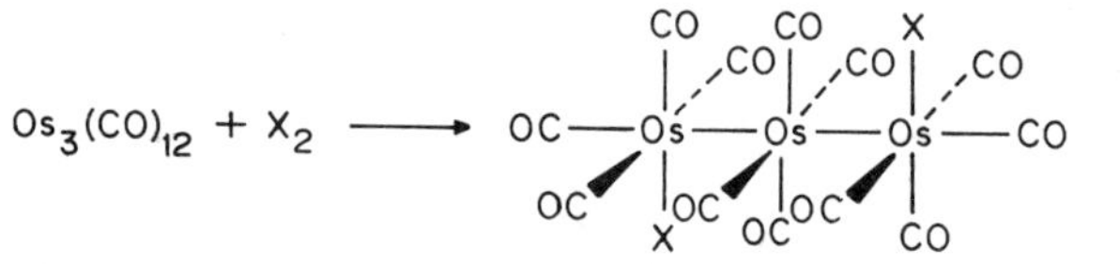

$$Os_3(CO)_{12} + X_2 \longrightarrow$$

p.e.c. 48

p.e.c. of $Os_3(CO)_{12}X_2 =$
$(3 \times 8) + (12 \times 2) + (2 \times 1) = 50$ or
$(34 + 34 - 18) = 50$

The latter can be considered as arising from two binary groups (p.e.c. 34) linked through one atom (p.e.c. 18).

When $Os_3(CO)_{12}$ is heated in a sealed tube to 190°C, $Os_6(CO)_{18}$ is formed, together with Os_5, Os_7 and Os_8 species (Fig. 11.11). The pentaosmium cluster $Os_5(CO)_{16}$ and the anion $Os_5(CO)_{15}^{2-}$ which is derived from it by reduction possess trigonal bipyramidal structures (p.e.c. 72). This figure can be viewed either as a *closo*-deltahedron $(14t + 2)$ or as a capped tetrahedron (2 tetrahedra sharing one face; p.e.c. $60 + 60 - 48$). $Os_5(CO)_{16}$ reacts with carbon monoxide to form $Os_5(CO)_{19}$ which has the fascinating 'bow-tie' arrangement of five metal atoms all lying in a plane.

In $Os_6(CO)_{18}$ (p.e.c. 84), the main product from heating $Os_3(CO)_{12}$, two faces of a central Os_4 tetrahedron are capped $(3 \times 60 - 2 \times 48)$. Reduction gives octahedral $Os_6(CO)_{18}^{2-}$ (p.e.c. 86, $14t + 2$) and action of CO followed by tertiary phosphite, $Os_7(CO)_{17}L_4$, in which the planar bow-tie arrangement of osmium atoms is extended still further.

11.6 Clusters and catalysis

Cluster complexes are being studied in the hope that they might provide novel catalytic systems. In a mononuclear complex potential coordination sites must

352

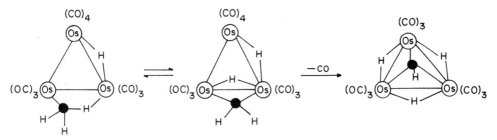

Fig. 11.12 Equilibrium between methyl and μ-hydrido-μ-methylene structures in an osmium carbonyl complex.

either be perpendicular or opposite to each other. In clusters there is the possibility that they might be on adjacent metal atoms and parallel, allowing an unsaturated substrate molecule to bridge two or more sites. Such a substrate could thus, for example, coordinate at one metal centre and undergo reaction at another. Activation of C—H bonds in this way has been demonstrated in the osmium cluster $HOs_3(CO)_{10}CH_3$. This has a formal electron count of 46, but one hydrogen atom of the methyl group is 'agostic'; in solution there is an equilibrium between the two structures shown in Fig. 11.12. $HOs_3(CO)_{10}CH_3$ loses CO to form the tetrahedral alkylidyne cluster $H_3Os_3(CO)_9CH$.

Some very large carbonyl cluster anions of the late transition elements have been prepared by condensation of carbonyl anions with polynuclear carbonyls. The first example of such a reaction was discovered by Hieber in 1965.

$$Fe_3(CO)_{11}^{2-} + Fe(CO)_5 \xrightarrow[\text{THF}]{25°C} Fe_4(CO)_{13}^{2-} + 3CO$$

These condensations are most favoured with elements of the Second and Third Transition Series e.g.

$$Rh_4(CO)_{12} + Rh(CO)_4^- \xrightarrow[\text{1 atm CO}]{25°C/\text{THF}} Rh_5(CO)_{15}^- + CO$$

$$Rh_5(CO)_{15}^- + Rh(CO)_4^- \underset{-70°C}{\overset{25°C/\text{THF}}{\rightleftharpoons}} Rh_6(CO)_{15}^{2-} + 4CO$$

octahedron, p.e.c. 86

$$Rh_6(CO)_{16}^{2-} + Rh(CO)_4^- \underset{-70°C}{\overset{25°C/\text{THF}}{\rightleftharpoons}} Rh_7(CO)_{16}^{3-} + 3CO$$

capped octahedron
p.e.c. 98

Reactions of this type may be endothermic. While $\bar{D}(M—CO)$ $160\,kJ\,mol^{-1}$, $\bar{D}(M—M)$ is only about $100\,kJ\,mol^{-1}$. Expulsion of CO is favoured on entropy grounds at higher temperatures. In this way large clusters can be built up:

$$Rh_4(CO)_{12} \xrightarrow[\text{(H}_3\text{C)}_2\text{CHOH}]{80°C/\text{OH}^-} \begin{array}{c} \xrightarrow{H_2} Rh_{13}(CO)_{24}H_3^{2-} \quad (50\%) \\ \xrightarrow{N_2} Rh_{15}(CO)_{27}^{3-} \quad (50\%) \end{array}$$

Cluster compounds

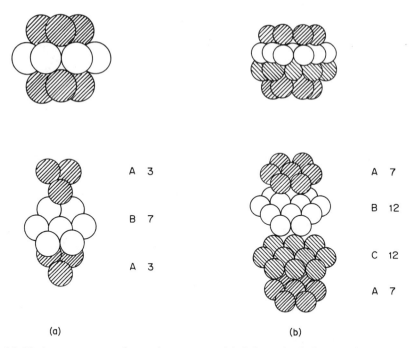

A 3

B 7

A 3

A 7

B 12

C 12

A 7

(a)   (b)

Fig. 11.13 Arrangements of metal atoms in (a) $[Rh_{13}H_2(CO)_{24}]^{3-}$, a fragment of hexagonal close packing and (b) $[Pt_{38}(CO)_{44}H]^{2-}$, a fragment of cubic close packing. The letters ABC refer to the arrangement of the planes in close packing and the figures to the number of metal atoms in each plane.

In the Rh_{13} cluster, the metal atoms describe a fragment of hexagonal close packing. There is a centre plane of seven rhodium atoms with hexagonal symmetry; above and below this plane are planes of three rhodium atoms in eclipsed configuration (ABA) (Fig. 11.13). In $Pt_{26}(CO)_{32}^{2-}$ there is an even larger fragment of hexagonal close packing. Above and below a central plane of 12 close packed platinum atoms there are planes of seven metal atoms in hexagonal symmetry, giving overall D_{3h} symmetry for the whole framework. In $Pt_{38}(CO)_{44}H_x^{2-}$ there is a set of four planes containing 7, 12, 12 and 7 platinum atoms arranged in the ABCA sequence of cubic close packing. The overall structure is that of a cube octahedron (O_h symmetry). Electron counting rules have been proposed even for such large clusters. Cluster compounds formed by platinum and gold, however, tend to conform to 16- and 14-electron rules respectively rather than to the 18-electron rule. Consequently the counting schemes which apply differ in detail from those presented above.

The structural results suggest that large clusters can be viewed as fragments of bulk metals and that reactions occurring on them might mimic those which take place on metal surfaces. Another view is that the two types of system bear little

354

relation to each other. Discussion of this area is outside the scope of this book; some relevant literature is cited.

11.7 References

Adams, R.D. and Horvath, I.T. (1985) Novel reactions of metal carbonyl cluster compounds. *Prog. Inorg. Chem.*, **33**, 127.

Bradley, J.S. (1983) Carbido carbonyl metal complexes. *Adv. Organomet. Chem.*, **22**, 1.

Chini, P. (1980) Large metal carbonyl clusters. *J. Organomet. Chem.*, **200**, 37.

Grimes, R.N. (1983) Metals in borane clusters. *Acc. Chem. Res.*, **16**, 22.

Hall, K.P. and Mingos, D.M.P. (1984) Cluster compounds of gold. *Prog. Inorg. Chem.*, **32**, 237.

Johnson, B.F.G. (Ed.) (1980) *Transition Metal Clusters*, Wiley, Chichester.

Lewis, J. and Johnson, B.F.G. (1981) Carbonyl clusters. *Adv. Inorg. Chem. Radiochem.*, **24**, 225.

Mingos, D.M.P. (1984) The polyhedral electron pair approach. *Acc. Chem. Res.*, **17**, 311.

Mingos, D.M.P. (1986) Bonding in molecular clusters and their relationship to bulk metals. *Chem. Soc. Revs.*, **15**, 31.

Muetterties, E.L. (1982) Metal clusters: bridges between molecular and solid-state chemistry. *Chem. & Eng. News*, Aug. 30, p. 28.

Wade, K. (1976) Structural and bonding patterns in cluster chemistry. *Adv. Inorg. Chem. Radiochem.*, **18**, 1.

Problems

1. Give an account of the basis and use of the skeletal electron counting rules for clusters.

Derive the polyhedral electron count for each of the following species and use this number to predict the structure of the molecule or ion.

    (a) B_5H_{11} (b) $C_3B_3H_7$ (c) $B_{10}H_{10}AsSb$ (d) $(Ph_3P)_2PtC_2B_9H_{11}$
    (e) $[SnC_5Me_5]^+$ (f) $Co_3(CO)_9SnMe$ (g) B_9H_{15} (h) $Fe_5(CO)_{15}C$
    (i) $[Pb_5]^{2-}$ (j) $B_{10}H_{11}PC$ (k) $Me_2C_2Co_4(CO)_{10}$ (l) $C_6H_6Cr(CO)_3$.

2. Show how the molecular structures of the following pairs are related:

(a) $B_6H_6^{2-}$ and $Ru_6(CO)_{18}^{2-}$

(b) B_5H_9 and $Fe(B_4H_8)(CO)_3$

(c) B_4H_{10} and $(Ph_3P)_2Cu(B_3H_8)$

Derive the p.e.c. for each cluster and discuss the isolobal relationships between the groups which are interchanged in the clusters.

3. Discuss the isolobal relationship between CO and $Os(CO)_3$ groups in the context of the structures of $Rh_6(CO)_{16}$ and of $[Os_{10}(CO)_{24}C]^{2-}$. Show how the p.e.c. for the latter is consistent with the replacement of the triply bridging CO groups in an octahedral cluster $[Os_6(CO)_{16}C]^{2-}$ by $Os(CO)_3$ groups.

4. Show how the electron counting rules can be extended to account for the

Cluster compounds

structures of the species $Os_6(CO)_{18}$ and $[Rh_7(CO)_{16}]^{3-}$. Reaction of the osmium compound with $NaBH_4$ affords $[HOs_6(CO)_{18}]^-$. What prediction can be made about the arrangement of the core atoms in the last ion?

5. Propose structures for the species given below, giving reasons for your choice. Discuss the measurements which could be made to confirm your proposal for one of these clusters.

(a) $Fe_3(CO)_9Te_2$ (b) $Fe_4C(CO)_{13}$ (c) $[Co_6N(CO)_{15}]^-$ (d) $[Rh_9As(CO)_{21}]^{2-}$
(*University of Southampton*).

6. A mixture of pentaborane (9) and ethyne was heated strongly for several hours in a sealed tube. The main volatile product, Y (Found B, 57.40%, C, 31.89%; H, 10.71%; r.m.m., 75.3) was separated from other components by g.l.c. The ^{11}B n.m.r. spectrum of Y consists of a broad doublet at low field and another at much higher field, relative areas 3:1 respectively. The low field doublet shows some resolved splitting through coupling to bridge hydrogen atoms. The ^1H n.m.r. spectrum of Y shows a C—H proton signal at $\delta - 6.32$ ppm relative to Me_4Si.
Identify Y. What is its most likely molecular structure? Comment on its ^{11}B n.m.r. spectrum.
(*University of Warwick*).

7. The reaction of the carborane $C_2B_6H_8$, A, with $Me_4N^+BH_4^-$ produced a number of degradation products including a volatile carborane B. This had maximum m/z in its mass spectrum of 76, while a relative molecular mass determination using a molecular weight bulb gave a value of 75.1. The ^{11}B n.m.r. spectrum showed two low-field doublets (X, Y) and one doublet at much higher field (Z), of relative areas X:Y:Z::2:2:1. X and Y showed additional fine structure. The ^1H n.m.r. spectrum showed one well-defined peak attributable to hydrogen bound to carbon, but the signals expected from the remaining hydrogens were broad (due to quadrupolar effects) and could not be resolved.
Identify B and comment on the structures of A and B and the ^{11}B n.m.r. spectrum of B. If the fine structure in the doublets X and Y in this spectrum could be resolved, what pattern would you expect it to show?
(*University of Warwick*).

12 Mechanisms of industrial processes clarified by studies of homogeneous catalysis by complexes of transition elements

12.1 Introduction

In this Chapter a number of organic reactions are discussed which are catalysed under homogeneous conditions by complexes of the transition elements. A few of these, notably the carbonylation of methanol to yield acetic acid and the conversion of alkenes into aldehydes or alcohols with carbon monoxide and hydrogen (hydroformylation) are used industrially on a large scale. Most industrial catalysts, however, are heterogeneous, that is, they do not belong to the same phase as the reactants. Normally a catalyst is a solid, often a highly dispersed metal or a metal oxide with a high surface area and reactions take place on adsorption of reactant molecules at its surface.

Table 12.1 represents an attempt to summarize some of the characteristics of homogeneous and heterogeneous catalysts and their advantages and disadvantages. In an industrial context the first three factors, *activity*, *selectivity* and *life* are of paramount importance. As far as activity is concerned, homogeneous systems might appear to make the most efficient use of the catalyst material as all metal atoms are involved in the catalytic process. Only sites on the surface of a heterogeneous catalyst, including those in pores, are available to the reactants so that those metal atoms in the interior of the crystallites are not utilized. By using catalysts of high surface area, however, this wastage can be minimized.

A catalyst cannot alter the *position* of equilibrium. It merely lowers the activation energy barrier between reactants and products so that equilibrium is attained more rapidly than in its absence. Even with the simplest starting materials, many reaction paths leading to different products are possible. A catalyst should be able to select one of these paths to the exclusion of all others. High selectivity of this type is desirable in any industrial process. Different

Table 12.1. Homogeneous and heterogeneous catalysts

Property	Homogeneous	Heterogeneous
Activity	All metal atoms active	Only surface site active, including those in pores
Selectivity	Usually good. Systematic control sometimes possible by varying ligands.	Structure of active site often not known. Selectivity less easily controlled, but good using correct promoters.
Life	Can be limited (e.g. lab system) but not in applied catalysts.[a]	Months or even years[b]
Ease of study	Mechanism and structures of intermediates can sometimes be determined (spectroscopic methods)	Difficult to establish clear mechanism
Mass transfer	Not restricted (one phase). Adequate stirring required if some reactants partitioned between gas and liquid phases (e.g. H_2, CO in hydroformylation)	May be restricted (more than one phase)
Heat transfer	Not restricted. Adequate stirring required.	May be restricted (more than one phase).
Temperature of operation	Usually below 200°C; often ambient.	Elevated temperatures above 200°C usually required.
Stability	Temperature range limited by decomposition of active complex.	Stable over wide temperature range. Catalyst may be inactivated by poisons in feed.
Corrosion of plant	Possible problem	No problem from catalyst
Ease of separation of catalyst from product	Difficult unless products volatile.	Generally easy

[a]Lifetime of Monsanto acetic acid catalyst (p. 385) over ly.
[b]Can be short only if regeneration is easy. Catalyst for cat. cracking, for example, must be continuously removed from main reactor to burn off coke.

catalysts may promote different reactions. A mixture of carbon monoxide and hydrogen (synthesis gas), for example, is converted respectively into hydrocarbons, methane or methanol depending on whether iron, nickel or copper is used (p. 383).

Catalysts are intimately involved in the chemical changes which are taking place. Deactivation can occur by the action of foreign substances (poisons) which may bind too strongly to the catalyst or convert it into inactive materials. The lifetime of a catalyst and its ease of regeneration should it become deactivated are major considerations in deciding whether it can be used commercially.

The structures of intermediates present in homogeneously catalysed systems can often be determined using solution techniques such as n.m.r. and infrared spectroscopy. Taken in conjunction with kinetic data a fairly detailed picture of the mechanism can, in favourable cases, be assembled. Good examples of this approach include the carbonylation of methanol (p. 385) and the oligomerization of butadiene (p. 367). In the latter case the selectivity to different products is dramatically changed by addition of phosphine or phosphite ligands. It is much more difficult to discover the nature of reacting species on metal surfaces. Different sites on the surface may have different coordination numbers which in turn exhibit different catalytic activities. The techniques available for studying metal surfaces such as LEED, Auger electron spectroscopy etc. do not permit the chemical characterization of adsorbed species except in fairly simple cases. Solid state n.m.r. is very promising, but as yet few results have been reported. Consequently rather little is known about the mechanisms of heterogeneously catalysed processes and the divising of such catalysts is still essentially an empirical art. Some homogeneous systems, however, behave very similarly to their heterogeneous counterparts (e.g. olefin metathesis, p. 373). Detailed study of the former can thus provide indirect information about the way the latter operate. This is a major theme in the Chapter. While heterogeneous systems are often preferred in an industrial context, it is only by studying the analogous homogeneous system that real clues to the mechanism of the process can be discovered.

One factor which influences the preference for heterogeneous catalysts in industry is the relative ease of separation of products from catalyst. Homogeneous catalysts often require a solvent. If this is an organic chemical, additional cost is incurred. There are also the problems of product separation and solvent recovery. The homogeneously catalysed Wacker process for oxidation of ethene (p. 380) does not suffer from these disadvantages because the product, acetaldehyde, is very volatile and water (containing some acetic acid) can be used as the solvent.

12.1.1 *Supported Catalysts*

Homogeneous catalysts for hydrogenation, hydroformylation etc. often contain ligands such as tertiary phosphines. If this phosphine group is bonded to a solid support such as silica or an organic polymer it is possible to anchor the active complex on to its surface. The high selectivity and activity of a homogeneous system are thus combined with the ease of separation of products which is characteristic of heterogeneous catalysts.

The silicon esters $Ph_2P(CH_2)_nSi(OEt)_3$, for example, have been used to anchor rhodium complexes to silica. In this way successful hydrogenation and hydroformylation (p. 387) catalysts have been prepared. Gradual loss of metal through leaching, however, does present a problem.

Table 12.2.

Process	Feedstock	Catalyst	Conditions	Reaction type	Products
PRODUCTION OF FEEDSTOCKS FOR LARGE SCALE CHEMICAL MANUFACTURE					
Steam cracking	Naphtha b.p. 30–190°C: steam ~ 1:1	None	750–900°C	Free radical chain, C–C cleavage.	Ethene, Propene Butenes, Butadiene
	Ethane	None	750–900°C	Dehydrogenation $C_2H_6 \rightarrow C_2H_4 + H_2$	Ethene
Synthesis gas $(CO + H_2)$:					
Gasification (Lurgi)	Coal, steam, oxygen	None	Fixed bed reactor ~ 1000°C, 30 atm	Pyrolysis, steam reforming, partial oxidation	CH_4, H_2, CO, CO_2
Steam reforming	Natural gas; (naphtha)	Supported Ni, alkali promoter	700–830°C 15–40 atm	$CH_4 + H_2O \rightleftharpoons CO + 3H_2$	CO, H_2 (CH_4, CO_2)
Partial oxidation	Hydrocarbons including heavy fuel residues; H_2O. O_2.	None?	1200–1500°C 30–80 atm		CO, H_2 (CH_4, CO_2)
ADDITIONAL PROCESSES FOR FUEL OILS					
Catalytic reforming	Naphtha b.p. 30–190°C: H_2 (6:1)	0.3–0.5% Pt on γ-Al_2O_3 promoted by 1% chlorine.	500–525°C 23.3 atm	Dehydrogenation Isomerization (Hydrocracking) Carbonium ion reactions (acidic sites)	Aromatics Alkanes
Catalytic cracking	Fractions b.p. 220–540°C	Zeolite or Al_2O_3 + SiO_2	500–550°C 1 atm	Carbonium ion reactions as in cat. cracking.	Gasoline (high olefin content)
Hydrocracking	Fractions b.p. 165–575°C/H_2.	e.g. Ni/Mo or Co/Mo as sulphides on silica/alumina. Pd/zeolite for hydrocracking.	316–482°C 68–204 atm	i) Carbonium ion reactions as in cat. cracking. ii) Hydrogenation of aromatics and alkenes.	Wide range of hydrocarbon fractions available.

12.2 Resources

Before 1950, about 80% of all organic chemicals were obtained either by fermentation or from coal. Since then there has been a shift to oil, which now provides the raw material for about 90% of such products. The primary use for oil is as a fuel, particularly for transport. Only 5–10% of crude oil is used for chemical manufacture. The lighter oil fractions provide the main feedstocks. These comprise light petroleum gases (ethane, propane and butanes) and naphtha (b.p. 30–190°C) which consists of paraffins, alicyclics (naphthenes) and aromatics.

The reactions discussed below can be divided into two very broad groups. The first group is based on alkenes which are produced by steam cracking of hydrocarbons. The second starts out from synthesis gas, which is essentially a mixture of carbon monoxide and hydrogen. The main methods used to obtain alkenes and synthesis gas are summarized in Table 12.2.

In the production of fuel oils, heavy petroleum fractions may be converted into lower molecular weight products by catalytic cracking over a zeolite catalyst. When this process is carried out under a high pressure of hydrogen (hydrocracking) saturated hydrocarbons are produced from the intermediate aromatics and alkenes. On the other hand, motor gasoline of high octane number contains a high proportion of aromatics. The content of aromatics may be increased by catalytic reforming over supported platinum catalysts. These processes are included in Table 12.2, but it is not economic to apply them to naphtha or heavy petroleum fractions to produce lighter fractions for chemical feedstocks.

12.3 Catalysis of reactions of alkenes by transition metal complexes

12.3.1 *Production of alkenes—steam cracking*

Only a limited number of building blocks are required by the large scale chemical industry. Of these ethene occupies a prime position. It is the principal feedstock for the manufacture of about 30% of all petrochemicals. In 1985 about 13.8 m tonnes were produced in the USA alone. Of this 44% was converted into polyethene, 18% into ethene oxide and 14% into vinyl chloride (for PVC). Ethene oxide is the precursor to 1, 2-ethane diol (ethylene glycol) which is used as an antifreeze and as a component for polyester production. These figures emphasize the dominant position of polymers in the petrochemical industry. Some of the processes, described below, which employ homogeneous catalysts are carried out on a much smaller scale and may represent only a small fraction of the total usage of chemical feedstocks.

Ethene is manufactured by steam cracking. Ethane is the feedstock of choice in the United States. ($C_2H_6 \rightarrow C_2H_4 + H_2$). The Fife ethene plant at Mossmorran which came into production in 1986 uses ethane piped from the Brent and

Mechanisms of industrial processes

Fulmer fields. In W. Europe and Japan, however, ethane is not available in sufficient quantity and naphtha has to be used. In the steam cracking of naphtha alkyl radicals are first formed by homolysis of C—C bonds and these then decompose to alkenes and smaller radicals.

Initiation $\quad RCH_2CH_2CH_2CH_2R' \longrightarrow RCH_2\dot{C}H_2 + R'CH_2\dot{C}H_2$

Propagation $\quad RCH_2\dot{C}H_2 \longrightarrow R^\cdot + H_2C{=}CH_2$

$\quad H_3\dot{C}CHCH_2CH_2CH_2R \longrightarrow H_3CCH{=}CH_2 + RCH_2\dot{C}H_2$

Termination

(Radical combination) $\quad R^\cdot + R'' \longrightarrow R{-}R'$

(Disproportionation) $\quad 2RCH_2\dot{C}H_2 \longrightarrow RCH_2CH_3 + RCH{=}CH_2$

Typically steam cracking of naphtha produces ethene (32%), propene (13%) and butadiene (4.5%). By-products consist of aromatics (13.5%), methane, hydrogen, gasoline and fuel oil (total 37%). The exact proportions are a function of temperature. A process by which the olefins are separated is described on p. 89.

12.3.2 Isomerization of alkenes

Many transition metal complexes which catalyse reactions such as hydrogenation of alkenes are also able to bring about their isomerization. This can sometimes be used to good effect. It can also be a nuisance, leading to a mixture of products. The possibility of isomerization must be borne in mind whenever transition metal catalysts are employed.

Two general mechanisms have been recognized. In the first, a metal hydride is an essential intermediate. This can, in principle, add to the alkene either

$$[M]{-}H + RCH_2CH{=}CH_2 \; \underset{}{\overset{\substack{\text{Markovnikov}\\\text{addition}}}{\rightleftharpoons}} \; \underset{\underset{[M]}{|}}{RCH_2CHCH_3} \; \underset{}{\overset{\substack{\beta\text{-H}\\\text{transfer}}}{\rightleftharpoons}} \; RCH{=}CHCH_3$$

$$+ [M]{-}H$$

$$\underset{}{\overset{\substack{\text{anti-Markovnikov}\\\text{addition}}}{\rightleftharpoons}} \; RCH_2CH_2CH_2\text{-}[M]$$

according to Markovnikov's rule, or in the opposite sense. Markovnikov addition is favoured where the metal—hydrogen bond is polarized $\overset{\delta-}{M}{-}\overset{\delta+}{H}$ and where steric hindrance from bulky ligands is low. Isomerization of a terminal alkene to a thermodynamically more stable mixture of internal alkenes can then occur. Note that for a terminal alkene, anti-Markovnikov addition cannot lead to isomerization as the only possible reversal regenerates starting material. This does not apply for an internal alkene. Firm evidence that this mechanism operates in the case of NiL_4/CF_3COOH, $L = P(OEt)_3$, comes from studies on deuteriated 1-pentene. The active catalyst is probably the 16-electron species $HNiL_3^+$. This has a vacant site which can coordinate alkene. The cycle (Fig. 12.1) is drawn on the basis of Tolman's 16/18-electron rule (p. 183). In such cycles it is not necessary

362

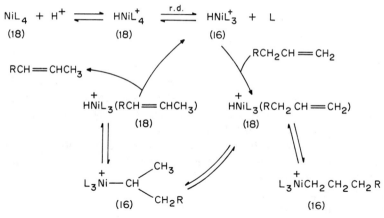

Fig. 12.1 Catalytic cycle for isomerization of a terminal alkene by a 1,2-hydrogen shift. The figures in parentheses (16), (18) indicate the electronic configuration of the metal. r.d. = rate determining step.

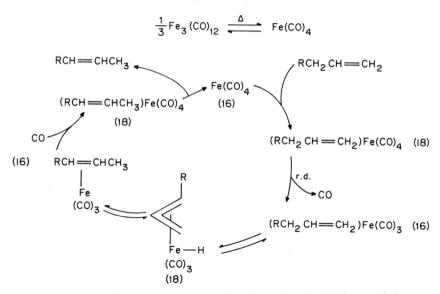

Fig. 12.2 Catalytic cycle for isomerization of an alkene by a 1,3-hydrogen shift.

that all species have an independent existence. Sometimes two consecutive steps could be linked by a synchronous process so that the species shown may represent a transition state rather than an identifiable intermediate.

The second mechanism is followed, for example, by $Fe_3(CO)_{12}$. It involves an η^3-allyl metal hydride (Fig. 12.2). The rate determining step is thought to be loss of CO from the 18-electron species $Fe(CO)_4$(alkene).

The η^3-allyl mechanism probably operates also in alkene isomerizations catalysed by bis(benzonitrile)dichloropalladium.

Mechanisms of industrial processes

12.3.3 *Hydrocyanation of ethene and of butadiene*

The 16-electron complex NiL_3 (L = tris-*o*-tolylphosphite) catalyses the addition of hydrogen cyanide to ethene and to 1, 3-butadiene below 25°C. A catalytic cycle for ethene based on the alternant formation of 16- and 18-electron intermediates is shown in Fig. 12.3. The inner route is considered to be the more likely, but the outer one cannot be excluded. If DCN is used instead of HCN, deuterium appears at both the α and β positions in the resulting CH_3CH_2CN. This shows that the addition of Ni—H (Ni—D) to C_2H_4 is reversible.

The Du Pont process for the manufacture of adiponitrile (1, 4-dicyanobutane) from butadiene depends on nickel catalysed hydrocyanation. Adiponitrile is the precursor to 1, 6-diaminohexane which is used to make Nylon 6, 6. The other component, adipic acid (1, 6-hexanedioic acid) is usually made by oxidation of cyclohexane, although in principle it could also be obtained from the dinitrile. The

$$HO_2C(CH_2)_4CO_2H + H_2N(CH_2)_6NH_2$$
$$\longrightarrow \cdots \bar{O}_2C(CH_2)_4\bar{C}O_2 \cdots H_3\overset{+}{N}(CH_2)_6\overset{+}{N}H_3 \cdots \bar{O}_2C(CH_2)_4CO_2 \cdots$$
Salt

$$\xrightarrow[-H_2O]{270°C} \sim -\overset{O}{\overset{\|}{C}}(CH_2)_4\overset{O}{\overset{\|}{C}}NH(CH_2)_6\overset{O}{\overset{\|}{N}}HC(CH_2)_4\overset{O}{\overset{\|}{C}}NH \sim \sim \quad \text{(Nylon 6, 6).}$$

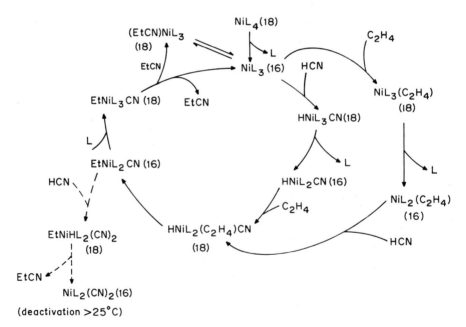

Fig. 12.3 Catalytic cycle for hydrocyanation of ethene.

364

commercial process, which has been in operation for about 15 years, occurs in three stages.

(i) Formation of mononitriles

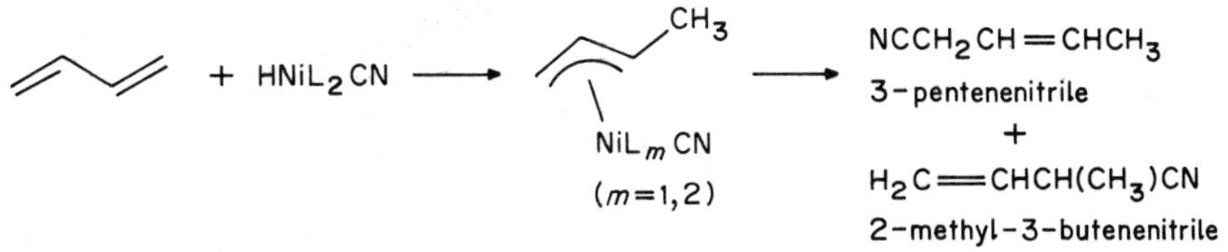

NCCH$_2$CH=CHCH$_3$
3-pentenenitrile
+
H$_2$C=CHCH(CH$_3$)CN
2-methyl-3-butenenitrile

(ii) Isomerization of the branched mononitrile into the linear 3-pentenenitrile using a similar NiL$_4$⇌NiL$_3$ system.
(iii) Hydrocyanation of 3-pentenenitrile.

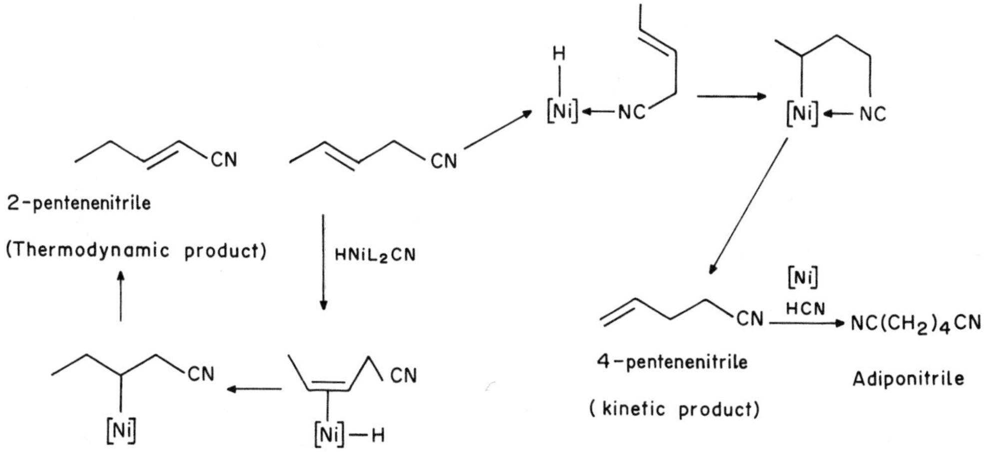

([Ni] means nickel + unspecified ligands)

3-Pentenenitrile is first rapidly isomerized by the nickel catalyst to 4-pentenenitrile. Addition of hydrogen cyanide then affords adiponitrile as the major product. This is fortunate, as the unwanted conjugated 2-pentenenitrile is thermodynamically more stable than either the 3- or the 4-isomer. The favourable initial course of the isomerization is thought to be controlled by the formation of a cyclic intermediate or transition state in which the nitrile group coordinates to nickel.

12.3.4 Dimerization of ethene

While investigating the oligomerization of ethene by aluminium alkyls (p. 80), Ziegler and coworkers noticed that in the presence of traces of nickel dimerization to 1-butene occurs instead. Very active catalysts can be generated by adding an alkylaluminium halide to [(η^3-C$_3$H$_5$)NiCl]$_2$. The phosphine complexes NiL$_4$ are also effective in the presence of acids. A nickel hydrido complex is first formed which adds to ethene to give an ethyl intermediate which can either eliminate 1-

365

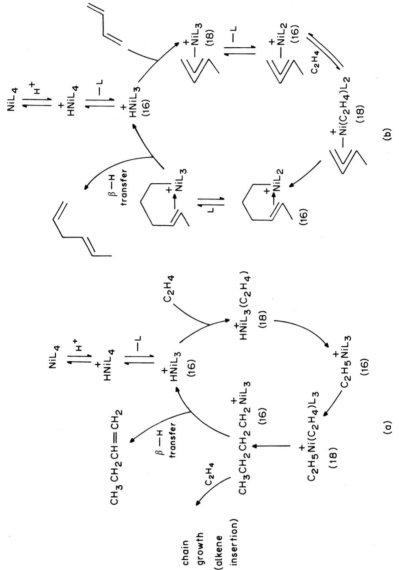

Fig. 12.4 Postulated catalytic cycles for (a) dimerization of ethene to 1-butene and (b) codimerization of 1, 3-butadiene and ethene to give *trans*-1, 4-hexadiene.

butene or add a second ethane molecule leading to chain growth. The relative importance of these two paths can be controlled by changing the phosphine ligands. In the system $[(\eta^3\text{-}C_3H_5)NiCl]_2/EtAlCl_2/L$ at $-20°C/1$ atm, for instance, only 1-butene is formed when $L = Me_3P$. Bulky phosphines hinder β-hydrogen transfer more than alkene insertion. When $L = P(\text{cyclohexyl})_3$, 70% butene, 25% hexene and 5% higher oligomers are formed. In the extreme case of $L = PBu^t_3$, olefin elimination is essentially repressed, so that polyethene is produced.

Selective codimerization of ethene and 1, 3-butadiene is also possible because the diene first forms an η^3-allyl complex with the metal centre (Fig. 12.4). An industrial process along these lines has been developed in the USA by Du Pont. A rhodium catalyst is employed; rhodium trichloride in ethanol itself gives 80% selectivity towards the desired product, *trans*-1, 4-hexadiene, which is used in the manufacture of an ethene–propene–hexadiene synthetic rubber. The diene introduces some double bonds into the polymer chains which are required for vulcanization.

12.3.5 *Oligomerization of butadiene*

Wilke and his coworkers at the Max Planck Institut für Kohlenforschung at Mülheim, Ruhr, have extensively studied oligomerization reactions of butadiene using a wide range of transition metal catalysts. Those based on nickel and on palladium are the most useful. Nickel systems generally afford cyclic products, whereas linear materials are obtained using palladium.

The nickel catalyst is obtained by reduction of a nickel(II) compound such as the acetylacetonate in the presence of butadiene, or simply by exchange of butadiene and bis(η^3-allyl)nickel or bis(cycloocta-1, 5-diene)nickel. In the absence of added ligands, butadiene is rapidly and catalytically converted at low temperatures into 1, 5, 9-cyclododecatriene, mainly the *trans, trans, trans* isomer. The trimerization bears the mark of a 'template' process, three butadiene molecules linking very specifically around a 'naked nickel' centre. Detailed studies show that the process is not synchronous but occurs in a stepwise fashion (Fig. 12.5).

The $C_{12}\eta^2$, η^3, η^3-intermediate has been isolated from the reaction mixture at $-40°C$. The ^1H n.m.r. spectrum shows that in solution two isomers are present in which the allyl groups bear *anti* substituents. With PMe_3, an adduct is formed in which isomerization to *syn* substituents has occurred, presumably *via* an η^1-intermediate. This isomerization explains why the *trans, trans, trans* isomer of cyclododecatriene is the major product. Cyclododecatriene is used as a precursor to Nylon 12.

When a tertiary phosphine or phosphite L is added in the ratio 1:1 (Ni:L), coordination of a third butadiene molecule is prevented, so that dimers rather than trimers result. The first key intermediate is an η^1, η^3-octadienyl complex. This gives rise to vinylcyclohexene (VCH). If L is weakly basic, X rearranges to an $\eta^3, \eta^3\text{-}C_8H_{12}$ species Y which leads to *cis*-1, 5-cyclooctadiene (COD) (Fig. 12.6).

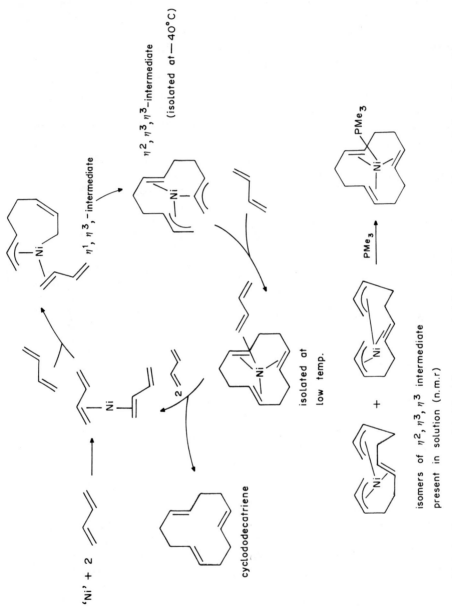

Fig. 12.5 Catalytic conversion of 1,3-butadiene into 1,5,9-cyclododecatriene using a 'naked nickel' catalyst.

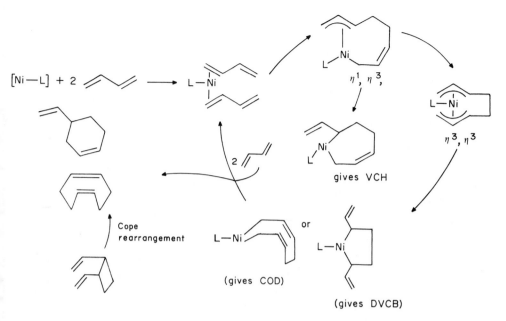

Fig. 12.6 Catalytic conversion of 1,3-butadiene into cyclic dimers. VCH = 4-vinylcyclohexene; COD = cis-1,5-cyclooctadiene; DVCB = cis-1,2-divinylcyclobutane. When L = P(OC$_6$H$_4$PH$_{-o}$)$_3$ at 60°C, COD = 94.5%; L = PPhEt$_2$, COD = 38.2%; DVCB = 12.6%; VCH = 46.0% of total products.

cis-Divinylcyclobutane is also formed with basic phosphines at low temperatures (≤ 60 °C). At higher temperature a nickel catalysed Cope rearrangement to 1,5-COD occurs.

Whereas nickel complexes catalyse the trimerization or dimerization of butadiene to cyclic products, palladium catalysts produce linear oligomers. When bis(dibenzoylacetone)palladium is treated with butadiene at low temperature, a yellow air sensitive complex (η^3, η^3-dodecatrienediyl)palladium is formed initially. It decomposes above − 20°C in the absence of butadiene and in its presence actively catalyses its trimerization to linear 1,3,6,10-dodecatetraene. In contrast to the corresponding nickel intermediate, the substituents on the allyl groups are *syn*. Treatment with a diphosphine affords a metallacycle which contains a huge 13-membered ring. If (A) or (B) is allowed to warm above − 20°C or is treated with ligands (Ph$_3$P or CO), dodecatetraene isomers are released (Fig. 12.7).

In the presence of added ligand (Pd:L = 1:1), 1,3,7-octatriene isomers are obtained (Fig. 12.8). The conformation of the C$_8$ chain in the intermediate (C) has been determined in solution by n.m.r. spectroscopy and also in the solid state (L = Me$_3$P) by X-ray diffraction. While the analogous nickel intermediate rearranges at room temperature, leading to cyclic products, the palladium compound does not. If the dimerization is carried out in the presence of active hydrogen

369

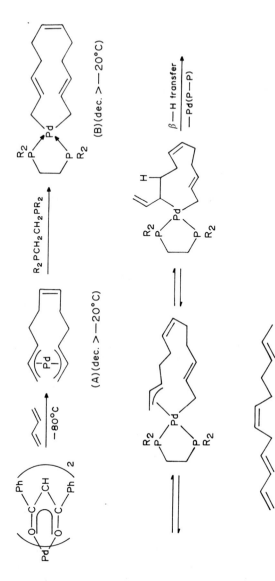

Fig. 12.7 Linear trimerization of 1,3-butadiene catalysed by palladium complexes.

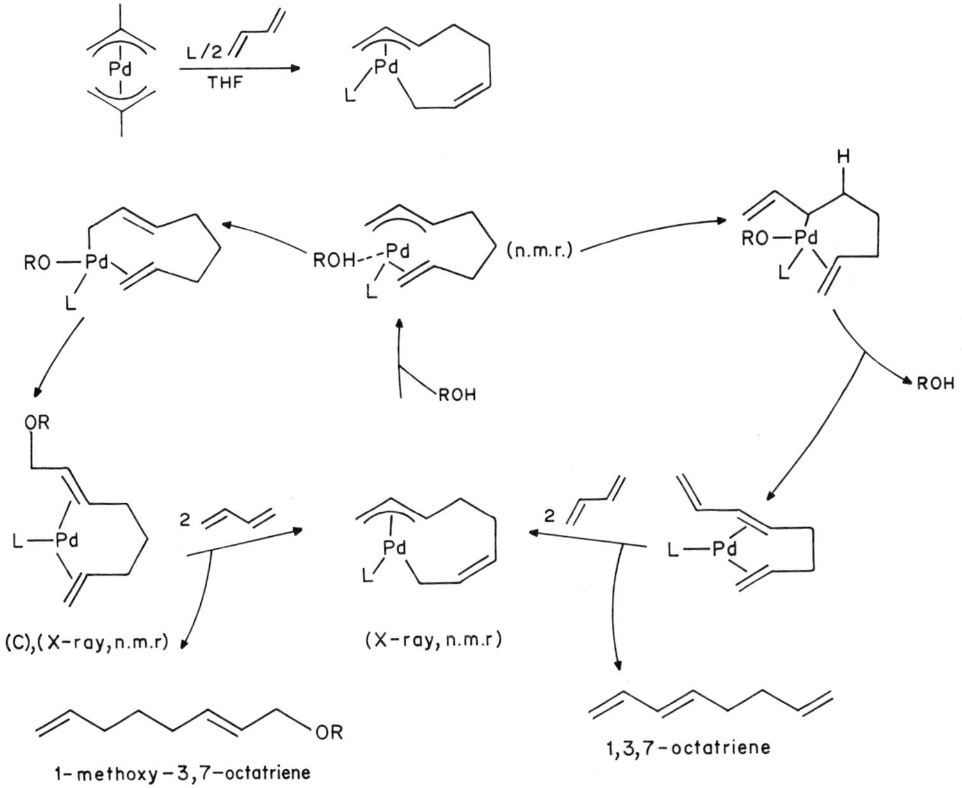

Fig. 12.8 Linear telomerization of 1, 3-butadiene, catalysed by palladium complexes.

compounds HX such as ROH, R_2NH, RCO_2H or $CH_2(COOEt)_2$, the major products are linear telomers. It does not appear that these reactions have been used industrially, although they have been included in some syntheses of natural products. The cycle shown in Fig. 12.8 accounts for the catalytic production of 1-methoxy-3, 7-octadiene from methanol and butadiene. Some of the intermediates have been characterized by n.m.r. (^{1}H, ^{13}C, ^{31}P) and where indicated, by X-ray diffraction.

12.3.6 Ziegler-Natta catalysis

While aluminium alkyls on their own convert ethene into oligomers (p. 80), in the presence of traces of nickel termination occurs after only one step, so that the dimer, butene, is produced. This observation led Ziegler to study the effect of compounds of other transition metals as cocatalysts. With titanium or zirconium compounds in particular, rapid polymerization of ethene occurs at 50°C/10 atm leading to a crystalline, high density polyethene (HDPE) which consists of long

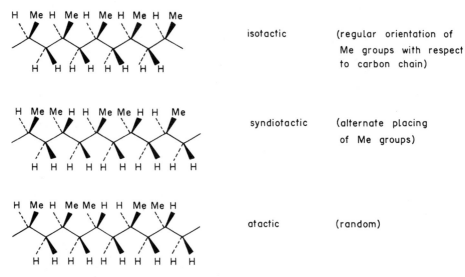

Fig. 12.9 Stereochemical arrangement of groups in polypropene.

chains with very few branches. This contrasts with low density polyethene (LDPE) produced by free radical initiated polymerization in a high pressure process. LDPE has a low degree of crystallinity because of extensive branching, typically 20–30 branches for every 1000 carbon atoms.

Natta, in collaboration with the Italian company, Montecatini, extended Ziegler's method to propene. He found that polymerization occurs almost entirely head to tail and that crystalline, stereoregular products can result. Using Et_3Al in conjunction with $TiCl_4$ (active species $TiCl_3$) isotactic polypropene is formed in which all the methyl groups are all on the same side of the carbon backbone (Fig. 12.9). With Et_2AlCl/VCl_4, syndiotactic polypropene results in which the configurations at the carbon atoms alternate along the chain.

Table 12.3. Consumption of low density polyethene (LDPE), high density polyethene (HDPE) and polypropene (PP) in 1985.

Polymer	Consumption (10⁶ tonnes)		Uses
	Continental W. Europe	USA	
LDPE	3.97	4.01	Bags, packaging
HDPE	1.76	2.90	Bottles, drums, film, pipes, cable insulation.
PP	2.02	2.32	Injection moulding and extrusion. Packaging, domestic appliances, automobiles, fibres.

The scale of consumption of polyethene and polypropene plastics in Western countries and some of the uses of these materials are listed in Table 12.3. For their contributions which have made such a mark on the plastics industry Karl Ziegler and Giulio Natta were jointly awarded the Nobel Prize for Chemistry in 1963.

All commercial catalysts for the production of HDPE and of PP are heterogeneous. Homogeneous systems are applied, however, in stereospecific polymerization of 1, 3-butadiene to give cis-1, 4-polybutadiene, which has properties similar to natural rubber. There have also been many laboratory studies of such homogeneous catalysts. Some of the most active are organolanthanides (p. 401). Cyclopentadienyl titanium complexes such as Cp_2TiCl_2 activated by aluminium alkyls are also effective. This work has contributed information about the mechanism of the polymerization. Even so, there is still much controversy about this; probably different pathways are followed, maybe simultaneously, in different systems. The most widely accepted mechanism is based on one put forward by Cossee in 1964 (Fig. 12.10). The early commercial Ziegler-Natta catalysts were prepared by treating $TiCl_4$ or $TiCl_3$ with an aluminium alkyl. Alkyl transfer to titanium and if necessary reduction to Ti(III) occurs initially. Polymerization apparently takes place at the edges of small $TiCl_3$ crystallites or at defects in the $TiCl_3$ structure. This may be responsible for producing a stereoregular product.

Catalysts more than a thousand times as active as the original Ziegler-Natta systems have now been developed. In one form magnesium chloride, which has a similar layer lattice to α-$TiCl_3$, is milled with ethyl benzoate and then treated with $TiCl_4$. The resulting solid retains about 1% titanium and ester; it is activated with Et_3Al. Polymerization of propene is carried out in the liquid phase under pressure (55°C/20 atm) and of ethene, in the gas phase. The reaction is strongly exothermic (p. 124). Such low concentrations of these modern supported catalysts are required that no removal of catalyst residues from the product is necessary. Moreover such highly crystalline polypropene is produced that there is no need to separate unwanted atactic material.

12.3.7 Olefin metathesis

Olefin metathesis or dismutation was discovered in the laboratories of Phillips Petroleum in the late 1950s. In one of the early experiments propene was passed over an alumina catalyst which had been pretreated with $Mo(CO)_6$. Ethene and 2-butene (cis and trans isomers) were obtained with negligible by-product formation. Without a catalyst, such reactions require temperatures above 700°C and do not occur cleanly.

$$2MeCH{=}CH_2 \underset{\text{catalyst}}{\overset{125°C/}{\rightleftharpoons}} MeCH{=}CHMe + H_2C{=}CH_2$$

Rapid development followed and the first commercial plant, the now defunct Triolefin Process, was opened in the mid 1960s near Montreal to deal with a local

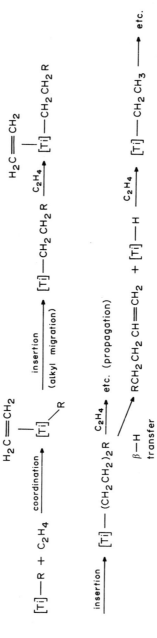

Fig. 12.10 Proposed mechanism for Ziegler-Natta polymerization of ethene (Cossee mechanism).

surplus of propene and a deficiency of polymerization grade ethene. Conversely a plant opened in 1985 in Channelview, Texas combines dimerization and metathesis to convert ethene into propene. The Shell Higher Olefins Process (p. 81) for generating C_{12}–C_{16} olefins from higher and lower fractions also involves a metathesis step. Neohexene, an intermediate in the manufacture of musk perfumes for soap, is obtained by co-metathesis of commercial diisobutene with ethene. The catalyst WO_3 on SiO_2/MgO functions both for isomerization and metathesis.

'Di isobutene'

$$Me_3CCH = CMe_2 \; + \; H_2C = CH_2 \; \rightleftharpoons \; Me_3CCH = CH_2 \; + \; H_2C = CMe_2$$

'neohexene' 'isobutene'

Isomerization   metathesis

$$Me_3CCHC(Me) = CH_2 \; + \; H_2C = CH_2 \; \rightleftharpoons \; \text{no net reaction}$$

Analogous homogeneous systems, however, give very similar results. Studies on the latter have thus shed much light on the mechanism of metathesis and incidentally have revealed some intriguing novel chemistry. Homogeneous metathesis catalysts resemble Ziegler-Natta systems in that they can be prepared from a transition metal halide (e.g. WCl_6, $MoCl_5$ or $ReCl_5$) and a Main Group alkyl (e.g. RLi, R_3Al or R_4Sn). A mechanism involving alkylidene intermediates, first proposed by Chauvin in 1970, is now well established. The more obvious but rather vague four centre mechanism has been effectively disproved.

Alkylidene mechanism:

RCH = M

R'CH = CHR" $\rightleftharpoons$ RCH — M

R'CH — CHR" $\rightleftharpoons$ RCH   M

R'CH   CHR" $\rightleftharpoons$

metallacyclobutane

Four-centre mechanism:

RCH   CHR"

M

R'CH   CHR''' $\rightleftharpoons$ RCH---CHR"

M

R'CH---CHR''' $\rightleftharpoons$ RCH = CHR"

M

R'CH = CHR'''

Cycloalkenes, such as cyclooctene, afford *linear* polymers on metathesis. This is in accord with the carbene mechanism, but not with the four-centre mechanism which predicts the formation of large rings (Fig. 12.11).

Cyclooctene is obtained by partial reduction of 1, 5-cyclooctadiene, which in turn is available from butadiene (p. 369). Co-metathesis between the polyoctene and *cis*-1, 4-polybutadiene gives a rubber which is very stable to heat, oxygen and light, is easily moulded and is readily combined with other rubbers by vulcanization.

375

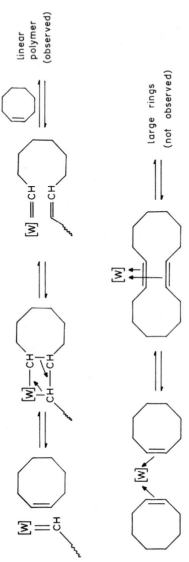

Fig. 12.11 Metathesis of cyclooctene to give a linear polymer (alkylidene mechanism) rather than large rings (four-centre mechanism).

Catalysis of reactions of alkenes by transition metal complexes

The evidence for alkylidene and metallacyclobutane intermediates in olefin metathesis is now overwhelming. Alkylidenes can arise by α-hydrogen transfer (p. 220). The tantalum alkylidenes $(Me_3CCH_2)_3Ta=CHCMe_3$ and $(Me_3CCH_2)TaCl_3(PMe_3)_2$ do react with alkenes, but their catalytic activity is short lived and there are side reactions. With ethene the former gives a metallacyclobutane, but this then rearranges by β-hydrogen transfer.

$$(Me_3CCH_2)_3Ta=CHCMe_3 \xrightarrow{\ C_2H_4\ } (Me_3CCH_2)_3Ta \overset{\displaystyle \underset{H}{\overset{CMe_3}{\diagdown C \diagup}}}{\underset{CH_2}{\diagdown \diagup CH_2}}$$

More successful are some oxo- and alkoxy derivatives of tungsten. One of these, $W(O)(=CHCMe_3)(PMe_3)_2Cl_2$, catalyses the metathesis of terminal and internal alkenes. Alkylidene complexes formed in the course of metathesis have been isolated (Fig. 12.12).

The tungsten alkylidyne complex $W(\equiv CBu^t)(OBu^t)_3$, which, if monomeric, is only 12-electron, catalyses the metathesis of 3-heptyne at a spectacular rate. Note that the presence of strongly π-donating alkoxy groups destabilizes some of the metal d-orbitals (p. 151) so that the 18-electron rule is no longer expected to hold. In this case the reactions are too fast to obtain evidence for tungstenacyclobutadiene intermediates. Another route to the catalysts, however, is provided by the remarkable reaction between the triply bonded tungsten alkoxides $W_2(OR)_6$ and alkynes.

$$(RO)_3W\equiv W(OR)_3 + RC\equiv CR \longrightarrow 2(RO)_3W\equiv CR$$

When R = 2, 6-$Me_2C_6H_3$—, tungstenacyclobutadiene complexes in which the metal atom adopts an approximately trigonal bipyramidal geometry, have been isolated from the reaction with 3-heptyne. X-ray crystallography shows that the four membered WC_3 ring is essentially planar. On account of the low coordination number (five) and low electron count (12), vacant sites are available at the metal centre for attachment of alkyne molecules. These properties are consistent with the high activity of these catalysts.

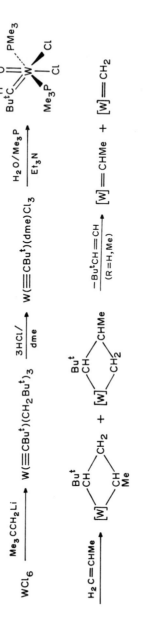

Fig. 12.12 Preparation of an oxo-tungsten complex, an active catalyst for metathesis of alkenes.

A related tungsten complex $(R'O)_2W(=NR)(=CHBu^t)$ is an exceptionally active catalyst for alkene metathesis. Its activity is unchanged over periods of at least 1 day and the turnover is in hundreds of conversions per minute. A trigonal bipyramidal tungstenacyclobutane intermediate is again involved.

Some alkene metathesis catalysts, notably WCl_6 activated by Me_4Sn, are tolerant of many functional groups such as —CO_2R, —COR, —CN, —halogen or —$OSiMe_3$. Fischer carbenes such as $Ph(MeO)C=W(CO)_5$ in the presence of Lewis acids are also useful. This has been exploited in the synthesis of long chain dicarboxylic esters from, for example, methyl oleate.

$$CH_3(CH_2)_7CH=CH(CH_2)_7CO_2Me \xrightleftharpoons[110°C]{WCl_6/Me_4Sn} CH_3(CH_2)_7CH=CH(CH_2)_7CH_3$$

$$+ MeO_2C(CH_2)_7CH=CH(CH_2)_7CO_2Me$$

Dicarboxylic esters are used to prepare polyesters and polyamides. The residual double bonds permit cross-linking of the polymer chains. Cross metathesis of methyl oleate with excess ethene affords methyl 9-decenoate, a key intermediate in the synthesis of 'queen bee substance', a honey bee pheromone.

$$CH_3(CH_2)_7CH=CH(CH_2)_7CO_2Me \rightleftharpoons H_2C=CH(CH_2)_7CH_3$$
$$+$$
$$+ H_2C=CH_2 \qquad H_2C=CH(CH_2)_7CO_2Me \rightarrow$$
$$\rightarrow MeCO(CH_2)_5CH\overset{tr}{=}CHCO_2Me$$

The compositions of classical Ziegler-Natta and metathesis catalysts are very similar. It has therefore been suggested by Green and also by Schrock that Ziegler-Natta polymerization might occur *via* alkylidene intermediates. The tantalum complex $Ta(CHCMe_3)H(PMe_3)_3I_2$, for example, in toluene converts ethene into a living polymer.

Mechanisms of industrial processes

When this mechanism was proposed, there was no clear example of insertion of an alkene into a transition metal alkyl bond, which is one of the key steps in the Cossee mechanism. Many complexes CpNiR(alkene) have been prepared, in which alkyl and alkene substituents coexist. When R = Ph, however, insertion does occur.

$$Cp_2Ni + RLi + C_2H_4 \xrightarrow[-30°C]{-70° \text{ to}} CpNiR(C_2H_4) \xrightarrow[R=Ph]{C_2H_4} CpNi \begin{smallmatrix} CH_2CH_2Ph \\ \| \end{smallmatrix}$$

Further evidence for insertion has also come from studies on a similar cobalt complex using deuterium labelling.

12.3.8 *Acetaldehyde and acetic acid*

Acetic acid is produced on a very large scale. In Western Europe alone the capacity is about 1.5 m tonnes per annum. Of this about one third is used to make polyvinyl acetate and its hydrolysis product, polyvinyl alcohol. Another 20% gives acetic anhydride which is required in the manufacture of cellulose acetate fibres and plastics. Acetate esters are important industrial solvents. The methods adopted for acetic acid manufacture have changed considerably over the past 100 years, reflecting the cost and availability of raw materials and developments in technical knowledge (Table 12.4). During the 1950s the chemical industry changed from coal to oil as its primary resource. To meet these changes the Celanese process, based on oxidation of butane, was developed. Although cheap it showed low selectivity. The Wacker process, widely used in the 1960s and 1970s, but now obsolescent, depends on the oxidation of ethene by palladium compounds. It has now been largely superseded by the Monsanto process which takes methanol as feedstock. The main reasons for the decline in the Wacker process are the increased cost of naphtha and the extra oxidation step, acetaldehyde to acetic acid, which is required. If the two routes are compared starting from hydrocarbon feedstocks, however, they have the same number of steps.

$$\text{Wacker: } C_2H_6 \longrightarrow C_2H_4 \longrightarrow CH_3CHO \longrightarrow CH_3CO_2H$$
$$\text{Monsanto: } CH_4 \longrightarrow CO + H_2 \longrightarrow CH_3OH \longrightarrow CH_3CO_2H$$

(a) THE WACKER PROCESS. The overall reaction $C_2H_4(g) + O_2(g) \rightarrow CH_3CHO(g)$ $\Delta H^\ominus = -58.2 \text{ kJ mol}^{-1}$ can be split into three parts.

(i) Product formation: $C_2H_4 + H_2O + PdCl_2 \rightarrow CH_3CHO + Pd + 2HCl$
(ii) Oxidation of palladium(0) by copper(II): $Pd + 2CuCl_2 \rightarrow PdCl_2 + 2CuCl^*$
(iii) Reoxidation of copper(I) to copper(II) by oxygen: $2CuCl + 2HCl + \frac{1}{2}O_2$ $\rightarrow CuCl_2 + H_2O$.

*PdCl$_2$, CuCl and CuCl$_2$ are present as complex anions.

380

Table 12.4. Manufacture of acetic acid over the past 100 years

Year	Raw material	Route	Comments
19th Century	Carbohydrate	$\xrightarrow{enzyme}$ CH$_3$CH$_2$OH $\xrightarrow[O_2]{enzyme}$ CH$_3$CO$_2$H	Still used to produce vinegar for food.
1930	Coal $\longrightarrow$ CaC$_2$	$\xrightarrow{H_2O}$ HC≡CH $\xrightarrow[Hg^{2+}]{H_2O/}$ CH$_3$CHO $\xrightarrow{O_2}$ CH$_3$CO$_2$H	First synthetic route
	Oil $\longrightarrow$ C$_2$H$_4$	$\xrightarrow{H_2O}$ C$_2$H$_5$OH $\xrightarrow[Ag/250°C]{Cu\ or}$ CH$_3$CHO $\xrightarrow{O_2}$ CH$_3$CO$_2$H	
1955	Oil $\longrightarrow$ Butane	$\xrightarrow{O_2}$ CH$_3$CO$_2$H + by-products	Celanese process
1962	Oil $\longrightarrow$ C$_2$H$_4$	$\xrightarrow[Pd, Cu]{O_2}$ CH$_3$CHO $\xrightarrow{O_2}$ CH$_3$CO$_2$H	Wacker process
1966	Synthesis gas(H$_2$ + CO)	$\xrightarrow[cat]{Cu}$ CH$_3$OH $\xrightarrow[Co, 500\ atm]{CO/}$ CH$_3$CO$_2$H	BASF process
1970	"	$\xrightarrow[20\ atm]{CO/Rh}$ CH$_3$CO$_2$H	Monsanto
1990?	Synthesis gas (H$_2$ + CO)	$\xrightarrow[40\ atm]{Rh\ cat.}$ CH$_3$CO$_2$H	

Mechanisms of industrial processes

Table 12.5. Conditions for single stage and two stage Wacker oxidations.

	Temp (°C)	Press (atm)	Oxidant	% Yield	% Selectivity	% C_2H_4 Conversion	Recycle
Single stage	130	3	O_2	95	98	40	C_2H_4 Liquid phase
Two stage	130	8–10	Air	95	?	100	

Composition of solution: $CuCl_2.2H_2O$, 200 g dm^{-3}; $PdCl_2$, 4 g dm^{-3}; CH_3CO_2H, 50 g dm^{-3} (solvent for C_2H_4); rest water.

The process can be operated either in one or two stages (Table 12.5). The first method requires pure oxygen which has to be separated from air by liquefaction. Although this is expensive, it is exclusively adopted in the West because very pure ethene is available, permitting recycling of the gas phase. The palladium/copper catalyst is continuously regenerated by oxygen while acetaldehyde is removed. One reason why this homogeneous process has been so viable commercially is that the product acetaldehyde is very volatile and is therefore easily separated from the solution of catalyst.

A mechanism has to account for the following observations.

(i) The rate is given by the expression

$$\frac{-d[\text{alkene}]}{dt} = \frac{k[PdCl_4^{2-}][\text{alkene}]}{[Cl^-]^2[H_3O^+]}$$

(ii) 1 mol of ethene is absorbed for each mol of palladium(II) which is reduced to palladium(0).

(iii) Ethanol is not an intermediate, as it is oxidized more slowly than ethene under the conditions of the Wacker reaction.

(iv) Vinyl alcohol is not an intermediate. If the reaction is carried out in D_2O, the CH_3CHO produced contains no deuterium.

(v) The rates of oxidation of C_2H_4 and C_2D_4 are very similar. This means that C—H bond breaking is unimportant in the rate determining step.

The following mechanism has been proposed:

$$PdCl_4^{2-} + C_2H_4 \rightleftharpoons Pd(C_2H_4)Cl_3^- + Cl^-$$
$$Pd(C_2H_4)Cl_3^- + H_2O \rightleftharpoons trans\,Pd(C_2H_4)(H_2O)Cl_2 + Cl^-$$

The anion $Pd(C_2H_4)Cl_3^-$ is the palladium analogue of Zeise's anion (p. 196). Substitution of Cl^- by water occurs *trans* to ethene rather than to chlorine, consistent with the greater *trans* effect of the former. The next step is thought to be nucleophilic attack of water on the ethene complex. Hydroxide cannot be the nucleophile as its concentration is too low under the prevailing acid conditions to account for the observed rate. There is still some controversy whether the

382

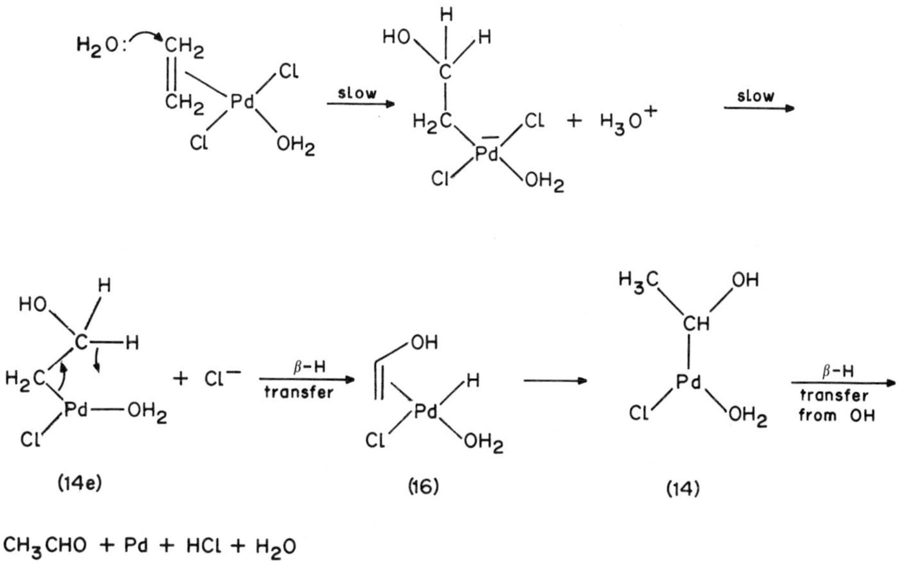

$CH_3CHO + Pd + HCl + H_2O$

Fig. 12.13 Suggested mechanism for the oxidation of ethene to acetaldehyde (Wacker process).

nucleophile attacks from outside the coordination sphere or whether initial attachment to palladium occurs. It has been clearly demonstrated, however, that at low temperatures amines or carbon nucleophiles attack η^2-alkenes *trans* to palladium. The weight of evidence seems to support the interpretation given here. Following loss of chloride ion to give a 14-electron intermediate, a series of fast rearrangements lead to the formation of acetaldehyde and palladium(0).

The major route for the industrial production of vinyl acetate, the monomer of polyvinyl acetate (emulsion paints, adhesives) and its hydrolysis product, polyvinyl alcohol (textiles, food packaging) is closely related to the Wacker acetaldehyde process, but the industrial catalysts are heterogeneous. A mixture of ethene, oxygen and acetic acid is passed over a palladium catalyst supported on alumina at 100–200°C. The overall reaction is $H_2C{=}CH_2 + CH_3CO_2H + \frac{1}{2}O_2$ $\rightarrow H_2C{=}CHCO_2CH_3 + H_2O$. Ethene is no longer cheap, so that work is being pursued to make vinyl acetate from synthesis gas (p. 384).

12.4 Chemistry based on synthesis gas

'Synthesis gas' (syn. gas) is a mixture of hydrogen and carbon monoxide. It may also include some carbon dioxide. Natural gas (methane) can be converted into syn. gas by reaction with water (steam reforming) and/or by partial oxidation. There are also several commercially proven processes for the gasification of coal using steam and oxygen (air?), which yield mixtures of H_2, CO and CO_2, sometimes with some methane. Of these the Lurgi process uses a fixed bed reactor

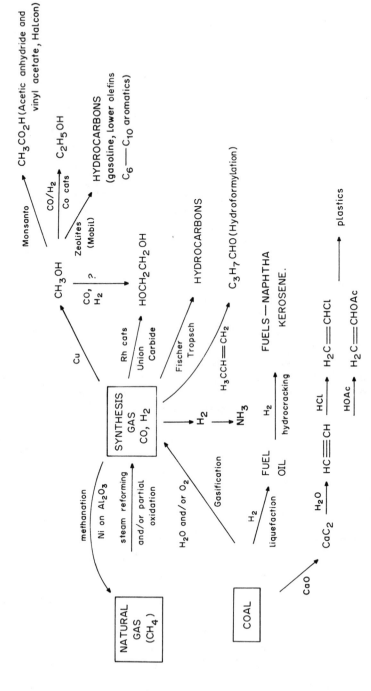

Fig. 12.14 Chemicals from synthesis gas.

at 1000°C and 30 atm. In Britain steam reforming of naphtha preceded that of natural gas but since the development of North Sea supplies the latter has become the preferred feedstock.

There is not space to discuss the details of syn. gas production here. The first reaction (steam reforming) is highly endothermic and involves an increase in entropy. It is carried out at high temperatures and moderate pressures (700–830°C/15–40 atm) over a nickel catalyst supported on α-alumina or calcium aluminate and promoted by alkali metal oxides, which help to prevent the deposition of carbon.

$$CH_4(g) + H_2O(g) \rightleftharpoons CO(g) + 3H_2(g) \quad \Delta H^\ominus = +206 \, kJ \, mol^{-1}$$
$$\Delta S^\ominus = +214 \, JK^{-1} \, mol^{-1}$$

It is important to be able to control the ratio of carbon monoxide to hydrogen. This may be done via the 'water gas shift' equilibrium

$$CO(g) + H_2O(g) \rightleftharpoons CO_2(g) + H_2(g) \quad \Delta H^\ominus = -41 \, kJ \, mol^{-1}$$
$$\Delta S^\ominus = -31 \, JK^{-1} \, mol^{-1}$$

This moves to the right as the temperature is lowered. Catalysts based on iron and chromium oxides (Fe_3O_4 on Cr_2O_3) require temperatures of 400–500°C, leaving 2–4% CO in the reactant gas. More active catalysts (copper and zinc oxides) can be operated at 190–260°C but are easily poisoned.

A summary of some of the chemicals which can be manufactured from syn. gas is given in Fig. 12.14.

12.4.1 The Monsanto acetic acid process

Methanol is manufactured from syn. gas by passing the latter at 230–270°C/50–100 atm over a copper oxide catalyst (CuO.ZnO or CuO.ZnO.Al_2O_3) (ICI Low Pressure Methanol process). (The copper oxide is reduced to metal.)

$$CO(g) + 2H_2(g) \rightleftharpoons CH_3OH(g) \quad \Delta H^\ominus = -92 \, kJ \, mol^{-1}$$

When syn. gas rich in hydrogen is used (the ratio H_2:CO from the steam reforming of methane is 3:1), the balance is restored by adding carbon dioxide.

$$CO_2(g) + 3H_2(g) \rightleftharpoons CH_3OH(g) + H_2O(g) \quad \Delta H^\ominus = -50 \, kJ \, mol^{-1}$$

Alternatively CO_2 can be introduced, along with steam, at the reforming stage.

$$CH_4 + CO_2 \rightleftharpoons 2CO + 2H_2.$$

Methanol is now the feedstock for production of acetic acid. The reaction $CH_3OH + CO \rightarrow CH_3COOH$ is homogeneously catalysed by soluble rhodium complexes. The development of the process from the laboratory bench to full industrial production was amazingly rapid. Only three years, from 1967 to 1970, were required. Very thorough studies were made of the reaction mechanism during this time and the information thus gleaned was used to optimize conditions during development of the process. This was a new approach for industrial laboratories in 1967, but has now become standard practice.

The process is operated at 175°C/13–25 atm. Methanol dissolved in aqueous acetic acid is treated with carbon monoxide in the presence of a soluble rhodium compound and an iodide. The nature of the rhodium compound is not critical; after an induction period in which conversion to the active complex occurs, similar activity is attained. The selectivity to acetic acid is 99% based on methanol and 90% on CO. The major side reaction is the water gas shift $CO + H_2O \rightleftharpoons CO_2 + H_2$. On thermodynamic grounds the reaction $CH_3OH + CO + 2H_2 \rightarrow CH_3CH_2OH$ should be favoured, but the catalyst is very specific in promoting only the desired conversion into acetic acid.

Kinetic studies produced some surprising results. The rate of carbonylation is *independent* not only of the concentration of methanol but also of the pressure of carbon monoxide. The reaction, however, is first order in each of the species rhodium and iodide (rate α [Rh][I$^-$]). Bands at 2064 and 1984 cm^{-1} in the infrared spectrum, which correspond to the known ion [Rh(CO)$_2$I$_2$]$^-$ shift to 2062 and 1711 cm^{-1} on addition of methyl iodide. The band at 1711 cm^{-1} suggests the formation of an acetyl complex by methyl migration (p. 225). Under

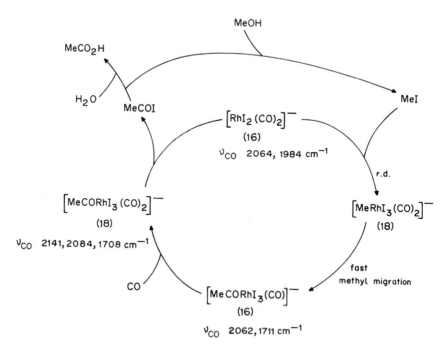

Fig. 12.15 Catalytic cycle for carbonylation of methanol (Monsanto process).

CO these bands change to peaks at 2141(w), 2084(vs) and 1708(s) which are assigned to an acetyldicarbonyl species. A cycle which accommodates these findings is given in Fig. 12.15. The rate determining step is the addition of methyl iodide to $[Rh(CO)_2I_2]^-$. The immediate product of this addition has never been observed, on account of its rapid conversion into the acetyl. The final step is elimination of acetyl iodide. This reacts with water to give acetic acid or with methanol forming methyl acetate. Hydrolysis of the ester takes place as methanol is consumed.

The process has been modified to yield acetic anhydride. Here the last step is

$$[MeCORhI_3(CO)_2]^- + MeCOOH \rightarrow (MeCO)_2O + [RhI_2(CO)_2]^- + HI.$$

An important goal is a single stage process for making acetic acid directly from syn. gas. There are indications that this is a real possibility in the near future. A homogeneous process to make ethylene glycol (ethane 1, 2-diol) has been developed by Union Carbide. Rhodium carbonyl clusters are probably implicated. Unfortunately very high pressures, 1000 atm, are required. The construction of C_2 and longer units directly from syn. gas tends to be unselective. The very first step

$$MH(CO) \rightleftharpoons MCHO \xrightarrow{H_2} \begin{matrix} H \\ | \\ M-O \\ \;\;\; \| \\ C \\ / \backslash \\ H\;\;H \end{matrix}$$

is usually strongly disfavoured thermodynamically and formation of CH_2O (methanal) becomes rate limiting. Even at high temperatures and pressures commercially useful rates are not normally achieved. Under such forcing conditions the reaction becomes unselective, following many concurrent pathways.

The Monsanto process affords acetic acid even in the presence of hydrogen. Modification of ligands or introduction of a second catalytic centre could in principle permit hydrogenation of the acyl intermediate rather than its solvolysis. In this way ethanal or ethanol are possible products.

12.4.2 Hydroformylation

(a) COBALT CATALYSTS. Production of aldehydes by catalytic addition of carbon monoxide and hydrogen to alkenes is known as hydroformylation. Formally, hydrogen and a formyl group (H and CHO) are added across the double bond. Aldehydes are used to make acids by oxidation, amines by reductive amination, polyols by reaction with methanal and alcohols by reduction. Butanal (n-butyraldehyde) is a precursor to 2-ethylhexanol and hence to di-iso-octyl phthalate, which is used as a plasticizer for PVC.

$$H_3CCH_2CH_2CHO \xrightarrow{OH^-} H_3CCH_2CH_2\overset{\overset{\displaystyle OH}{|}}{C}HCHCHO \xrightarrow{-H_2O} H_3CCH_2CH_2CH=CCHO \xrightarrow{H_2}$$
$$\underset{|}{\qquad\qquad\qquad\qquad\qquad C_2H_5} \qquad\qquad\qquad \underset{|}{\qquad C_2H_5}$$

$$C_4H_9\overset{\overset{\displaystyle }{|}}{C}HCH_2OH \longrightarrow \qquad\qquad (R=C_4H_9\,CH(C_2H_5)CH_2-)$$
$$\underset{|}{\;C_2H_5}$$

The world capacity for its production from butyraldehydes is at least 3 m tonnes per annum.

Hydroformylation was discovered in 1938 by Otto Roehlen (Ruhrchemie) during investigations on the Fischer-Tropsch process (p. 393). An alkene (propene to produce butyraldehydes), carbon monoxide and hydrogen are heated under pressure in the presence of a cobalt catalyst which operates homogeneously. Typical conditions used today are 140–180°C/190–210 atm. The active catalyst is $HCo(CO)_4$ which is formed *in situ via* $Co_2(CO)_8$ from a cobalt salt such as the oleate or the acetate, present in 0.1–1% concentration. The active species is stable only under at least 100 atm of carbon monoxide. Catalyst life depends on maintaining a high pressure of CO. The products consist mainly of aldehydes (Table 12.6). From propene, butanal and 2-methylpropanal (n- and iso-butyraldehydes) are formed in a ratio of about 3:1. As only the former is useful for making 2-ethylhexanol, the low selectivity towards the straight chain product is a disadvantage.

Hydrogenation of an alkene is thermodynamically favoured over hydroformylation and it is thus the function of the catalyst to direct the reaction into the latter course by providing an energetically favourable path.

$$C_2H_4(g) + H_2(g) + CO(g) \rightleftharpoons C_2H_5CHO(g) \quad \Delta G^\ominus = -61\,kJ\,mol^{-1}$$
$$C_2H_4(g) + H_2(g) \rightleftharpoons C_2H_6(g) \qquad\qquad \Delta G^\ominus = -95\,kJ\,mol^{-1}$$

The rate expression shows inhibition by carbon monoxide. As the catalyst is stable only above 100 atm CO, this sets a limit on the rates which can be achieved.

$$\frac{d[\text{aldehyde}]}{dt} = k\frac{[\text{alkene}][\text{Co}]p\text{H}_2}{p\text{CO}}$$

A conventional cycle is shown in Fig. 12.16. As the $\overset{\delta+}{H}-\overset{\delta-}{Co}$ bond is polarized as shown ($HCo(CO)_4$ is a strong acid), formal Markovnikov addition would be predicted. On steric grounds, however, the bulky metal centre prefers to add to the unsubstituted carbon atom of the alkene. Addition of CO and alkyl migration lead to $RCH_2CH_2COCo(CO)_3$. Acyl tetracarbonyls, however, have been detected by infrared spectroscopy under the conditions of hydroformylation and are therefore in equilibrium with the tricarbonyls. The rate determining step seems to be the hydrogenolysis. This can be thought of as proceeding either by addition of H_2 to cobalt followed by elimination of aldehyde, or as a concerted process.

388

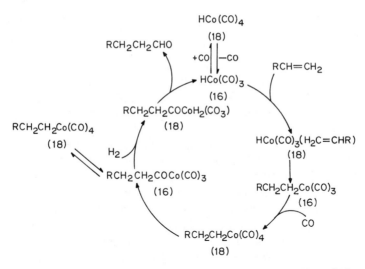

Fig. 12.16 Conventional mechanism for hydroformylation catalysed by cobalt carbonyl.

This cycle is probably an oversimplification. It has been suggested from kinetic studies that $HCo(CO)_3(RCH{=}CH_2)$ arises *via* attack of alkene on the 17-electron radical $Co(CO)_4$, which is labile to substitution. It is possible that $HCo(CO)_4$ is responsible for releasing the aldehyde in the final step, but there is no direct evidence for this yet. Probably several concurrent pathways are followed in actual hydroformylation.

Isomerization of alkenes by these cobalt catalysts takes place faster than hydroformylation itself. Mixtures of similar proportions of 1-hexanal and 2-hexanal are obtained irrespective of whether the starting material is 1-pentene or 2-pentene. 3-Heptene affords 1-octanal as the major product. Hydroformylation of internal alkenes such as 3-heptene is difficult, because of steric hindrance. Although 3-heptene is the most stable isomer and the formation of 1-heptene from it by isomerization is thermodynamically disfavoured, any 1-heptene formed can be hydroformylated very much faster than 2- or 3-heptene.

There are several problems with cobalt catalysed hydroformylation. The pressures required are high, leading to high capital costs. The selectivity (n:iso) is rather low and there are side reactions. Catalyst losses arise through its volatility and also by decomposition to metallic cobalt. The metal has to be removed with acids from time to time, causing corrosion. Fortunately, cobalt is cheap. As better processes are now available it is unlikely that new plants of this type will be built. It is economic, however, to keep plants running which are already in operation. This is still the major industrial route to butanal.

Steric hindrance to the addition of H—Co to a terminal alkene and also in the alkyl to acyl migration step favours the eventual formation of the required unbranched product. Replacement of $HCo(CO)_4$ by the more bulky phosphine derivative $HCo(CO)_3PBu_3$ leads to an increase in n:iso ratio to about 9:1. This

389

Mechanisms of industrial processes

catalyst has improved stability so that the process can be operated at lower pressures of carbon monoxide, but only about one fifth of the activity of the simple carbonyl. Consequently larger reactors and slightly higher temperatures are required. The aldehydes, which are formed initially are further catalytically hydrogenated to alcohols. Unfortunately some of the alkene feedstock is also converted into alkane. Nevertheless the process provides a valuable method of converting alkenes into unbranched alcohols in a single stage.

(b) RHODIUM CATALYSTS. Work at Union Carbide and concurrently in Wilkinson's laboratory at Imperial College, London in the 1960s led to the discovery that the rhodium complex HRh(CO)(PPh$_3$)$_3$ is a very active and selective catalyst for the hydroformylation of terminal alkenes. A pilot plant was set up in 1972 and the process first entered commercial production in Puerto Rico early in 1976. The great advantages of the rhodium over the cobalt catalysed processes are apparent from Table 12.6; lower operating pressures, lower temperatures, high n:iso ratio, lack of by-products and long catalyst life (> 1 year).

Several rhodium complexes including RhCl(CO)L$_2$, RhH(CO)L$_3$, RhClL$_3$ or even rhodium trichloride + L (L = Ph$_3$P) have been used as catalyst precursors. Under the conditions of hydroformylation they are all converted, apparently, into HRh(CO)$_2$L$_2$. Two mechanisms have been suggested. In the first initial loss of L from HRh(CO)$_2$L$_2$ gives a four-coordinate 16-electron intermediate which then adds alkene (not shown). In the second (Fig. 12.17) the Rh—H bond of HRh(CO)$_2$L$_2$ adds to alkene in a concerted step yielding an alkyl complex. Alkene cannot reasonably coordinate to HRh(CO)$_2$L$_2$, as a 20-electron intermediate would then arise. Two phosphine ligands remain coordinated to rhodium

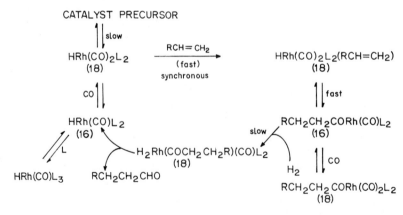

Fig. 12.17 Possible reaction mechanism (associative) for hydroformylation catalysed by rhodium carbonyl phosphine complexes. L = Ph$_3$P. Formation of n (unbranched) product only shown. 16ε complexes square planar, 18ε probably trigonal bipyramidal.

Table 12.6. Comparison of processes for hydroformylation.

	Cobalt carbonyl	Tributylphosphine modified cobalt carbonyl	Rhodium carbonyl phosphine
Catalyst	$HCo(CO)_4$	$HCo(CO)_3PBu_3$	$HRh(CO)_2(PPh_3)_2$
Temperature (°C)	140–150	180–200	80–120
Pressure (atm)	260–350	50–100	12–25
n:iso ratio	3 or 4:1	9:1	> 10:1
% n product	67	67	83
Main product	Aldehyde	Alcohol	Aldehyde
% Propane	2	10	2
By-products	Aldehyde trimer Formate ester Ethers Acetals	Alkane	Aldehyde trimer

throughout the cycle given here, whereas only one is present at some stages in the alternative mechanism. The mechanism in Fig. 12.17 is therefore the more sterically demanding and is therefore consistent with the high proportion of unbranched (n) product formed. Addition of L increases the selectivity, although it reduces the rate. It does have the advantage, however, of reducing the activity of the catalyst towards isomerization of the alkenes. There is thus an optimum L:Rh ratio. It has been shown by n.m.r. spectroscopy that the monocarbonyl $HRh(CO)L_2 \rightleftharpoons HRh(CO)L_3$ does not react with alkene at 25°C, 1 atm, whereas the dicarbonyl does. This excludes a mechanism in which dissociation of CO from $HRh(CO)_2L_2$ is the initial step.

12.4.3 Carbonylation of alkynes and of alkenes

During the period 1920–1960 ethyne was an important building block for a wide range of substances. It can be produced from coal via calcium carbide

$$CaO + 3C \xrightarrow{2200°C} CaC_2 + CO$$

$$CaC_2 + 2H_2O \longrightarrow C_2H_2 + Ca(OH)_2$$

It can also be obtained by steam cracking of ethane at temperatures above 1200°C. Nowadays acetylene has generally been replaced by alkenes as a chemical feedstock. India and East European countries however, continue to use some acetylene technology as they have adequate coal reserves but are oil importers.

A pioneer in the development of homogeneous transition metal catalysts was W. Reppe of the German firm I.G. Farben. Working during the late 1930s and early 1940s he discovered a series of carbonylation reactions of alkynes, alkenes

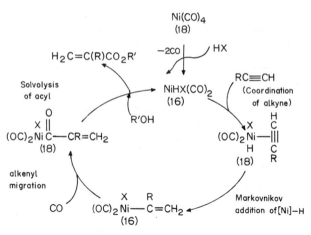

Fig. 12.18 Possible mechanism for carbonylation of an alkyne.

and alcohols which lead to carboxylic acids or their derivatives.

$$HC\equiv CH + CO + H_2O \longrightarrow H_2C=CHCO_2H$$
$$HC\equiv CH + CO + ROH \longrightarrow H_2C=CHCO_2R$$
$$R'CH=CH_2 + CO + H_2O \longrightarrow RCH_2CH_2CO_2H$$

One example, the formation of acrylic (propenoic) acid or its esters from acetylene which is catalysed by nickel carbonyl, is taken to illustrate this area. Details of related processes are to be found in literature cited at the end of the Chapter.

The reaction is carried out in aqueous tetrahydrofuran, if acrylic acid is the desired product, or in aqueous alcohol if the ester is required. Nickel is introduced as bromide or iodide and is converted into carbonyl complexes under the reaction conditions, typically $200°C/100$ atm. One catalytic cycle which has been postulated for this process is shown in Fig. 12.18. Selectivity in the formation of acrylic acid from acetylene is better than 90%. Even from propyne, where anti-Markovnikov addition of [Ni]—H competes with the desired pathway, selectivities of over 80% to methyl methacrylate $H_2C=C(Me)CO_2Me$ are achieved. The major by-product is methyl crotonate, $MeCH=CHCO_2Me$.

Palladium complexes catalyse similar reactions, usually under milder conditions than those required using nickel. Azelaic acid (nonanedioic acid), which is an intermediate in the synthesis of the speciality nylon 6, 9, is obtained from butadiene *via* 1, 5-cyclo-octadiene.

Some related carbonylations, useful in the laboratory, are mentioned on p. 230.

12.4.4 *The Fischer-Tropsch process*

The Institut für Kohlenforschung at Mülheim was founded in 1914 with the prime objective of converting coal into oil. From 1919 onwards its Director, F. Fischer, decided to approach this goal *via* synthesis gas. He and Tropsch soon found that mixtures of liquid hydrocarbons and oxygenated products can be produced using heterogeneous promoted iron or cobalt catalysts. By 1934 the process was ready for commercial introduction. During World War II nine plants each producing about 10^5 tonnes per annum were operated in Germany. These used cobalt catalysts and gas pressures of 5–30 atm.

From the end of the War until the 1970s, when the price of crude oil increased sharply, the only interest shown in this technology was in South Africa. The first plant (SASOL I) used syn. gas produced from coal by the Lurgi gasification process. It had a capacity of 2×10^5 tonnes yr^{-1} and employed alkali-treated iron catalysts. Fixed-bed operation (220–240°C/25 atm) produces predominantly straight chain hydrocarbons with a wide spread of molecular weights. A fluid bed, however, gives about 70% hydrocarbons, both alkanes and alkenes, in the C_5–C_{11} gasoline range, together with about 13% of alcohols and ketones. Two much larger plants SASOL II and SASOL III have since been introduced and together they produce about half of South Africa's liquid fuel needs.

New Zealand has extensive natural gas reserves but lacks oil resources. It is remote from major oil producers, so that transportation costs are high. An alternative to Fischer-Tropsch has been developed (Mobil process), by which syn. gas is converted first into methanol and then, by using a special Zeolite catalyst, into gasoline hydrocarbons (C_5–C_{10}). This is essentially a dehydration of methanol and unlike Fischer-Tropsch produces a high proportion of aromatic (28%) and high octane branched aliphatics. Water is formed in 56% yield by weight. These examples, together with others shown in Fig. 12.14, indicate that methods are already available for converting coal or natural gas, either *via* syn. gas or by other routes, into a very wide range of fuels and chemical feedstocks. Most of these employ heterogeneous catalysts, although in some cases homogeneous systems have also been investigated.

(a) MECHANISM OF THE FISCHER-TROPSCH REACTION　The Fischer-Tropsch reaction takes place at the surface of the metallic catalyst. It does not, therefore, necessarily relate to transformations which occur under homogeneous conditions in solution. There have been many speculations about the mechanism, but there is still little definite evidence about the nature of the species bound to the surface during the reaction. Much research has been directed towards making complexes which contain various groups postulated in these speculations and studying their reactions.

Carbon monoxide is dissociatively adsorbed on the surfaces of those metals, such as iron or cobalt, which are active in hydrocarbon synthesis. On nickel, dissociation also occurs, although at higher temperatures than on iron. Chain

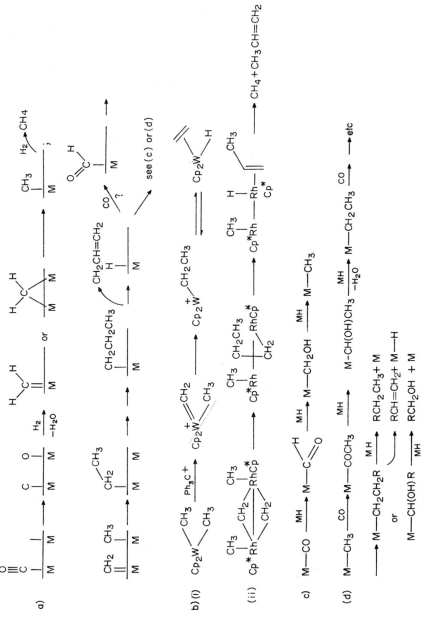

Fig. 12.19 Possible reaction paths in Fischer Tropsch reactions. (a) A suggested scheme involving surface methylene groups. (b) Analogues from chemistry of complexes. (c) (d) Further possible pathways. M = surface metal atom. All these steps have analogies in the chemistry of known complexes.

growth is suppressed and methane is produced selectively. Copper, however, adsorbs carbon monoxide without dissociation. Consequently it catalyses reactions of syn. gas, such as the production of methanol, in which the C—O bond remains intact.

Fischer and Tropsch suggested that surface bound CH_2 groups are formed initially, which then link together to form hydrocarbon chains. This has been tested by studying the decomposition of diazomethane on iron surfaces. Between 25° and 250°C, ethene is the main product. If hydrogen is present, however, a mixture of linear alkenes and alkanes, of molecular weight distribution very similar to that found in the Fischer-Tropsch reaction, are obtained. Methylene complexes in which the CH_2 group is bound either to one metal atom ($M{=}CH_2$) or bridges two centres are well characterized (p. 236). Alkyl migration to methylene, either at a single metal centre or between adjacent centres in a binuclear complex has been observed (Fig. 12.19).

Another suggested mechanism implicates formyl, hydroxyalkyl and acyl intermediates. All the key steps have analogies in solution chemistry. Reduction of coordinated carbon monoxide throughout all the intervening stages shown in Fig. 12.19 has been achieved, starting from $[CpRe(CO)(NO)]^+BF_4^-$ and one, two or three equivalents respectively of sodium borohydride in tetrahydrofuran.

Whereas intramolecular alkyl migration to coordinated carbon monoxide to form an acyl is well established (p. 225), the corresponding hydrogen migration to give a formyl complex was unknown until fairly recently. Metal—hydrogen bonds are generally stronger, it is thought, than metal—carbon bonds. Consequently the equilibrium $MH(CO){\rightleftharpoons}M{-}CHO$ lies further to the left than $MCH_3(CO){\rightleftharpoons}MCOCH_3$. The conversion of a thorium carbonyl hydride into a formyl complex, however, has now been demonstrated.

12.5 Conclusion

In this Chapter the emphasis has been placed on applications which are of interest to the chemical industry. In the laboratory many of the features of homogeneous catalysts which can make them unattractive on a large scale lose their significance. Laboratory preparations are carried out in batches. Work up using extraction or distillation, which can be uneconomic in industry, presents few problems. Recovery of catalyst is usually unimportant. Homogeneous catalysis by transition metal complexes is therefore fast becoming established in the organic chemist's synthetic armoury.

12.6 References

General texts

Atwood, J.D. (1985) *Mechanisms of Inorganic and Organometallic Reactions*, Brooks/Cole, CA.

Mechanisms of industrial processes

Bond, G.C. (1987) *Heterogeneous Catalysis, Principles and Applications*, 2nd Edn, Oxford University Press, Oxford.

Masters, C. (1981) *Homogeneous Transition-Metal Catalysis—a gentle art*, Chapman and Hall, London.

Parshall, G.W. (1980) *Homogeneous Catalysis*, Wiley-Interscience, New York.

Somorjai, G.A. (1981) *Chemistry in Two Dimensions*, Cornell University Press, Ithaca, NY.

Industrial applications

Pearce, C. and Patterson, W. (Eds) (1981) *Catalysis and Industrial Processes*, Leonard Hill, Glasgow.

Wiseman, P. (1979) *An Introduction to Industrial Organic Chemistry*, 2nd Edn, Applied Science, London.

Specific processes

HYDROCYANATION

Tolman, C.A. (1986) *J. Chem. Educ.*, **63**, 199.

DIMERIZATION

Su, A.C.L. (1979) *Adv. Organomet. Chem.*, **17**, 269.

Bogdanovic, B. (1979) *Adv. Organomet. Chem.*, **17**, 105.

OLIGOMERIZATION OF BUTADIENE

Jolly, P.W. (1985) *Angew. Chem. (Int. Edn)*, **24**, 283. Comprehensive Organometallic Chemistry, Ch. 56.4.

ZIEGLER-NATTA CATALYSIS

Eisch, J.J. (1983) *J. Chem. Educ.*, **60**, 1009.

Pino, P. and Mulhaupt, R. (1980) *Angew. Chem. (Int. Edn)*, **19**, 857.

Ziegler, K. (1967) *Adv. Organomet. Chem.*, **6**, 1.

OLEFIN METATHESIS

Banks, R.L. (1979) *Chem. Tech.*, 494; (1986), *Chem. Tech.*, 112.

Calderon, N., Lawrence, J.P., Ofstead, E.A. (1979) *Adv. Organomet. Chem.*, **17**, 449.

Katz, T.J. (1978) *Adv. Organomet. Chem.*, **16**, 283.

Mol, J.C. (1983) *Chem. Tech.*, 250.

Schrock, R.R. (1983) *Science*, **219**, 13; (1986) *J. Organomet. Chem.*, **300**, 249.

Streck, R. (1983) *Chem. Tech.*, 758.

WACKER

Backvall, J.E. (1983) *Acc. Chem. Res.*, **16**, 335.

HYDROFORMYLATION

Pruett, R.L. (1979) *Adv. Organomet. Chem.*, **17**, 1.

FISCHER-TROPSCH

Henrici-Olivé, G. and Olivé, S. (1984) *The Chemistry of the Catalyzed Hydrogenation of Carbon Monoxide*, Springer Verlag, Berlin.
Herrmann, W.A. (1982) *Angew. Chem. (Int. Edn)*, **21**, 117.

SYN. GAS

Forster, D. (1979) *Adv. Organomet. Chem.*, **17**, 255.
King, D.L. and Grate, J.H. (1985) *Chem. Tech.*, 244.
Maugh, T.H., II (1983) *Science*, **220**, 1032.

SUPPORTED METAL CATALYSTS

Bailey, D.C. and Langer, S.H. (1981) *Chem. Revs.*, **81**, 109.

Problems

1. 'The activity of a catalytic process is the major criterion for its large scale industrial application.' Discuss this statement and illustrate with examples from homogeneously catalysed processes.
(*Royal Holloway College and Bedford College*).

2. 'Homogeneous catalysis mediated by complexes of transition elements relies primarily on the ability of such complexes to undergo oxidative addition/reductive elimination reactions.'
Propose suitable mechanisms for the following processes:

(a) $RC{\equiv}CH + CO + MeOH \xrightarrow{\text{NiHCl(CO)}_2} H_2C{=}C\underset{COOMe}{\overset{R}{<}}$

(b) 1,5-cyclooctadiene $\xrightarrow{\text{Fe(CO)}_5}$ 1,3-cyclooctadiene

(*Heriot-Watt University*)

3. Complexes of the type $Ni\{P(OR)_3\}_4$ catalyse the isomerization of but-1-ene to *cis*- and *trans*-but-2-ene in benzene solution in the presence of CF_3CO_2H. Propose a cycle for this process, discussing the important steps.
(*University of Southampton*).

4. Both the selectivity and the activity of rhodium systems which catalyse alkene hydroformylation are improved by the addition of PPh_3. However, the rhodium complex so formed, $[HRh(CO)_2(PPh_3)_2]$, does not catalyse alkene isomerization.

By considering the mechanism of the hydroformylation reaction, account for these observations.
(*University of Southampton*).

397

Mechanisms of industrial processes

5. Consider the following reactions:

$$NiBr_2 + 2CH_2{=}CH{-}CH_2MgCl \xrightarrow{Et_2O} (C_3H_5)_2Ni + 2MgClBr$$

[1]

$$(C_3H_5)_2Ni + 3CH_2{=}CH{-}CH{=}CH_2 \longrightarrow C_{12}H_{18}Ni + C_6H_{10}$$
$$\text{[1]} \qquad\qquad\qquad\qquad\qquad\qquad \text{[2]}$$

$$C_{12}H_{18}Ni \xrightarrow{PPh_3} C_{12}H_{18}NiPPh_3$$
$$\text{[2]} \qquad\qquad\qquad \text{[3]}$$

$$\xrightarrow{5CO} C_{12}H_{18}CO + Ni(CO)_4$$
$$\text{[4]}$$

The ^1H n.m.r. spectrum of [1] in solution at 273 K shows resonances (relative to TMS) at δ 4.91 (multiplet), 3.78 (doublet) and 1.70 (doublet) with relative intensities 1:2:2; in addition there is another set of similar resonances at δ 4.43 (multiplet), 3.57 (doublet) and 2.17 (doublet) also with relative intensities of 1:2:2. The ratio of the total area of the first pattern to that of the second is temperature dependent.

Comment on the above reactions and identify [1], [2], [3] and [4], giving structures whenever possible. Interpret the n.m.r. data given for [1]. (*University of Warwick*).

13

Some complexes of the lanthanides and actinides

13.1 Introduction—lanthanides

The lanthanides are the fourteen elements between cerium (Z = 58) and lutetium (Z = 71) inclusive. The Group IIIA elements yttrium (Z = 45) and lanthanum (Z = 57), which are chemically very similar, are also included in this discussion. The 4f orbitals are being filled across the lanthanide series. These are inner orbitals, that is, they do not extend significantly beyond the filled $5s^2 5p^6$ orbitals in the xenon core. Strong interactions between metal (4f) and ligand orbitals would therefore not be expected. These elements are electropositive, forming quite ionic compounds and they possess the common property of forming the tripositive ions Ln^{3+} (4f^n). There is a gradual contraction in size across the series from La^{3+} ($r = 1.032$ Å) to Lu^{3+} (0.861 Å). Y^{3+} (0.90 Å) fits into the second half of the series, near Er^{3+} (0.89 Å).

The structures, stability and reactivity of organolanthanide compounds seem to be largely determined by electrostatic and steric effects. This is in marked contrast to d-block transition elements where interactions between metal and ligand orbitals are important. Nothing equivalent to the 18-electron rule is to be expected, but rather similar chemistry regardless of the 4f^n configuration. Size is of crucial importance. By selecting a metal of appropriate size it may in future be possible to tailor the steric effect so as to provide just the right environment to bring about a desired reaction. Stability of complexes is often enhanced by filling the coordination sphere with bulky ligands, up to a certain limit, so as to block decomposition pathways.

Organolanthanide chemistry is a new, rapidly developing area. There are considerable experimental difficulties. The complexes are very susceptible to hydrolysis and to oxidation in air; the metal centres are strongly electrophilic. The complexes are paramagnetic, except for those having f^0 (La^{3+}, Y^{3+}) or f^{14} (Lu^{3+}) configurations. The use of n.m.r. spectroscopy is therefore generally limited to compounds of these metals. Yttrium complexes are valuable in n.m.r. studies as not only are they diamagnetic, but the only naturally occurring isotope of this element, $_{45}^{89}Y$, has nuclear spin I = $\frac{1}{2}$. ^{89}Y spectra have been observed and $^{89}Y-^1H$ coupling appears in proton spectra.

Lanthanides and actinides

A few examples of reactions which occur at lanthanide centres are highlighted below and compared with those undergone by d-block transition metal complexes.

13.2 Introduction—actinides

The actinide elements comprise the fourteen elements Th to Lr which follow actinium ($Z = 89$) in the Periodic Table. Across the series the $5f$ orbitals are filled. These, like the $4f$ orbitals of the lanthanides, are inner orbitals, which do not extend significantly outside the radon core. Strong participation of these orbitals in covalent bonding is therefore not to be expected. At the beginning of the actinide series, however, the $5f$ and $6d$ orbitals are rather similar in energy. This leads to a greater range of oxidation states in actinide compounds than for the early lanthanides, for which (apart from Ce(IV)) the $+3$ ($4f^n$) state is established right from the start. Uranium, for example, forms fluorides UF_3, UF_4, UF_5 and UF_6. Moreover, while steric and electrostatic considerations are still thought to play the dominant role in determining structure and reactivity in the organometallic derivatives, a greater degree of covalent interaction with ligand molecules through metal $6d$ and $5f$ orbitals is likely compared with the lanthanides. Only a few aspects of this chemistry can be touched upon; a more complete perspective can be obtained from the articles cited.

13.3 Cyclopentadienyls

Tris-cyclopentadienyls of the lanthanides have been prepared from the trichlorides and sodium cyclopentadienide. They are isolated from tetrahydrofuran as adducts, $Cp_3Ln(THF)$, in which the metal atom is formally ten-coordinate (Cp formally occupies 3 coordination positions). These high coordination numbers (8–12) are characteristic of lanthanide chemistry. $Cp_3Ln(THF)$ are converted into dimeric halides by reacting with more $LnCl_3$. These species are useful starting materials for preparing other derivatives.

$$LnCl_3 + 3NaCp \xrightarrow{\text{THF}} Cp_3Ln.THF$$
$$2Cp_3Ln.THF + LnCl_3 \longrightarrow [Cp_2LnCl]_2$$

The dimer $Cp_2Y(\mu\text{-}H)_2YCp_2$ contains strongly hydridic hydrogen. With water it gives hydrogen (D_2O gives HD) and with iodomethane, methane. The ^1H and ^{89}Y

n.m.r. spectra each show a triplet, $J(^1H^{89}Y) = 27.2$ Hz, indicating that the dimeric structure is retained in solution. Hydride complexes also result from β-hydrogen transfer, as with d-block transition elements. The tertiary butyl group is useful, not only because it has nine β-hydrogen atoms but because it is too large to form a bridge like the methyl group.

$$[Cp_2YCl]_2 \xrightarrow{\text{LiCH}_3} Cp_2YCH_3.THF \xrightarrow[\text{benzene}]{\text{recyst}} Cp_2Y\underset{\underset{H}{\overset{C}{\big|}}}{\overset{\overset{H}{\underset{|}{C}}}{<>}}YCp_2$$

$$\Big\downarrow H_2$$

bridging methyls cf Al_2Me_6

$$Cp_2Y\underset{H}{\overset{H}{<>}}YCp_2$$

THF / THF

Insertion of CO, another reaction characteristic of transition metal alkyls, has also been observed. The dihapto acyl (cf. Zr, p. 291) reacts further to yield a product in which four CO molecules have been linked through carbon—carbon bonds.

$$[Cp_2LuCl]_2 \xrightarrow[\text{THF}]{\text{LiBu}^t} Cp_2Lu\{C(CH_3)_3\}\, THF \xrightarrow[-(H_3C)_2C=CH_2]{\Delta} \tfrac{1}{2}\left[Cp_2Lu\overset{H}{\underset{H}{<>}}LuCp_2\right]$$

$$\Big\downarrow CO$$

$$Cp_2Lu\overset{O}{<}{\overset{|}{\underset{}{C}}}-R \quad \longleftarrow \quad Cp_2Lu\cdots$$

By replacing cyclopentadienyl by pentamethylcyclopentadienyl, steric compression at the metal centre is greatly increased, especially at lutetium, the smallest lanthanide. The complex $Cp_2^*LuCH_3(OEt_2)$ polymerizes ethene and oligomerizes propene. ($Cp^* = \eta^5\text{-}C_5Me_5$). Its behaviour thus provides a good model for Ziegler-Natta catalysis. The insertion of propene into the Lu—CH_3 bond and some further propagation steps have been followed by n.m.r. spectroscopy. Clear examples of such insertions are rare in d-block chemistry (p. 380). Termination can occur either by β-H transfer or by β-methyl transfer; the latter is a novel process which has yet to be observed for d-block elements.

PROPAGATION $\quad Cp_2^*LuCH_3 + H_3CCH=CH_2 \longrightarrow Cp_2^*LuCH_2CH(CH_3)_2 \longrightarrow Cp_2^*LuCH_2CH\overset{\overset{CH_3}{|}}{}$

β–H TRANSFER $\qquad\qquad\qquad\qquad\qquad\qquad Cp_2^*LuH + H_2C=C(CH_3)_2$

$$Cp_2^*Lu-CH_2CH\overset{CH_3}{\underset{CH_3}{<}}$$

β–Me TRANSFER $\qquad\qquad\qquad\qquad\qquad Cp_2^*LuCH_3 + H_2C=CHCH_3$

Lanthanides and actinides

$Cp_2^*LuCH_3$ is in equilibrium with its dimer in solution. The monomer and one site in the dimer must therefore possess room for coordination of another molecule— they are 'sterically unsaturated'. $Cp_2^*LuCH_3$ and $Cp_2^*LuH_2$ effect metallation (p. 40) viz. C—H bond activation (p. 224) for weakly acidic hydrocarbons such as alkynes or arenes. Like $RLi(TMEDA)_2$ these complexes possess sites of high Lewis acidity and basicity in the same molecule. The lutetium centre appears to be even more fiercely electrophilic and can interact even with methane. The exchange

$$Cp_2^*LuCH_3 + {}^{13}CH_4 \rightleftharpoons Cp_2^*Lu^{13}CH_3 + CH_4$$

obeys a second order rate law and has been followed by n.m.r. spectroscopy.

Uranium forms tris and tetrakis-cyclopentadienyls Cp_3U and Cp_4U. In the latter there are four η^5-cyclopentadienyl groups so that the metal atom is formally 12-coordinate. This high coordination number is consistent with essentially ionic (electrostatic) bonding. This may also apply to the complexes MCp_4 ($M = Ti, Zr$) where the structures apparently depend on the size of the metal atom and, for Cp_4Zr, the 18-electron rule does not apply.

	Ti	Zr	U
$r/Å$ (M^{4+})	0.74	0.91	1.17
Formal Coord. No	8	10	12
Attachment	$2\eta^5, 2\eta^1$	$3\eta^5, \eta^1$	$4\eta^5$
Electron count	18	20	26

13.4 Cyclooctatetraene complexes

Cyclooctatetraene forms complexes with both lanthanide and actinide elements. They are prepared by reacting the anhydrous metal halides with $K_2(C_8H_8)$ in tetrahydrofuran. The anion $C_8H_8^{2-}$ is a planar ten-electron aromatic system and this planar structure is retained by the C_8H_8 ligand in the complexes.

$$LnCl_3 + 2K_2(C_8H_8) \xrightarrow{\text{THF}} K^+[Ln(C_8H_8)_2]^- + 3KCl$$

$$(Ln = La, Ce, Pr, Nd, Sm, Gd, Tb, Y)$$

$$MCl_4 + 2K_2(C_8H_8) \xrightarrow{\text{THF}} [M(C_8H_8)_2]$$

$$(M = Th, Pa, U, Np, Pu, Am)$$

Mixed cyclopentadienylcyclooctatetraene complexes have also been prepared:

$$LnCl_3 + K_2(C_8H_8) \longrightarrow [Ln(C_8H_8) Cl(THF)_2]_2$$

$$NaCp \diagdown THF$$

$$Ln(C_5H_5)(C_8H_8)(THF) \longleftarrow Ln(C_5H_5) Cl_2 (THF)_3$$
$$\text{(heavier lanthanides)}$$

In the salt $[K^+(\text{diglyme})][Ce(C_8H_8)_2^-]$ the anion has staggered D_{8d} symmetry. The cerium system is unusual for a lanthanide complex in showing two one-electron oxidation-reduction steps.

$$Ce(OPr^i)_4 + 2C_8H_8 \xrightarrow{\text{AlEt}_3} \underset{\text{Black needles}}{Ce(C_8H_8)_2} \xrightarrow[\text{diglyme}]{K} \underset{\text{Green}}{[K(\text{diglyme})^+][Ce(C_8H_8)_2]^-}$$

$$\xrightarrow{K} [K(\text{diglyme})^+]_2 [Ce(C_8H_8)_2]^{2-}$$

Streitwieser noticed that the HOMOs of $C_8H_8^{2-}$ have the correct symmetry to interact with the $5f$ orbitals of an actinide (or $4f$ of a lanthanide). This led to the preparation of the sandwich complex bis(cyclooctatetraene)uranium (uranocene). Uranocene is pyrophoric in air, but it thermally very robust. It forms deep green crystals which can be sublimed under vacuum. X-ray diffraction shows that the rings are parallel and eclipsed (D_{8h}).

The extent of electrostatic (ionic) and orbital (covalent) considerations in determining the structures of such compounds of lanthanides, actinides and early d-block transition elements is not clear. As with Cp_4M (M = Ti, Zr, Th, U) the complexes $M(C_8H_8)_2$ show structural trends apparently related to the size of the metal atom.

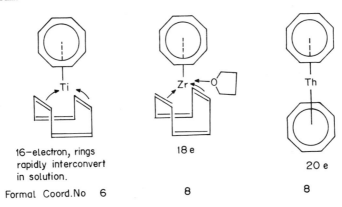

16−electron, rings
rapidly interconvert
in solution.

Formal Coord.No   6

18 e

8

20 e

8

13.5 References

Evans, W.J. (1985) Organolanthanide chemistry. *Adv. Organomet. Chem.*, **24**, 131.

Marks, T.J. (1978) *Prog. Inorg. Chem.*, **24**, 51; (1979) *Prog. Inorg. Chem.*, **25**, 223. Reviews of early work.

Watson, P.L. and Parshall, G. W. (1985) Organolanthanides in catalysis. *Acc. Chem. Res.*, **18**, 51.

Index

Organometallic compounds are indexed under the metal concerned. Where possible a major reference is indicated in **bold** type when several are given.

Index

Index

Index

Index

Index